Structural Fire Loads

About the International Code Council

The International Code Council (ICC), a membership association dedicated to building safety, fire prevention, and energy efficiency, develops the codes and standards used to construct residential and commercial buildings, including homes and schools. The mission of ICC is to provide the highest quality codes, standards, products, and services for all concerned with the safety and performance of the built environment. Most United States cities, counties, and states choose the International Codes, building safety codes developed by the International Code Council. The International Codes also serve as the basis for construction of federal properties around the world, and as a reference for many nations outside the United States. The Code Council is also dedicated to innovation and sustainability, and a Code Council subsidiary, ICC Evaluation Service, issues Evaluation Reports for innovative products and reports of Sustainable Attributes Verification and Evaluation (SAVE).

Headquarters: 500 New Jersey Avenue NW, 6th Floor, Washington, DC 20001-2070

District Offices: Birmingham, AL; Chicago, IL; Los Angeles, CA

1-888-422-7233; www.iccsafe.org

Structural Fire Loads

Theory and Principles

Leo Razdolsky, Ph.D., P.E., S.E.

New York Chicago San Francisco
Lisbon London Madrid Mexico City
Milan New Delhi San Juan
Seoul Singapore Sydney Toronto

The *McGraw·Hill* Companies

Library of Congress Cataloging-in-Publication Data

Razdolsky, Leo.
    Structural fire loads : theory and principles / Leo Razdolsky.
       p.   cm.
    Includes bibliographical references and index.
    ISBN 978-0-07-178973-8 (hardback)
    1. Building, Fireproof.   2. Live loads.   3. Structural failures—
Prevention.   I. Title.
    TH1065.R28   2012
    624.1'71—dc23   2012012409

ISBN 978-0-07-178973-8
MHID 0-07-178973-1

Sponsoring Editor	**Copy Editor**
Joy Evangeline Bramble	James K. Madru
Editing Supervisor	**Proofreader**
Stephen M. Smith	Barnali Ojha
Production Supervisor	**Indexer**
Pamela A. Pelton	Robert Swanson
Acquisitions Coordinator	**Art Director, Cover**
Molly T. Wyand	Jeff Weeks
Project Manager	**Composition**
Sheena Uprety,	Cenveo Publisher Services
Cenveo Publisher Services	

Printed and bound by RR Donnelley.

McGraw-Hill books are available at special quantity discounts to use as premi-
ums and sales promotions, or for use in corporate training programs. To contact
a representative, please e-mail us at bulksales@mcgraw-hill.com.

This book is printed on acid-free paper.

About the Author

Leo Razdolsky, Ph.D., P.E., S.E., has more than 45 years of experience in the field of structural engineering. He was the Chief Structural Engineer with the Department of Buildings, City of Chicago, for almost 20 years. His broad range of experience includes high-rise and mid-rise building design, field inspections, and construction management, as well as specialty projects including stadiums, cable structures, exhibition halls and pavilions, restoration and rehabilitation of buildings, power plants, cooling towers, and bridges. He also has experience in computer modeling, wind tunnel testing analysis, dynamic analysis of structures, seismic design, and complex foundation systems analysis and design. For the past 7 years, Dr. Razdolsky has been conducting research connected with analytical methods of obtaining the structural fire load for a high-rise building subjected to abnormal fire conditions. He has taught various structural engineering courses for over 15 years at the University of Illinois at Chicago and Northwestern University. He is currently a member of the Fire & Safety Working Group at the Council on Tall Buildings and Urban Habitat (CTBUH), and is a member of the American Society of Civil Engineers (ASCE), the American Concrete Institute (ACI), and the American Institute of Steel Construction (AISC). He is a licensed Structural Engineer in the State of Illinois and the president of LR Structural Engineering, Inc.

Contents

Foreword

In the design of buildings, loads caused by gravity, wind, and earthquake must be considered by the structural designer. Gravity loads, both dead and live, are always present but somewhat variable and hard to forecast because of lack of knowledge about extremes and are considered to be the most important loads. Wind load is less easy to predict and is subject to wide variation, including the extremes caused by hurricanes and tornados. Earthquake loads, also subject to extremes, are infrequent at any given location. However, the designer recognizes that all three of these load effects must be provided for in the design.

Fire is another loading condition that must be accommodated in the design of a building. Owing to the prescriptive way in which fire-resistant design has been handled historically, designers tend not to think of fire as a loading condition. Rather, they commonly handle fire design by prescribing fireproofing systems that have been calibrated by standard testing of full-size building components. These prescriptive rules for producing a fire-resistant structure tend to mask the effects of a major fire on a real building.

Although performance-based fire-resistant design has been available for many years, it tends to be used only by a few very knowledgeable designers. A lack of good reference books and knowledgeable design professionals has inhibited the broader use of performance-based design for fire resistance.

Historically, building codes have not considered fire as a load. Rather, it was considered as a hazard that could be mitigated by prescriptive rules. This masking of the real effects of fire has prevented many building departments from accepting performance-based design.

The *International Building Code*, first published in 2000, does have a procedure for performance-based design of fire-resistant structures. What has been lacking is appropriate reference materials to permit designers to understand and building officials to accept these procedures.

History shows that fire is a frequent and deadly event that strikes buildings. Historically, single-family homes have had few requirements for fire resistance. Often, low-rise buildings have been built without fire-resistive construction. As a result, many low-rise residential and commercial buildings collapse during fires.

High-rise buildings have an excellent record of protecting life when fires occur. Loss of life in high-rise office and apartment buildings has been extremely low.

Despite the excellent record, experience with the First Interstate Bank Building in Los Angeles, One Meridian Plaza in Philadelphia, and Buildings 5 and 7 at the World Trade Center after 9/11 shows that burnouts can occur in buildings. When a burnout occurs, there is a potential for partial or even complete collapse of the structure. Performance-based procedures can be used to help mitigate the risk of collapse and, at the same time, produce a cost-effective design.

This book is timely in that it provides both prescriptive and performance-based procedures for the design of fire-resistant buildings. In addition to providing a good understanding of how fire affects buildings and how to get the desired performance, the book provides guidance for both design professionals and building officials. In the performance-based portion of the book, a means for considering fire as another load is provided. Using this approach, all types of fires including burnout can be considered in the design procedures.

Through the use of numerous design examples, this book provides practical guidance to design professionals, educators, and building officials. It can be used as both a design guide and a resource for educators. It fills a gap in the technical literature on the subject of fire-resistive design.

W. Gene Corley, Ph.D., P.E., S.E.

Preface

S *tructural Fire Loads: Theory and Principles* is a practical book on structural fire loads for fire prevention engineers, structural engineers, architects, and educators. The goal of this book is to bridge the gap between prescriptive and performance-based methods and to simplify very complex and comprehensive computer analyses to the point that structural fire loads have a simple approximate analytical expression that can be used in structural analysis and design on a day-to-day basis. The main audience is practicing structural and fire prevention engineers. The scope of the work is broad enough to be useful to practicing and research engineers and to serve as a teaching tool in colleges and universities.

This book contains a large amount of original research material (substantially modified and increased) from the author's previously published articles. At the same time, the book contains many other results obtained by other research primarily reflecting the most important data in the area of defining and computing the structural fire load. It is worthwhile to underline here that the structural fire load in general as part of the performance-based method has been evolving very rapidly in recent years, and the author has limited himself to only very few research papers connected with the structural fire load.

The main portion of the book is devoted to the additional assumptions and simplification that are specifically tailored to the structural fire load problem only. The main results are compared with the current provisions from Eurocode.

This book is constructed in such a way that the research fire protection engineer will find the simplified versions of energy, mass, and momentum equations written in dimensionless form and their solutions in tabular form, and the fire protection and structural engineer will find the "best-to-fit" analytical formulas ready to be used just for practical computations. For emergency cases (e.g., an ongoing fire scenario), many sections of this book have the scaled graphical solutions that might help to do "on the back of the envelope"–type calculations.

This book has a large number of practical examples (for fire protection and structural engineering design) that are presented in a simple step-by-step computational form. The standard structural systems (beams, trusses, frames, arches, etc.) are used in all these examples.

Chapter 1 introduces the philosophy of structural fire load design and the assumptions that are made in this book in order to achieve the main goal: provide the temperature-time relationships that are based on conservation of energy, with mass and momentum equations on the one hand and practical and simple formulas for future structural engineering design on the other. It is indicated here that the burning process during fire development and the nonsteady combustion process have many similarities; therefore, the mathematical modeling of structural fire load also should be similar.

Chapter 2 presents an overview of the main simplified methods of obtaining the structural fire load at the present time: the time-equivalence method and the parametric design method.

The "traditional way" of structural fire design using the standard temperature-time curve in many cases results in a design on the safe side, causing unsatisfactory costs for fire protection measures. In some cases, the structural fire design with a standard temperature-time curve can result in underestimation of thermal exposure. The parametric natural fire model considers the actual boundary conditions of the fire compartment concerning fire load, ventilation conditions, geometry, and thermal properties of the enclosure. The parametric fire curves are derived by heat-balance simulations, assuming a great number of natural design fires by varying the above-mentioned parameters. These curves have been incorporated into a Swedish standard and also have served as the basis for the parametric temperature-time curves of Eurocode 1-1-2 and can be applied to the structural fire design of small to medium rooms where a fully developed fire is assumed. The parametric temperature-time curves of Eurocode 1-1-2, annex A, in some cases provide an unrealistic temperature increase or decrease.

Chapter 3 provides a review of computer simulations of a design fire: zone and field models. Zone models are relatively simple from a computational point of view and based on the assumption that the temperature in a fire compartment is uniformly distributed in each zone and the hydrodynamic portion of a burning process is practically omitted. There are many zone modeling packages available now on a market, and the summary of available current zone models is presented in this chapter. The field computer models (FDS) are very comprehensive on the one hand, but on the other hand they are very complex from a computational point of view and very sensitive to even small changes in input data or any boundary conditions. At the same time, it is a well-known fact that the physical properties of real burning materials are often unknown (or very difficult to obtain), and this is a very serious limitation in practical application of field models for design purposes. The FDS model is very reliable when the heat release rate (HRR) of the fire is specified (given), and it is much less reliable for fire scenarios where the HRR is predicted (unknown). This is why the field models are used primarily for investigation purposes of large projects but not for the design stages on a day-to-day basis. A summary of available current field models is also presented in this chapter.

The principal aim of Chap. 4 is to overcome the two major obstacles of simplifying (as much as possible) the conservation of energy, mass, and momentum equations in case of structural fire load only. What this means is that all other life safety issues of fire protection, such as modeling the transport of smoke, a detailed mixture fraction-based combustion model, the activation of sprinklers, heat and smoke detector modeling, and so on, are not part of this simplification process. In order to achieve this goal, the following assumptions are made: (1) All differential equations (thermal and mass transfer coupled with Navier-Stokes hydrodynamic flow equations) are written in dimensionless form. This drastically reduces the total number of parameters in the input data. (2) The equations are simplified to such a degree that solutions will be acceptable and easy to use by the structural engineer. (3) For all practical purposes, it will be assumed in this study that the structural system is robust enough and doesn't induce any measurable interior forces (moments, shears, etc.) when the maximum gas temperature in the compartment is below $T = 300°C$, and, therefore, the earliest stage of fire (initial burning) is not important (from a structural fire load point of view). (4) The chemical reaction of the burning process can be drastically simplified and presented as a first-order chemical reaction. (5) The dimensionless solution of differential equations should be verified against the standard fire test results as well as the natural fire test data results.

Further simplifications are contained in Chap. 5. First, the conservation of energy, mass, and momentum equations are broken up in two parts (similar to zone and FDS models): conservation of energy and mass on the one hand and the momentum equation (Navier-Stokes equations) on the other. Almost universally the Navier-Stokes equations are written for a simple class of fluids (which most liquids and all known gases belong to) known as *Newtonian fluids*. Second, the additional simplification has been made because the heat transfer due to radiation plays much more significant role than the heat transfer due to conduction and convection (in case of "small" fire compartments). For larger sizes of fire compartments the convection forces have been added in hydrodynamic equations (Navier-Stokes equations), and the velocity field (or flow field) has been obtained. These velocities then were included in the first approximation of the conservation of energy and mass equations, and corrected temperature-time functions have been obtained. This type of computational procedure was repeated for each case of fire severity (very fast, fast, medium, and slow fire) and different geometric parameters characterizing the size of a fire compartment. Since all differential equations have had dimensionless variables, the solutions are presented in a tabular form that is later substituted by the nonlinear analytical approximation.

Chapter 6 is dedicated to the application of computational procedures developed in the previous chapter to the functional relationships between SFL and opening factor, fuel load, fire duration, decay period, geometry of the fire compartment, etc. The comparison of Eurocode parametric curves and SFL curves is also presented here. Finally, the computational procedure for passive fire protection design is established in this chapter. The output of such computations is the temperature-time function that will be used for structural analyses and design in the next chapter.

Chapter 7 is devoted to the structural response factor (similar to the gust response factor for wind analyses) and structural analyses and design of various structural systems subjected to SFL. The application of general mechanical creep theory to the analysis of stiffness reduction owing to high-temperature load is also presented here. Traditional (standard) structural systems (such as beams, simple frames, trusses, and arches) subjected to SFL are analyzed and designed.

The website associated with this book, www.mhprofessional .com/sfltp, contains additional material the reader will find interesting.

Leo Razdolsky, Ph.D., P.E., S.E.

Acknowledgments

I would like to extend my appreciation and gratitude to John R. Henry, P.E., principal staff engineer, International Code Council, Inc.; Joy Evangeline Bramble, senior editor, McGraw-Hill Professional; Molly Wyand, acquisitions coordinator, McGraw-Hill Professional; and Stephen Smith, editing manager, McGraw-Hill Professional, for their guidelines, support, and expert advice.

My special thanks to Dr. Pravinray D. Gandhi, director, corporate research, Underwriters Laboratories, for review of this manuscript. He provided me with an ample amount of knowledge about the field of structural fire load and its role in the overall performance-based design method in fire protection engineering.

My special thanks also to Dr. W. Gene Corley, senior vice president of Construction Technology Laboratories, P.C., for writing the Foreword for this book and brilliantly defining the role of structural fire load in the overall structural design load spectrum.

CHAPTER 1

Introduction

1.1 Objectives and Goals

The main goal of this chapter is to establish the roadmap for defining the structural fire load (SFL) and its components. Obviously, this can be done only by using the performance-based design (PBD) method. Many structural engineering disciplines (such as structural building engineering, mechanical engineering, aircraft engineering, etc.) have used this method successfully for many years. Some principles, assumptions, and engineering judgments that are currently widely used in structural performance-design engineering could be used in structural fire protection performance-based design. The discussion below is intended to set the stage for this chapter's main emphasis on a methodology for performance-based structural fire protection. Structural fire engineering encompasses only a small but essential aspect of fire protection engineering. By definition, structural fire engineering should involve close collaboration between fire protection engineers and structural engineers. Establishing the lines of communication between the structural engineer and the fire protection engineer is the main goal of this book. In many respects, the challenges faced in performance-based fire engineering are similar to those of performance-based structural engineering. Fire effects on structures have to do primarily with structural response of materials, members, and systems under high temperatures that occur during a "structurally significant" (postflashover) fire. An accurate structural fire assessment generally will require information describing the time history of elevated temperatures in all the structural members and the applied gravity and environmental loads.

Specifically, this chapter aims to create a list of SFL components that are essential in successful application of such loads in a day-to-day structural engineering practice. Some areas of SFL practical application are only in a research and developmental (R&D) stage at present. Therefore, this type of information is presented in this chapter in discussion form. For example, this chapter provides a discussion of the design criteria regarding *structural fire load* (SFL), and the intent here is to present commentary and background information on the

1

"how" and the "why" of the design criteria, including, where appropriate, a discussion of how the criteria presented in this book differ from the existing recommendations contained in ASCE-7-05 [1] and AISC's *Manual of Steel Construction*, 13th ed. [2]. It is imperative that engineers exercise good judgment in the design of a building to resist SFL so that actual building performance falls within expected or desired ranges.

It was recognized a long time ago that the performance of structural members in a real fire can be very different from the fire resistance of a single element of the structural system in a standard furnace. This is important because structural analysis of the whole building (or part of it) subjected to a structural fire load can provide information regarding robustness of the structural design or point out weaknesses in the structural system. This is particularly important in innovative structural systems and iconic buildings, which generally are much taller and have longer spans and many vertical and horizontal irregularities.

The structural engineer is responsible to check the building structure subjected to the SFL and to quantify the response of the originally proposed structural system in realistic fire scenarios in order to determine if this response is acceptable. Strengths and weaknesses then can be clearly identified and addressed within the structural design, as appropriate. Behavior of the structural system under SFL should be considered an integral part of the structural design process. The role of a structural engineer today involves a significant understanding of both static and dynamic loading and the structures that are available to resist them. The complexity of modern structures often requires a great deal of creativity from the engineer in order to ensure that the structures support and resist the loads to which they are subjected. Structural building engineering is driven primarily by the creative manipulation of materials and forms and the underlying mathematical and scientific ideas to achieve an end that fulfills its functional requirements and is structurally safe when subjected to all the loads it reasonably could be expected to experience. The structural design for a building must ensure that the building is able to stand up safely and function without excessive deflections or movements that may cause fatigue of structural elements, cracking, or failure of fixtures, fittings, or partitions or discomfort for occupants. It must account for movements and forces owing to temperature, creep, cracking, and imposed loads. It also must ensure that the design is practically buildable within acceptable manufacturing tolerances of the materials. It must allow the architecture to work and the mechanical systems to function and fit within the building. The structural design of a modern building can be extremely complex and often requires a large team to complete.

The SFL is the transient temperature load that should be applied to the structural system in combination with other design loads such as dead load, live load, wind load, etc. The philosophy of establishing any structural design load has some commonsense logical steps. For example:

1. We assume that the live loads are uniformly distributed over the floor area. This assumption is in the code because it represents the best "engineering judgment" and provides substantial simplification in structural analyses and design. In relatively rare cases, if the location and distribution of live load are known, then the code requires the structural analysis and design for a given location and magnitude of live load. This uniform distribution assumption will be used with respect to SFL application: (1) All structural elements (beams, girders, columns, etc.) located inside the compartment will have a uniformly distributed temperature load, and (2) this will simplify all major three-dimensional differential equations (thermal and mass transfer coupled with Navier-Stokes hydrodynamic flow equations) because the goal is to determine the temperature-time relationship only. These approximate solutions allow checking complex computer analyses with more simple methods and reduce the computer model uncertainty and human error.

2. Most of the time, the live-load locations are not shown on structural floor plans (people, furniture, etc.). Nevertheless, the structural code has general recommendations with respect to how to locate the design live load on a floor plan (e.g., see Sec. 8.9 from ACI 318-05 [3], "Arrangement of Live Load"). For any structural element that is under design consideration, the live load shall be located "on two adjacent spans" or "on alternate spans." The ACI code has similar provisions with respect to the stiffness of a structural element (see Sec. 8.6: "Use of any set of reasonable assumptions shall be permitted for computing relative flexural and torsional stiffnesses"). The reasoning for this statement is given in the "Commentary": ". . . the complexities involved in selecting different stiffnesses for all members of a frame would make analyses inefficient in design offices." In structural fire design, the collaboration between structural and fire protection engineers should be as follows: The fire protection engineer and the owner shall be responsible for providing the structural fire load and fire scenarios (preferably in the form of a temperature-time relationship), and the structural engineer shall be responsible for locating this load on a structural floor plan and computing the structural system reaction to the application of such

design load and load combinations (such as dead load, live load, etc.).

3. ASCE 7-05 permits a reduction of the full live-load intensity over portions of the structure or member based on the area of influence. ASCE 7-05 also provides exceptions and limitations to this rule based on a maximum reduction coefficient and the type of occupancy. There are other cases where the live load can be reduced only when approved by the authority having jurisdiction. Similar engineering judgment shall be used in case of a structural fire load. For example, if the size of a fire compartment exceeds 400 ft², it should be allowed (with some exceptions) to reduce the fire load (and therefore the SFL). If the fire scenario is such that it is possible to have multiple fires on the same floor, then the SFL is the summation of all "simple" fires. If the fire scenario includes the possibility of having a fire on two adjacent floors, then the SFL shall be applied simultaneously to the top and bottom of a structural element or system. Any special fire scenarios can be assigned by the authority having jurisdiction.

4. *Importance factor I* should reflect the degree of fire hazard to life safety (human life) and damage of a building structure as a whole. The SFL in this case is presented as an extraordinary event. The occurrence of such an event is likely to lead to structural failure (or, possibly, to a progressive collapse). The structural system should be designed in such a way that in case of a fire event, the probability of structural damage is sufficiently small. The design philosophy to limit the spread of structural damage is quite different from the traditional approach to withstand the design load combinations. These low probabilities are reflected in the load combinations of dead load, live load, wind load, etc. and SFL. The importance factor adjusts the SFL based on annual probabilities. A structural engineer may use a limit state design (LSD) that incorporates catenary action of a structural system, large deformation theory, plastic deformations, etc.

5. Wind and seismic loads are dynamic structural loads. However, the structural code permits a static equivalence of such loads. In order to achieve this goal, the *structural response factor* is introduced. For example, in the case of wind load, the static portion of the load is multiplied by the gust-effect factor. The real SFL is a dynamic load owing to rapid temperature changes in the compartment. In general, it always has been assumed that the temperature-time effect on a structural system has two components: static and dynamic [4,5].

The static component is proportional to the difference between maximum average gas temperature in the compartment and the ambient temperature, and the dynamic component is proportional to the inertia force (product of mass and the acceleration of elongations owing to temperature-time changes, second derivative of the temperature-time function). Inertia force acts on a structural element as well as on a structural building system as a whole. The magnitude of the inertia force depends on the flexibility (or natural frequency) of a structural element or structural system as a whole. The structural response factor for the structural fire load is similar to the structural response factor for wind load.

Building design practice treats the force owing to seasonal climatic temperature change in a structural member as a static load and assumes that the corresponding inertia force is small enough to be neglected. This may not be true for all applications of an SFL. The static thermal load application induces internal forces (bending moments, shears, axial forces, etc.) only in statically indeterminate structural systems. It does not induce internal forces in statically determinate structural systems. The heat flux from "very fast" and "fast" growth fire scenarios can be large enough to create substantial dynamic force acting on a structural system. In this case, the inertia forces can induce internal stresses and forces even in statically determinate structural systems. To illustrate this, consider the weightless hanger supporting a weight P (see Fig. 1.1).

This system has one degree of freedom (ODOF) and will be used often throughout the text. The total displacement of mass $m = P/g$ is presented as follows:

$$y = Y_{1t} + y_d \tag{1.1}$$

where Y_{1t} = static component owing to temperature change
$\quad\quad y_d$ = dynamic component owing to inertia force

The dynamic component of displacement is

$$y_d = -\delta_{11} m \ddot{y} \tag{1.2}$$

where δ_{11} = unit displacement from force $P = 1\,k$
$\quad\quad \ddot{y}$ = second derivative of displacement y with respect to time t

After substituting Eq. (1.2) into Eq. (1.1), we have

$$\delta_{11} m \ddot{y} + y = Y_{1t} \tag{1.3}$$

Differentiating Eq. (1.1) twice with respect to time t, we have

$$\ddot{y} = \ddot{Y}_{1t} + \ddot{y}_d \tag{1.4}$$

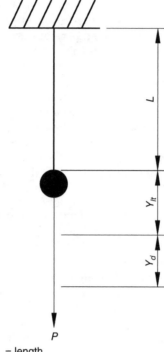

L = length
P = force
Y_{lt} = static displacement
Y_d = dynamic displacement

Figure 1.1   One-degree-of-freedom systems.

After substituting Eqs. (1.1) and (1.4) into Eq. (1.3), we have

$$\delta_{11} m(\ddot{Y}_{1t} + \ddot{y}_d) + (Y_{1t} + y_d) = Y_{1t} \tag{1.5}$$

After simplifying Eq. (1.5), we have

$$\ddot{y}_d + \omega^2 y_d = -\ddot{Y}_{1t} \tag{1.6}$$

where

$$\omega = \sqrt{\frac{1}{m\delta_{11}}} \qquad \text{Natural frequency of ODOF system} \tag{1.7}$$

and

$$\ddot{Y}_{1t} = \alpha L \ddot{T}(t) \tag{1.8}$$

where α = the coefficient of linear expansion
         L = member length
         $\ddot{T}(t)$ = second derivative of temperature-time function

Full analyses and applications to different fire scenarios will be provided in Chap. 7, where structural design examples of different structural system will be presented. For now, let's just underline here that the dynamic component of structural system reaction is important for impact (or close to impact) SFL applications ("very fast" and "fast" fires). On the other hand, one of the reasons the temperature-time curve from a standard fire test does not represent real-life fire is that the second derivative of this curve is presented by a rapidly decreasing function. From a practical point of view, this is an indication that the SFL is applied statically (only!) for all structural systems and all fire-load scenarios. However, the real-life fire has three major stages: (1) growth period, (2) fully developed fire, and (3) decay period (see Fig. 1.2).

Each stage is characterized by a different curvature sign (second derivative): 1 = positive; 2 = negative; and 3 = positive. The point where the second derivative is equal to zero is the flashover point.

Performance-based structural fire design methods have been gaining more and more recognition in recent years in Europe, New Zealand, and Australia. The effectiveness of this method was clearly demonstrated in the NIST report [6] (August 26, 2008) regarding investigation of the WTC 7 building collapse. The fires in this case were simulated using Fire Dynamic Simulator (FDS). This very complex and comprehensive software calculates the temperature-time relationship for any given fire model, which then is passed on to the structural engineer for structural analysis and design of the building. In general, the input data (in the *FDS* software) regarding physical properties (e.g., thermal conductivity, specific gravity, density, mechanism of chain chemical reaction, etc.) are very approximate. For example, the NIST report [6] makes the following comment: "Indeed,

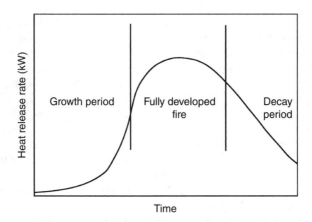

Figure 1.2 Three stages of real-life fire.

the mathematical modeling of the physical and chemical transformations of real materials as they burn is still in its infancy [and] … in order to make progress, the questions that are asked [are] having to be greatly simplified. To begin with, instead of seeking a methodology that can be applied to all fire problems, we begin by looking at a few scenarios that seem to be most amenable to analysis. Second, we must learn to live with idealized descriptions of fires and approximate solutions to our idealized equations." In this study, a few major steps are taken in the direction of simplification and approximation: (1) All differential equations (thermal and mass transfer coupled with Navier-Stokes hydrodynamic flow equations) are written in the dimensionless form. This drastically reduces the total number of parameters in the input data. (2) The equations are simplified to such a degree that solutions will be acceptable and easy to use by the structural engineer. (3) For all practical purposes, it will be assumed here that the structural system is robust enough and doesn't induce any measurable interior forces (i.e., moments, shears, etc.) when the maximum gas temperature in the compartment is below $T = 300°C$, and therefore, the earliest stage of fire (initial burning) is not important (from an SFL point of view), and the chemical reaction of the burning process can be drastically simplified and presented as a first-order chemical reaction. (4) Structural fire load should be viewed the same way as any other environmental load (e.g., wind load). This means that the design temperature load also should include the structural response factor (similar to the gust-effect factor for wind load) and force factor (similar to the turbulent effect of wind load). (5) Dimensionless solution of differential equations will allow the structural engineer to analyze and use the standard fire test results (e.g., by using the time equivalence method) as well as to apply the results to the real-life fires. This will allow the structural engineer to use the standard fire test data in applications of performance-based design methods.

The SFL has seven major characteristics from a structural design load point of view. First, the incident flux is very high; therefore, the dynamic impact on the structural system should be taken into consideration (e.g., in fires after earthquakes, when the reinforced-concrete structure has fully developed cracks). Second, the fire could be localized and act on individual structural members, as well as a fully developed fire that acts on the major part of the whole structural system. From a structural design load point of view, this means that the thermal load is a function of coordinates and time. Third, the duration of such a fire could be much longer than prescriptive recommendations given by the standard fire test, and the question is how to extrapolate the standard fire test data in this case to prevent the progressive collapse of the whole high-rise building structure. Fourth, much more elevated temperatures in this case cause a rapid decrease in concrete and steel strength, stiffness of structural elements, and strength of the system as a whole, which, in turns, require consideration of large deformations with the catenary's action of structural members in the case of progressive collapse prevention structural calculations. Fifth, the existing fire test facilities have some size limitations (3.7×2.7 m), therefore, the extrapolation of fire tests results of structural elements (beams, slabs, etc.) on the real-world high-rise building elements and systems raises a potential concern. Sixth, in a high-rise building, the incident flux on the structural elements or major portions of the whole structural system is expected to fluctuate with time, causing dynamic stresses on top of the static stresses created by temperature load application. Seventh, sustainability of a high-rise building at the design stage under abnormal structural fire loading conditions is a very important subject in structural design.

There are two primary means to address global stability of a compromised high-rise building structure: direct and indirect design approaches. The indirect design approach provides general statements to enhance the structural system as a whole by increasing robustness, ductility, etc. without specific consideration of an abnormal thermal load owing to the fire event [7]. The direct design approach considers abnormal thermal design loading combination and develops a structural system sufficient to arrest a progressive collapse. Structural analyses in this case are sophisticated, complex, and costly [8]. However, they are very sensitive to small changes in assumptions. For this reason, the approximate analysis of global stability of a compromised structure has been developed [9,10]. In order to achieve all these goals, the general theory of creep deformations has been employed [11] because it allows a structural engineer to analyze the structural problem from the very beginning of a fire development to the very end. Obviously, in today's computer-oriented world, most of these problems should be solved by using very sophisticated and complex structural engineering software. Approximate structural analyses in the case of thermal load are very useful in weeding out the less important parameters required for structural design, and on the other hand, they are very helpful in establishing the group of parameters that are critical for structural analysis and design. Any structural system in this study will be substituted by ODOF dynamic analysis. The temperature load in case of fire is presented by a nondimensional approximation of the temperature-time curve.

From a mathematical point of view, both the thermodynamic process of multiple fire propagation in tall buildings and the theory of nonsteady combustion (part of explosion theory) are described by a similar system of differential equations. For this purpose, first, let's examine some differences and amalgamations between combined effects on a structural system of multiple fires and "local" explosions. They are as follows:

1. Both are thermopositive chemical reactions that can be described by similar differential equations.

2. Both have periods of ignition ("growth period" in case of fire). However, nondimensional parameters are different.

3. Both have a self-ignition period ("flashover" in case of fire). However, nondimensional parameters characterizing self-ignition are different.

4. Thermodynamics (combination of conduction, radiation, and convection) can be described by similar parameters in both cases.

5. Hydrodynamics of both processes are described by using so-called opening factor F in the case of fire and a similar parameter K_v used in this book. This is the most important parameter in both cases.

6. The type of fire that may occur is defined by the amount of combustible materials and the size and locations of the windows in the building. Based on the heat release rate, the fire can be classified as slow, medium, fast, and very fast.

7. The total energy released during a "local" explosion or a building fire has a quasi-dynamic effect on the structural system that depends on the period of ignition or the flashover period in case of fire.

8. The temperature-time curves as a function of the opening factor K_v (F in the case of fire) had been developed.

The results of approximate structural analysis are presented in a compact analytical form that can be used later on in establishing a set of goals or rules, that is, codes or standards. The final results of this study are presented in simple form, and practical examples are provided.

1.2 Structural Design Requirements Review

The structural design of a building must ensure that the building is able to stand up safely and is able to function without excessive deflections or movements that may cause excessive cracking or failure of structural elements, fittings or partitions or discomfort for occupants. It must account for deformations and forces owing to temperature, creep, cracking, and imposed SFLs. The structural design of a modern supertall building can be extremely complex and often requires a large team to complete.

Any design process involves a number of assumptions. The design loads to which a structural system will be subjected must be estimated, sizes of members to check must be chosen, and design criteria must be selected. All engineering design criteria have a common goal—that of ensuring a safe and functional structure.

Limit state design (LSD) requires the structure to satisfy two major criteria: the *ultimate limit state* (ULS) and the *serviceability limit state* (SLS). A *limit state* is a set of performance criteria that must be met when the structure is subject to all design load combinations [1], including the SFL. To satisfy the ULS, the structure must not collapse when subjected to the peak design load for which it was designed. A structure is deemed to satisfy the ULS criteria if all factored bending, shear, and tensile or compressive stresses are below the factored resistance calculated for the section under consideration. The limit-state criteria also can be set in terms of stress rather than load. Thus, the structural element being analyzed (e.g., a beam or a column or other load-bearing element such as a wall) is shown to be safe when the factored "magnified" loads are less than their factored "reduced" resistance.

Currently, the standard fire curve, E119 Standard [12], is prescribed by a series of points rather than an equation but is almost identical to the British standard (BS) curve. Both the BS temperature time curve and the ASTM curve are illustrated in Fig. 1.3.

The BS temperature-time curve is given by Eq. (1.9), first published in 1932 [13].

$$T = T_0 + 345 \log(0.133t + 1) \tag{1.9}$$

where t = time (s) and T = temperature of the furnace atmosphere next to the specimen (°C). As was stated earlier, the standard fire

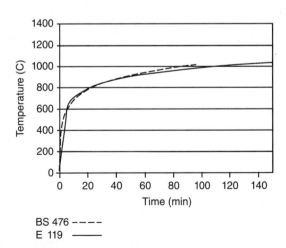

BS 476 −−−−
E 119 ———

Figure 1.3 Standard temperature-time curves.

resistance test methods are lacking the engineering data required for using the full spectrum of structural analyses and design. Relatively new methods (time-equivalence and parametric methods) have been developed in the past few decades to provide more specific and definite information about temperature-time relationships for different fire scenarios and different geometric and physical compartment parameters. Chapter 2 contains a brief review of these methods.

References

1. ASCE. *Minimum Design Loads for Buildings and Other Structures*, ASCE 7-05. ASCE, New York, 2005.
2. American Institute of Steel Construction (AISC). *Manual of Steel Construction*, 13th ed. AISC, Chicago, 2005.
3. American Concrete Institute (ACI). *Building Code Requirements for Structural Concrete*, ACI 318-05. ACI, Farmington Hills, MI, 2005.
4. Nowacki, W. *Dynamic Problems of Thermoelasticity*, edited by Philip H. Francis. Noordhoff International Publisher, London, 1975.
5. Truesdell, C. A., and Noll, W. "The Nonlinear Field Theory of Mechanics," in *Hand buch der Physik*. Springer, Berlin, 1965.
6. National Institute of Standards and Technology (NIST). *Fire Dynamics Simulator* (Version 5), Special Publication 1018-5. NIST, U.S. Government Printing Office, Washington, DC, 2007.
7. O'Donald, Dusenberry. "Review of Existing Guidelines and Provisions Related to Progressive Collapse," Progressive Collapse Workshop, Arlington, MA, 2004.
8. Kranthammer, T., Hall, R. L., Woodson, S. C., Baylot, J. T., Hayes, J. R., and Shon, Y. "Development of Progressive Collapse Analysis Procedure and Condition Assessment for Structures." *AISC Structural Journal*, 3rd Quarter, pp. 201–215, 2006.
9. Khan, F. R., and Sbarounis, T. A. "Interaction of Sheer Walls and Frames." *Proc. ASCE* 90(3), 285–335, 1964.
10. Razdolsky, L. "Local Explosions in a High-Rise Building," in *Proceedings of the 2004 Structural Engineering Congress*, ASCE-SEI, New York, 2004.
11. Rabotnov, Y. N. *Some Problems of the Theory of Creep*. National Advisory Committee for Aeronautics (NACA), Washington, DC, 1953.
12. American Society for Testing and Materials (ASTM). *Standard Test Method for Fire Tests of Building Construction and Materials*. ASTM E 119. ASTM, West Conshohocken, PA, 2007.
13. British Standards Institution (BSI). *Fire Tests on Building Materials and Structures*, BS 476, Parts 20–23. BSI, London, 1987.

CHAPTER **2**

Overview of Current Practice

2.1 Introduction

This chapter provides an overview of the performance-based method for the fire resistance of structural systems in building design. A real fire can be distinguished by three different phases (Fig. 2.1). First, during the *preflashover* or *growth phase* (*A*), the fire load begins to burn; temperature within the compartment varies from one point to another, with important gradients, and there is a gradual propagation of the fire. The average temperature in the compartment grows; if it reaches about 300 to 500°C, the upper layer is subjected to a sudden ignition called *flashover*, and the fire develops fully. In the *second phase* (*B*), after flashover, the gas temperature increases very rapidly from about 500°C to a peak value, often in excess of 1000°C, and becomes practically uniform throughout the compartment. After this phase, the available fire load begins to decrease, and the gas temperature decays (*C*), the *cooling phase*. The fire severity and duration of these phases depend on the amount and distribution of combustible materials (*fire load*), the *burning rate* of these materials, the *ventilation conditions* (openings), the compartment *geometry*, and the *thermal properties* of surrounding walls.

The design fire exposure and the structural data for the structure, the thermal properties of the structural materials, and the coefficients of heat transfer for various surfaces of the structure give the necessary information to determine the temperature development in the fire-exposed structure. Together with the mechanical properties of structural materials, the load characteristics, and the variation of resistant forces and moments, it is possible to determine thermal stresses and load bearing capacity in fire conditions. The *Natural Fire Safety Concept* (NFSC) or the *Global Fire Safety Concept* is a more realistic and credible approach to the analysis and design of structural fire safety. It accounts for active firefighting measures and real fire characteristics. In June 1994, European research titled *Natural Fire*

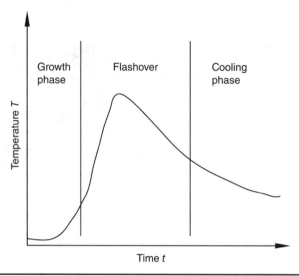

FIGURE **2.1**   Real fire phases.

Safety Concept (NFSC) started. It had been undertaken by 11 European partners and was coordinated by Profilarbed Research, Luxembourg. The research project ended in June 1998 [1]. The NFSC (1) takes into account the building characteristics relevant to fire growth (i.e., fire scenario, fire load, compartment type, and ventilation conditions), (2) quantifies the risk of fire initiation and considers the influence of active firefighting measures and egress time (this risk analysis is based on probabilities deduced from European databases of real fires), (3) deduces from the previous step design values for the main parameters such as the fire load, (4) determines the design exposure curve as a function of the design fire load that takes into account the fire risk and therefore the firefighting measures, (5) simulates the global behavior of the structure submitted to the design exposure curve and the static load in case of fire, (6) deduces the fire resistance time, and (7) verifies the safety of the structure by comparing the fire resistance time with the required time depending on the evacuation time and the consequences of failure. The European research on the NFSC [1] analyzed fire models based on more than 100 new natural fire tests, consequently permitting consideration of natural fire models in the European standard. Furthermore, these natural fire models allow, through design fire load, consideration of the beneficial effect of the active fire safety measures (i.e., by safe escape ways, proper smoke venting, or conveniently designed and maintained sprinkler systems). Also, the danger of fire activation is taken into account. Thus the so-called global fire safety concept produces real safety for people and at the same time guarantees the required structural fire resistance in real life.

The probability of the combined occurrence of a fire in a building and a high level of environmental and live loads is very small. Therefore, actions on structures from fire exposure are classified as extraordinary events and should be combined by using the design load combinations presented in ASCE 7-05 [2]. The concept of design performance levels is addressed in terms of the different performance groups and serviceability levels currently used for earthquake design. This concept is modified to address fire, which requires a different treatment from other environmental hazards.

What performance is really needed from structural systems in case of fire? The answer to this question depends on the performance objectives. As noted by Buchanan [3],

> The fundamental step in designing structures for fire safety is to verify that the fire resistance of the structure (or each part of the structure) is greater than the severity of the fire to which the structure is exposed. This verification requires that the following design equation be satisfied:
>
> $$\text{Fire resistance} \geq \text{fire severity} \qquad (2.1)$$
>
> where fire resistance is a measure of the ability of the structure to resist collapse, fire spread, or other failure during exposure to a fire of specified severity, and fire severity is a measure of the destructive impact of a fire, or a measure of the forces or temperatures which could cause collapse or other failure as a result of the fire.

The design fire resistance of structures depends on thermal actions and material properties at elevated temperatures. To determine the gas temperatures in the compartment, the appropriate temperature-time curve has to be defined first. Then the temperature increase in the structural member can be calculated using standard or advanced calculation methods. The thermal actions are given by the net heat flux $\dot{h}_{net}$ (W/m^2) to the surface of the member. On fire-exposed surfaces, the net heat flux should be determined by considering heat transfer by convection and radiation. The design resistance is determined by the temperature and mechanical properties at elevated temperatures. Typically, the process involves the following steps:

1. Define the temperature-time curve.
2. Calculate the heat transfer to the structure from exposure to the thermal environment.
3. Conduct structural analysis and design based on degraded material properties at elevated temperatures.

In EN 1991-1-2: 2002 [4], three nominal temperature-time curves are given:

1. The standard temperature-time curve (or ISO fire curve) is given by

$$T(t) = 20 + 345 \log_{10}(8t + 1) \ (°C) \qquad (2.2)$$

   where t is the time (min). This fire curve does not represent realistic fire conditions in a compartment. The temperature is always increasing; the cooling phase and the real fire load of the compartment are not considered. This curve has no probabilistic background. The real fire evolution is not considered, and it is not possible to calculate realistic temperatures with this curve. But this curve is simple to handle (only one parameter, the time t).

2. For a large hydrocarbon pool fire,

$$T(t) = 1,100(1 - 0.325e^{-0.167t} - 0.204e^{-1.417t} - 0.47e^{-15.833t}) + 20$$

$$(2.3)$$

3. For a smoldering fire:

$$T(t) = 154t^{0.25} + 20 \qquad \text{if } 0 < t < 21 \text{ minutes}$$

$$T(t) = 345\log_{10}[8(t - 20) + 1] + 20 \qquad \text{if } t > 21 \text{ minutes}$$

$$(2.4)$$

In the last few decades, modern design models to describe real fire behavior were developed. These natural fire models take into account the main parameters that influence the growth and development of fires. Natural fires depend substantially on fire loads, area of openings, and thermal properties of the surrounding structure. The gas temperature in the compartment can be determined using these parameters with parametric temperature-time curves. The parametric fire curves account for compartment ventilation by use of an opening factor. The fire load density design value may be determined by survey of existing conditions or by using representative values from EN 1991-1-2, Annex A [4]. In addition, advanced fire models taking into account the gas properties, the mass, and the energy exchange are given in EN 1991-1-2, Annex D [4]. An essential parameter in advanced fire models is the *heat release rate* (HRR) Q in kilowatts. It is the source of the gas temperature rise and the driving force behind the spread of heat and smoke. The *heat of combustion,* or *enthalpy of combustion,* is the energy released as heat when 1 mole of a compound undergoes complete combustion with oxygen under standard conditions. The chemical reaction typically is a hydrocarbon reacting with oxygen to form carbon dioxide, water, and heat. The calorific value of the heat of

Fuel	kJ/g	kcal/g	Btu/lb
Hydrogen	141.9	33.9	61,000
Gasoline	47.0	11.3	20,000
Diesel	45.0	10.7	19,300
Ethanol	29.8	7.1	12,000
Propane	49.9	11.9	21,000
Butane	49.2	11.8	21,200
Wood	15.0	3.6	6,000
Coal (lignite)	15.0	4.4	8,000
Coal (anthracite)	27.0	7.8	14,000
Natural gas	54.0	13.0	23,000

TABLE 2.1 Heat of Combustion for Some Common Fuels

combustion is a characteristic of each substance. It is measured in energy per unit of mass of the substance. Heat of combustion values are commonly determined by use of a bomb calorimeter. Many materials used in building construction are synthetics or hydrocarbon-based materials such as plastics. These materials have much greater energy potential than traditional building materials such as wood. The hydrocarbon-based fuels have approximately twice the energy potential of ordinary combustibles. Data regarding the heat of combustion for most common combustibles are presented in Table 2.1.

The HRR is determined by multiplying the mass burning rate (mass/time) by the heat of combustion (energy/mass) and a combustion efficiency coefficient to account for the portion of the fuel mass actually converted to the energy. A fire that has an increase in energy output over time is classified as a *growing fire*. When the HRR becomes close to a constant value over time, the fire is considered to be in a *steady state*. When the HRR starts to decrease over time, the fire is considered to be in a *decay stage*. The mass burning rate in a compartment fire has been defined by Kawagoe and Sekine [5].

According to the British Standards Institution (BSI) [6,7], the design of structural members in fire situations has to be carried out at the ultimate limit state. For many tall steel or concrete building structures, this is the most common performance objective. The left side of inequality (2.1) is the result of structural analyses and design of a building based on the limit-state design (LSD) method, which includes ultimate limit state design (ULS) and serviceability limit state (SLS) design. To satisfy the ultimate limit state, the structure must not collapse when subjected to the structural fire load (SFL). A structural system is deemed to satisfy the ultimate limit-state criteria if all factored bending moments and shear and tensile or compressive stresses are below the factored resistance calculated for the section under consideration. The right side of inequality (2.1) presents fire severity based on Society of Fire Protection Engineers (SFPE) 2004 [8] or the so-called Swedish curves method [9]. The quantitative measure of *fire severity* is defined by the temperature-time curve's parameters, such as maximum gas temperature, total duration time of a real compartment fire, time period of rising temperature, decay period, and the velocity of rising temperature that is connected with the HRR and the flashover point (the second derivation of the temperature-time function is zero). All these parameters combined are extremely important in characterizing the specifics of a future SFL that afterwards will be used as an input in structural engineering analyses and design.

Historically, fire protection engineers and combustion research scientists were trying to obtain all these parameters from real compartment fire experiments and standard furnace testing data. However, it was recognized a long time ago that [10] "the lack of engineering data from standard fire resistance test methods requires that performance-based design utilize data obtained from ad hoc test methods performed outside of the scope of standard test methodologies. This process is lacking in both standardization and efficiency. In addition to other limitations with respect to test procedures, measurements, and reporting, reproducibility of standard furnace testing always has been a serious issue. Fire resistance tests are unique within the fire test world in that the apparatus is only generally specified in the test standard. Fuels, burners, furnace linings, furnace dimensions, loading levels, and loading mechanisms are either unspecified or only generally specified. This has led to the situation that test results cannot be reproduced from laboratory to laboratory. This situation causes significant problems in a performance-based design environment." Similar difficulties and uncertainties are described in mathematical modeling of a real compartment fire by the National Institute of Standards and Technology (NIST) [11]: "However, for fire scenarios where the heat release rate is predicted rather than specified, the uncertainty of the model is higher." There are several reasons for this: (1) Properties of real materials and real fuels are often unknown or difficult to obtain, (2) the physical processes of combustion, radiation, and solid-phase heat transfer are more complicated than their mathematical representations in Fire Dynamics Simulator (FDS), and (3) the results of calculations are sensitive to both the numerical and physical parameters. Current research is aimed at improving this situation, but it is safe to say that modeling fire growth and spread will always require a higher level of user skill and judgment than that required for modeling the transport of smoke and heat from specified fires. Structural engineers make many decisions during the design of a structural system. Most of these decisions are performed under uncertainty, although we often do not think about that uncertainty because we have techniques for dealing with it. Uncertainty can be separated into two broad categories: aleatory and epistemic.

Aleatory means depending on luck or chance; so aleatory uncertainty is the uncertainty that arises from the randomness inherent in nature. The fact that any building that has a source of ignition and flammable materials can catch a fire is an example of an aleatory uncertainty. The fact that a fire can occur in the building is the possibility, and the quantitative measurement of that fact is the probability. If a building's fire rating is more than zero, then the probability of an aleatory uncertainty is equal to one. *Epistemic* means depending on human knowledge. Thus epistemic uncertainty is uncertainty that could, in theory, be reduced by increasing the profession's knowledge about the area of interest. The uncertainties in SFL design can come from a range of sources. Since a fire in a building is a very complex phenomenon, the sources of epistemic uncertainty are also very complex. Indeed, Dr. V. Babrauskas is correct in his writing about uncertainties measuring flame temperatures [12]: "Even careful laboratory reconstructions of fires cannot bring in the kind of painstaking temperature measuring technologies which are used by combustion scientists doing fundamental research studies. Thus, it must be kept in mind that fire temperatures, when applied to the context of measurement of building fires, may be quite imprecise, and their errors not well characterized." Lord Kelvin once said, "It's no trick to get the answers when you have all the data. The trick is to get the answers when you only have half the data and half that is wrong, and you don't know which half." Although he was talking about science, Lord Kelvin's statement fairly well summarizes the problems associated with uncertainty in SFL engineering. To move forward, we have to answer two major questions: (1) How can we reduce the level of uncertainty? (2) What is the acceptable level of uncertainty for defining the SFL?

Potential ways to reduce uncertainty include the following:

- Develop and improve the technical basis for changes and additions to American Society for Testing and Materials (ASTM) E119 so that measurements and results can be used in performance-based design.

- Use dimensionless forms of the energy, mass, and momentum equations which reduces the total number of unknown parameters (see Chaps. 4 and 5).

- Use scaling factors to compare the heating effect of real fires and a standard fire (*t*-equivalent approach to define fire severity; see Chap. 6).

- Make appropriately conservative analysis assumptions in complex computer modeling. FDS provides time and spatial temperature distribution in a compartment, but for all practical reasons in structural design and analysis, the equivalent uniformly distributed temperature load will be used in the same way as is done for live load distribution [2].

- Check complex analyses with simple approximate methods where possible to reduce model uncertainty and human error.

- Use engineering judgment.

- Use a heuristic approach to simplify mathematical modeling of real fire scenarios.

- Recognize that heuristics are used everywhere in design, and think about their limits.

- Unknown parameters in dimensionless differential equations can be obtained by using the general mathematical optimal control theory [13,14] when additional data from real fire tests are available.

2.2   *t*-Equivalence Method

The basic underlying concept of performance-based fire analysis is that a building structure should be designed for the fire severity that actually might occur in the building. Using factors such as fuel load and ventilation, the maximum credible fire in different locations in the building is calculated, and the structural response to such fires is determined. Analysis of room fire tests revealed that fire load was an important factor in determining fire severity. Ingberg [15] suggested that fire severity could be related to the fire load of a room expressed as an areas under the two temperature-time curve. The severity of two fires was equal if the areas under the two temperature-time curves were equal (above a baseline of 300°C). As was described later by Law [16], "The term *t*-equivalent is usually taken to be the exposure time in the standard fire resistance test which gives the same heating effect on a structure as a given compartment fire." Thus, any fire temperature-time history could be compared with the standard curve. Ingberg [15] related fire load to an equivalent time in the standard furnace. The results are provided in Table 2.2.

The *t*-equivalence method is used in Chap. 6 to obtain the uncertain parameters in a real-fire mathematical model.

Ingberg also provided a comparison of standard fire and real temperature-time curves including fire load and ventilation-opening factor. This comparison is presented by Fig. 2.2.

The fire load is in kilograms per square meter, and the ventilation is a fraction of one wall [e.g., 15(1/2) corresponds to a fire load of 15 kg/m² and ventilation equal to half of one wall] [17]. After Ingberg's publications, many other researchers developed similar but more sophisticated time-equivalent relationships. For example, Law's [16] concept of the *t*-equivalence method relates the actual maximum temperature of a structural member from an anticipated fire severity

Combustible Content (Wood Equivalent)		Equivalent (kJ/m² × 10⁻⁶)	Standard Fire Duration (h)
lb/ft²	kg/m²		
10	49	0.90	1
15	73	1.34	1.5
20	98	1.80	2
30	146	2.69	3
40	195	3.59	4.5
50	244	4.49	6
60	293	5.39	7.5

TABLE 2.2 Ingberg Fuel Load Fire Severity Relationship [17]

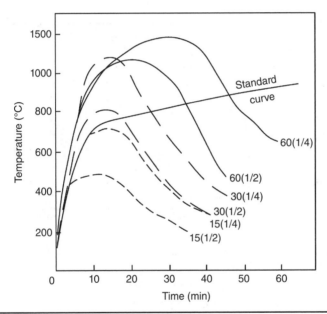

FIGURE 2.2 Comparison of the standard fire curve and real temperature-time histories.

to the time taken for the same member to attain the same temperature when subjected to a standard fire (Fig. 2.3).

Time Equivalence by Law [16]

Law developed a time-equivalence relationship to include the effect of ventilation using data gathered from fully developed

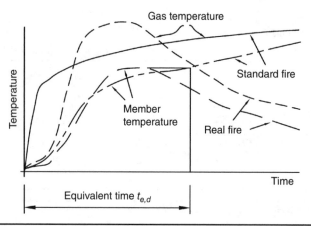

FIGURE 2.3   Law's concept of time equivalence.

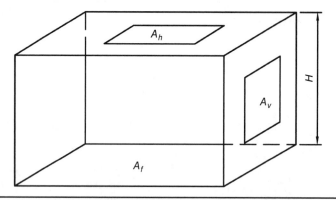

FIGURE 2.4   A fire compartment with horizontal and vertical openings.

compartment fires. This relationship is described by Eq. (2.5) and Fig. 2.4.

$$\tau_e = \frac{L}{\sqrt{A_v(A_t - A_v)}} \tag{2.5}$$

where τ_e = equivalent fire resistance (h)
A_f = floor area (m^2)
A_t = total area of the compartment boundaries, including the compartment opening (m^2)
A_v = area of the wall ventilation opening (m^2)
A_h = area of the roof ventilation opening (m^2)
L = fire load (kg/m^2)

Formulating equivalent fire exposures traditionally has been achieved by gathering data from real (natural) fire compartment experiments where protected steel temperatures were recorded and variables relating to the fire severity were changed systematically (e.g., ventilation, fire load, compartment shape). In 1993, British Steel (now Corus), in collaboration with the Building Research Establishment (BRE), carried out a test program of nine natural fire simulations in large-scale compartments to validate the time-equivalence method of Eurocode 1. The tests were conducted in a compartment 23 m long × 6 m wide × 3 m high constructed within the BRE ex-airship hanger testing facility at Cardington in Bedfordshire, UK. The test program examined the effects of fire loads and ventilation on fire severity and involved growing fires, simultaneous ignition, and changes in lining material and compartment geometry.

Time Equivalence by Pettersson et al. [18]

Pettersson and colleagues had adopted Law's method of t-equivalence but developed a new expression using the family of calculated temperature-time curves for particular compartments derived by Magnusson and Thelandersson [19]. Pettersson's t-equivalence approach takes into consideration the effect of the thermal inertia of the compartment wall lining [see Eq. (2.6)].

$$ \tau_e = \frac{1.21L}{\sqrt{A_v \sqrt{h} A_t}} \tag{2.6} $$

where h is the height of the ventilation opening (m).

Time Equivalence by Harmathy [20,21]

The normalized heat load concept is one of the most recent developments in this area and was introduced by Harmathy. He proposed that the total heat load incident on the enclosure surfaces per unit area was a measure of the maximum temperature that a load-bearing element would be expected to obtain during a fire. Recognizing that not all compartments are the same by virtue of the construction of the boundaries, it was necessary to develop an approach that could compare fires in dissimilar enclosures. With regard to calculating the normalized heat load, the only factors that are variable are the ventilation and fuel load factors. The other factors are a function of the compartment geometry being analyzed.

Harmathy proposed that the fuel load should be calculated based on the 80th or 95th percentile, similar to what had been proposed previously. The effective multiplier to the mean value ranges from 1.25 for the 80th percentile value to 1.6 for the 95th percentile value depending on occupancy. The *normalized heat load, H'*, is defined as the heat absorbed by the element per unit surface area during fire exposure. Harmathy presents an equation based on room burn experiments for compartments with cellulose fire loads and vertical openings only as follows:

$$H' = 10^6 \frac{11.0\delta + 1.6}{A_t\sqrt{k\rho c_p} + 935\sqrt{\Phi L A_f}}(LA_f) \qquad (2.7)$$

where A_f = floor area of the compartment (m^2)
        A_t = total area of compartment boundaries (m^2)
        H_c = height of compartment (m)
        $k\rho c$ = surface averaged thermal inertia of compartment boundaries (J/m^2 s$^{1/2}$ K^1)
        Φ = ventilation parameter (kg/s)
        L = specific fuel load per unit floor area (kg/m^2)
        k = thermal conductivity of the compartment boundaries (W/mK)
        ρ = density of the compartment boundaries (kg/m^3)
        c_p = heat capacity of the compartment boundaries (J/kg K)

For the ventilation factor, Harmathy proposes the following:

$$\Phi_{min} = \rho A_v \sqrt{g H_v} \qquad (2.8)$$

The minimum value for ventilation factor yields the highest value for normalized heat load and is therefore conservative. The premise is that the minimum value is represented by airflow introduced to the compartment through the openings in the absence of drafts or winds.

Parameter δ is presented as follows:

$$\delta = \left\{ 0.79\sqrt{H_c^3 / \Phi} \right\} \qquad \text{or} \qquad \delta = 1 \quad \text{whichever is less} \qquad (2.9)$$

Harmathy further proposes a relationship between the normalized heat load in the standard test and the duration of the test (fire resistance rating) as follows:

$$\tau_e = 0.11 + 0.16 \times 10^{-4} H' + 0.13 \times 10^{-9} (H')^2 \qquad (2.10)$$

Time Equivalence: Eurocode BSEN 1991-1-2 [22]

The t-equivalent method is described as follows:

$$t_{e,d} = q_{f,d} k_b w_f k_c \qquad (2.11)$$

where $t_{e,d}$ = time equivalent of exposure (minutes)
        $q_{f,d}$ = design fire load density (MJ/m^2)

Cross-sectional Material	k_c
Reinforced concrete	1.0
Protected steel	1.0
Unprotected steel	13.7*0

TABLE 2.3 Correction Factor k_c for Various Materials According to BSEN 1991-1-2 [22]

k_b = conversion factor related to the thermal inertia of the enclosure (min·m²/MJ)

k_c = correction factor of the member material as given in Table 2.3

w_f = ventilation factor as given in Eq. (2.12)

The ventilation factor w_f for a compartment with openings as shown in Fig. 2.5 is given by

$$w_f = \left(\frac{6.0}{H}\right)^{0.3}\left[0.62 + \frac{90(0.4 - \alpha_v)^4}{1 + b_v\alpha_h}\right] \geq 0.5 \qquad (2.12)$$

where $\alpha_h = A_h/A_f$
$\alpha_v = A_v/A_f$ but $0.025 \leq \alpha_v \leq 0.25$
$b_v = 12.5$ $(1 + 10\alpha_v - \alpha_v^2) \geq 10.0$
A_f = floor area of the compartment (m²)
A_h = area of horizontal openings in the roof (m²)
A_v = area of vertical openings in the facade (m²)
H = height of the fire compartment (m)

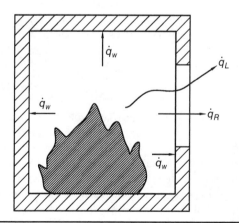

FIGURE 2.5 Compartment with openings. $\dot{q}_c$ = heat released by combustion, $\dot{q}_L$ = convective heat loss of hot gases and smoke through openings, $\dot{q}_R$ = radiative heat loss through openings, $\dot{q}_W$ = conduction heat loss to the walls.

Thermal Inertia b [J/m²s$^{1/2}$ K]	K_b in BSEN 1991-1-2 (min m²/MJ)	K_b in PD7974-3 (min m²/MJ)
> 2,500	0.04	0.05
720 < b < 2,500	0.055	0.07
b < 720	0.07	0.09

TABLE 2.4 Values of Conversion Factor k_b

In Table 2.3, O is the opening factor as given in Eq. (2.13).

$$O = (A_v \sqrt{h_{eq}})/A_t \qquad \text{but } 0.02 < O < 0.2 \ (\text{m}^{1/2}) \qquad (2.13)$$

where A_t = total area of enclosure (walls, ceiling, and floor including openings) (m²)
   A_v = total area of vertical openings on all walls (m²)
   h_{eq} = height of vertical openings (m)

The conversion factor k_b is related to the thermal inertia b of the enclosure, as given in Table 2.4. It is noteworthy that the values assigned to k_b in BSEN 1991-1-2 may be replaced nationally by the values given in PD 7974-3: 2003 for use in the United Kingdom, which have been validated by a test program of natural fires in large compartments by British Steel (now Corus) and the Building Research Establishment.

For compartments bounded with a typical building surfaces (e.g., masonry and gypsum plaster), k_b has a value of 0.07 according to PD 7974-3. For compartments with high levels of insulation (e.g., proprietary wall insulation systems with mineral wools), k_b has a value of 0.09. Table 2.5 shows the values of thermal inertia b for some typical compartment lining materials.

For a small fire compartment with a floor area $A_f < 100$ m² and without openings in the roof, the ventilation factor w_f can be calculated as

$$w_f = O^{-1/2} A_f / A_t \qquad (2.14)$$

where A_t is the total area of enclosure (i.e., walls, ceiling, and floor including openings) (m²), c is the specific heat capacity of the enclosure boundary at ambient temperature (J/kg L), ρ is the density of the enclosure boundary at ambient temperature (kg/m³), and $\lambda = k$ is the thermal conductivity of the enclosure boundary at ambient temperature (W/m K).

Law's general conclusion from his review of the t-equivalence formulas [16] is that the models may not be the most appropriate design parameters when the importance of fire temperature and duration is to be assessed. The concern is that t-equivalence formulas provide a general "feel" for the total heating effect but do not allow for the difference between short, hot fires and longer, cooler fires with

Boundary Materials	Thermal Inertia b $(J/m^2 s^{1/2} K)$
Aerated concrete	386
Wood (pine)	426
Mineral wool	426
Vermiculate plaster	650
Gypsum plaster	761
Clay brick	961
Glass	1312
Fireclay brick	1432
Ordinary concrete	1650
Stone	2423
Steel	12747

Source: Table A.2 of PD 7974-3: 2003.

TABLE 2.5   Thermal Inertia b for Typical Compartment-Lining Materials

the same value for t equivalence. This concern is supported by Buchanan [3], who suggests that t-equivalence models provide only a crude approximation of real fire behavior and that first principles, such as those used to develop parametric design fires, are more appropriate for estimating the effects of postflashover fires.

2.3   Parametric Fire Curves

General Information

The concept of parametric fires provides a simple approximation of a postflashover compartment fire. It is assumed that the temperature is uniform within the fire compartment. A parametric fire curve takes into account the compartment size, fuel load, ventilation conditions, and thermal properties of compartment walls and ceilings. The validity of the theory, assumptions, and limitations for parametric fire curves has been investigated in an extensive research work (titled "Natural Fire Safety Concept") supported by the European Coal and Steel Community (ECSC) between 1994 and 1998. A database of more than 100 natural fire tests made in Australia, France, the Netherlands, and the United Kingdom between 1973 and 1997 had been compiled. In 1999 and 2000, more full-scale large-compartment fires were carried out by Building Research Establishment (BRE) at Cardington in the United Kingdom to further supplement the database. As stated in the Society of Fire Protection Engineers (SFPE) standard [23], currently we have "… a group of twenty-three different methods or method variations. All methods considered had been documented in published material and did not involve the use of computer simulations. The methods included simplistic approaches such as a constant temperature exposure, correlations of particular data sets, generalized parametric approaches, and correlations of computer generated data." The SFPE standard [23] also underlines that "careful evaluation of the potential fire exposure scenarios must be considered to ensure adequate levels of conservatism are provided. Post flashover scenarios are typically of interest when considering structural design, although, in some cases, localized exposures may be more severe." In addition, the standard states: "This standard provides methodologies to predict thermal boundary conditions for fully developed fires to a structure over time. Information developed using this standard will provide input to thermal response and structural response calculations undertaken as part of an engineered structural fire resistance design." The major portion of this standard is devoted to thermal boundary conditions for fully developed fires. Even a brief review of all 23 methods or variations of parametric approaches would take up too much space; therefore, just a few of them will be reviewed. The first theoretical temperature-time relationship for a compartment fire based on a series of full and small-scale compartment fire experiments has been developed by Kawagoe and Sekine [24]. This model was further refined later by Kawagoe [25]. Figure 2.6 shows a typical parametric fire curve.

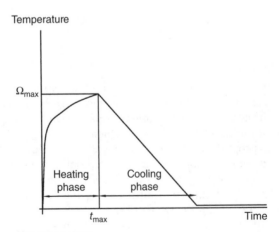

Key parameters:
1. Fire load
2. Opening factor
3. Boundary properties

FIGURE 2.6   Typical parametric fire curve.

A complete fire curve comprises a heating phase represented by an exponential curve until a maximum temperature T_{max} is achieved, followed by a linearly decreasing cooling phase until a residual temperature is reached, which is usually the ambient temperature. The maximum temperature T_{max} and fire duration t_{max} are two primary parameters that are adopted as the governing parameters in the design formulas for parametric fires. This theoretical model is based on the fundamental heat balance of a compartment fire, as indicated in the following equation:

$$q_c = q_l + q_w + q_r \tag{2.15}$$

In the sections that follow, the various compartment time-temperature curves models are described, and it is demonstrated that it is possible from an engineering design standpoint to use these curves. In all the models described, several fundamental simplifying assumptions are necessary, including the following:

- Combustion is complete and takes place exclusively inside the compartment.

- The compartment is well stirred so that the temperature is uniform throughout.

- The heat-transfer coefficient of the compartment surfaces is a constant and uniform throughout the compartment.

- The heat loss through the compartment boundaries is uniformly distributed.

In order to examine the key variables in the fundamental heat-balance equation and their related significance, each of the terms will be looked at separately.

q_l, Rate of Radiative Heat Loss Through the Ventilation Opening

The general form of this term, which is a direct derivation from the Steffan-Boltzman law [26], is as follows:

$$q_l = A_v \varepsilon_f \sigma \left(T_f^{\,4} - T_0^4 \right) \tag{2.16}$$

The gas emissivity, ε_f, typically is taken as 0.7 and usually is in the range of 0.6 to 0.9 [26]. However, the SFPE standard [23] recommends 1.0 for structural fire load design use.

q_w, Rate of Heat Loss Through Compartment Boundaries

Determination of the rate of heat transfer through the compartment boundaries is fairly complicated. The general calculation technique requires that the boundary surface be broken down into multiple layers and that a numerical technique be used to determine conduction as a function of time from one layer to the next. The more layers that are assumed, the more accurate is the resulting calculation. A real-world problem often involves a compartment constructed of different wall, ceiling, and floor types. This potentially complicates the calculation because each surface must be treated separately.

The general form of the term to be used is as follows:

$$q_w = (A_t - A_v)\left[1/\left(\frac{1}{\alpha_1} + \frac{\Delta x_1}{2k}\right)\right](T_t - T_1) \qquad (2.17)$$

where α_1 is the surface coefficient of heat transfer in the boundary layer between the combustion gases and the suspended ceiling, and Δx_1 is the thickness of layer being assessed.

Rate of Convective Heat Loss Through the Opening

The general form of the equation is as follows:

$$\dot{q}_l = \dot{m}_f c_p (T_f - T_*) \qquad (2.18)$$

One of the more significant outcomes of Kawagoe's research was the development of a term for the mass burning rate in a compartment fire, which is

$$\dot{m} = 5.5 A_v H_v^{0.5} \; (\text{kg}/\text{min}) \qquad (2.19)$$

This term is significant because it represents the rate at which the fuel in the compartment is releasing volatile gases into the compartment atmosphere, which are then burned as fuel by the fire. Numerous other experiments have followed the original work by Kawagoe to refine the relationship with the following concerns:

1. The burning rate can be predicted by this expression only over a limited range.

2. The expression implies that the burning rate is influenced only by the ventilation rate when the radiative contribution to the burning rate in a compartment is known to be significant because the radiative influence is a function of T^4 [27].

Swedish Curves

The most often cited time-temperature curves for compartment fires are the Swedish curves, which are described in detail by Pettersson and colleagues [28]. Based on the fundamental heat-balance equation and Kawagoe's burning-rate equation, a series of time-temperature curves has been developed for different ventilation and fuel-load values. These curves are shown in Fig. 2.7. The applicable mathematical model is

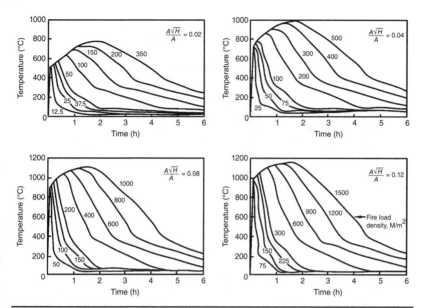

FIGURE 2.7   Analytical time-temperature curves: Swedish method [28].

$$T_t = \frac{\dot{q}_c + 0.09c_p A_v H_v^{0.5} T_0 + (A_t - A_v)\left(\dfrac{1}{\alpha_i} + \dfrac{\Delta x}{2k}\right)^{-1}(T_t - T_1) - \dot{q}_l}{0.09c_p A_v H_v^{0.5} + (A_t - A_v)\left(\dfrac{1}{\alpha_i} + \dfrac{\Delta x}{2k}\right)^{-1}} \qquad (2.20)$$

where

$$\alpha_i = \frac{\varepsilon_r \sigma}{T_t - T_i}(T_t^4 - T_i^4) + 0.023 \text{ (kW/m}^2\text{K)} \qquad (2.21)$$

$$\varepsilon_r = \left(\frac{1}{\varepsilon_f} + \frac{1}{\varepsilon_i} - 1\right)^{-1} \qquad (2.22)$$

$$\dot{q}_l = A_v \varepsilon_f \sigma\left(T_f^4 - T_0^4\right)\text{(kW)} \qquad (2.23)$$

$$\dot{q}_c = 0.09 A_v H_v^{0.5} H_{ui}\text{(kW) (kW based on the combustion} \\ \text{of wood} = 18.8 \text{ MJ/kg)} \qquad (2.24)$$

The solution is complicated and requires numerical integration that does not lend itself easily to hand calculations. For this reason, the series of curves shown in Fig. 2.8 has been developed by designers in Sweden.

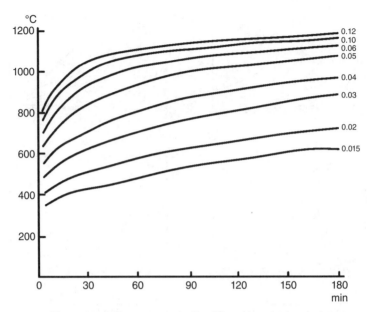

Temperature-time curve under the different temperature factors
(where k = 1.0, C = 0.3, p = 2,400)

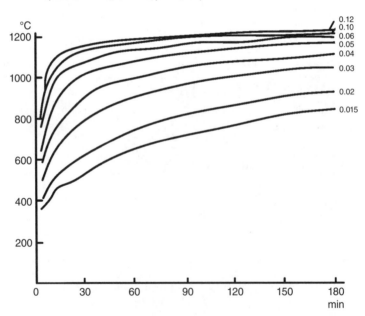

Temperature-time curve under the different temperature factors
(where k = 0.5, C = 0.24, p = 1,700)

FIGURE **2.8** Time-temperature curves for compartments with different bounding surfaces.

The designer simply has to match the physical characteristics of the actual compartment to be modeled with the closest curve to establish a fire time-temperature curve. The curves shown in Fig. 2.8 are currently the basis for the design of fire resistance requirements in Sweden and form the basis for the Eurocode time-temperature curves. Some of the assumptions of the model are as follows [28]:

- The mass burning rate is $330A\sqrt{h}$ kg/h.
- The curves are based on wood crib fires with the energy content of wood = 18,800 kJ/kg.
- The decay phase assumes a rate of cooling of 10°C/min.
- The fire is assumed to be ventilation controlled.

Furthermore, the curves shown in Fig. 2.8 are based on a predefined type A compartment, which is a compartment with surrounding structures that have thermal properties similar to concrete, brick, and lightweight concrete, where the thermal conductivity is $\sqrt{k\rho c_p}$ = 1,160 J/m² s$^{1/2}$ K. Multipliers are provided for other compartment types that normally might be found in buildings.

Babrauskas and Williamson Model

This theoretical model is also based on the heat balance equation for the compartment and some of the original assumptions developed by Kawagoe [24,25]. It diverges from Kawagoe's work in that it treats the burning rate in a theoretical manner rather than an empirical manner, as is done by Kawagoe and by Pettersson and colleagues, presenting a final heat-balance equation:

$$h''_c - (\dot{m}_{\text{air}} + \dot{m}_p) \int_{298}^{T_t} c_p d = TA_t \sigma \left[\frac{T_t^4 - T_w^4}{(1/\varepsilon_f) + (1/\varepsilon_w) - 1} \right]$$

$$+ A_t h_c (T_t - T_w) + A_v \sigma (T_t - T_0^4) \qquad (2.25)$$

where the combustion enthalpy $\dot{h}_c$, infiltration airflow rate $\dot{m}_{\text{air}}$, and mass flow rate of the products of combustion $\dot{m}_p$ are defined in Babrauskas and Williamson [29]. Specifically, the model discusses the difficulty in defining the actual combustion efficiency of the compartment fire and proposes that the enthalpy release rate is the lesser of the potential enthalpy of gas released from the fuel or the enthalpy release rate from perfect burning. This is different from Kawagoe's suggestion, which coupled the mass burning rate with the ventilation factor, as shown in Eq. (2.21). Furthermore, the model offers a comparison of the pyrolysis rates of plastic fuels with those of wood fuels, and the difference is significant. Given the proliferation of plastics in typical residential, commercial, or institutional occupancies, this is cause for concern. Unfortunately, the model does not specifically address the actual impact of these issues on the results of calculated time-temperature curves based on Kawagoe's burning rate. The models developed by Pettersson and colleagues and Babrauskas and Williamson are very mathematically involved and do not lend themselves to reasonable computation times required for a day-to-day structural engineering practice. In addition, in all the models just described, it is necessary to determine some of "uncertain" parameters. Drysdale [27] suggests that owing to the uncertainties associated with compartment fires, Lie's approach [30] may be used to obtain a "rough sketch" of the compartment fire time-temperature curve. Lie's approach is to eliminate the need to determine these parameters, suggesting that it is not important to predict a time-temperature curve that is representative of the fire scenario but rather a time-temperature curve that with reasonable probability will not be exceeded. Lie also suggests that the importance of correctly modeling the decay period of the fire is minor because the impact of the decay phase on the maximum room temperature is small, as determined by Kawagoe. Lie proposed that "a characteristic temperature-time curve that, with reasonable likelihood, will not be exceeded during the lifetime of the building" [30] should be developed. Based on the theoretical approach developed by Kawagoe [24,25], Lie developed an expression that approximately described the theoretical curves for any value of opening factors. This development was based on two distinct compartment types: those constructed from light materials and those constructed from heavy materials. The defining density is 1,600 kg/m³. Lie argues that owing to the lack of sensitivity of the heat-balance model to small changes in this variable, his approach represents a reasonable simplification. The expression that Lie proposes is as follows:

$$T_t = 250(10F_v)^{0.1/F_v^{0.3}} e^{-F_v^2 t}[3(1-e^{-0.6t})-(1-e^{-3t})$$

$$+ 4(1-e^{-12t})]+C\left(\frac{600}{F_v}\right)^{0.5} \tag{2.26}$$

where
$$F_v = \frac{A_v(H_v)^{1/2}}{A_t} \tag{2.27}$$

and C is a constant taking into account the properties of the boundary material ($C = 0$ for heavy materials and $C = 1$ for light materials).

Figures 2.9 and 2.10 compare this expression to Kawagoe's theoretical model for various opening factors.

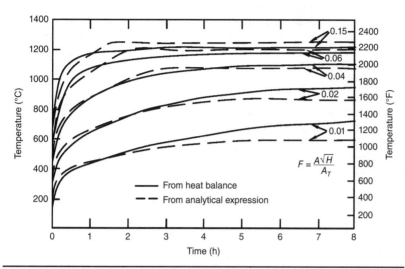

FIGURE 2.9   Theoretical vs. experimental time-temperature curves: heavyweight construction [30].

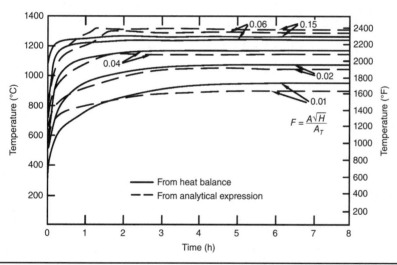

FIGURE 2.10   Theoretical vs. experimental time-temperature curves: lightweight construction [30].

To model the decay phase of the fire that must be applied to the curves generated by the primary expression, Lie proposed the following:

$$T_t = -600\left(\frac{t}{\tau} - 1\right) + T_\tau \tag{2.28}$$

where

$$\tau = \frac{L_t A_t}{330 A_v (H_v)^{1/2}} \tag{2.29}$$

Recognizing that this equation is based on the expression for burning rate developed by Kawagoe, it can be shown that the expression proposed by Lie reasonably approximates both experimental data and the Swedish approach. The benefit is that the expression proposed by Lie is simplistic enough that it may be applied to real-life day-to-day structural design calculations with a hand calculator or spreadsheet. It is important to remember that Lie's expression is based on curves developed with the heat-balance approach and that Lie has developed an expression that allows the designer to avoid the significant calculations necessary to perform a heat balance in order to develop a reasonable time-temperature curve for design purposes. One concern raised by Buchanan [3] is that Lie's curves are unrealistic for rooms with small openings because the calculated compartment temperatures are not sufficient for the occurrence of flashover.

Eurocode Model (EC1)

The parametric temperature-time curve in EC1 [4] is designed to predict the $T(t)$ function of postflashover compartment fires for any combination of fuel load, ventilation, and wall lining materials. The time-temperature curve is as follows (the parametric curve valid for compartments up to a floor area of 100 m² and compartment height of 4.5 m):

$$T_t = 20 + 1{,}325(1 - 0.324e^{-0.2t^*} - 0.204e^{-1.7t^*} - 0.472e^{-19t^*}) \quad (2.30)$$

where

$$t^* = t(\Gamma)$$

$$\Gamma = \frac{(F_v/0.04)^2}{(b/1160)^2} \quad (2.31)$$

The decay rates are

$$T_t = T_{max} - 625[t^* - t^*_{max}(x)] \qquad \text{for } t^*_{max} < 0.5 \quad (2.32)$$

$$T_t = T_{max} - 250(3 - t^*_{max})(t^* - t^*_{max}x) \qquad \text{for } 0.5 < t^*_{max} < 2.0 \quad (2.33)$$

$$T_t = T_{max} - 250(t^* - t^*_{max}x) \qquad \text{for } t^*_{max} > 2.0 \quad (2.34)$$

where

$$t^*_{max} = 0.2(10^{-3})\frac{L_{t,d}}{F_v}\Gamma \quad (2.35)$$

and

$$x = 1.0 \qquad \text{if } t^*_{max} > t_{lim} \quad \text{or}$$

$$x = t_{lim}\frac{\Gamma}{t^*_{max}} \qquad \text{if } t^*_{max} = t_{lim} \quad (2.36)$$

where t_{lim} = 25 minutes for a slow-growth fire

t_{lim} = 20 minutes for a medium-growth fire

t_{lim} = 15 minutes for a fast-growth fire

The duration of the fire is determined by the fire load in Eq. (2.37):

$$t^* = 0.13(10^{-3})q_{t,d}\Gamma / K_v \qquad (2.37)$$

where

$$q_{t,d} = q_{f,d}A_f / A_t \qquad (2.38)$$

and

$$50 \leq q_{t,d} \leq 1,000$$

$q_{t,d}$ is the design value of the fire-load density (MJ/m^2) related to the surface area A_t (m^2) of the enclosure, and $q_{f,d}$ is the design value of the fire load density related to the surface area of the floor (MJ/m^2).

SFPE Standard, 2011 [23]

SFPE standard recommends two different models for the parametric fire analysis. These methods allow calculating the thermal boundary conditions to a structural system resulting from a fully developed fire. The basis for selecting these methods is clearly identified in the commentary to Sec. 5.1.2:

The methods presented in this Standard for computing a time-temperature profile in an enclosure were selected from a group of twenty-three different methods or method variations. All methods considered had been documented in published material and did not involve the use of computer simulations. The methods included simplistic approaches such as a constant temperature exposure, correlations of particular data sets, generalized parametric approaches, and correlations of computer generated data. The selection process involved assessing the performance of all twenty-three methods against a database containing about 130 fully-developed single compartment fire tests. The database was compiled largely from four decades of published enclosure fire test results. Most of the tests were conducted using full-scale compartments that represented a wide range of parameters that would have some influence on the time-temperature development. Such parameters included the absolute enclosure dimensions, the absolute opening dimensions, the number and location of the openings, the type of boundary materials, the ventilation factor, and the type of fuel burning. The dominant criterion for selecting a method or methods was the need to produce reliably conservative results when applied as intended. Other factors considered included the accuracy, the correlation factor, the prediction trend, the ease of use, the generality, and the method's technical basis. To objectively assess the performance of the methods against the data, a set of four metrics were developed that reflect both raw time-temperature predictions and the manner in which the time-temperature predictions would be used. These metrics were as follows:

- The time-temperature profile;
- The thickness of bare steel required to prevent it from reaching 538°C, a common threshold temperature for structural design;
- The thickness of concrete required to prevent a steel plate from reaching 538°C;
- The thickness of a mineral-based insulation required to prevent a steel plate from reaching 538°C.

The time-temperature profile comparisons were based directly on the measured test data and the correlation predictions. In contrast, the thickness comparisons were effectively integrated average heat flux comparisons for different types of building materials. The integrated average comparisons involved the use of an iterative heat transfer model to compute and compare a material thickness given the measured test data time-temperature exposure profile v. a thickness computed using the predicted time-temperature exposure profile. The integrated average computations used the thermal radiation and convection boundary conditions as recommended in this Standard though it is noted that because the computations were relative to one another, there was not a great deal of sensitivity to the specific boundary conditions assumed.

A statistical analysis was also conducted on the comparison results for each metric to quantify trends, accuracy, and correlation. It was clear from both a direct comparison of the results and the statistical analysis that there was a great deal of scatter and therefore marked uncertainty in the predictive capability of nearly all methods. Some methods had a high correlation factor but were not consistently conservative even to within a reasonable percentage; some methods appeared only to perform well for a small subset of data and generate very conservative or very non-conservative results when outside this range. Still, others were generally conservative for all tests but were sometimes overly conservative predictions.

Given these observations, it was deemed prudent to select two methods that were nearly always conservative while recognizing that there may be situations where the predictions are substantially over-conservative. An important aspect of this selection is the recognition by the user that the results are defined with two parameters: the temperature and the time. A raw comparison between the predicted time-temperature and measured time-temperature profiles readily shows deviations in one or both parameters for any method. However, there are many paths available to reach a correct result for the integrated average metrics. Because the information in this Standard is intended for use as input for defining the boundary conditions that will subsequently be used to compute a thermal response, a greater emphasis was placed on the integrated average metrics, which temper the degree of over-conservativeness.

Thus, while the methods presented in this Standard can have a tendency to over-predict the compartment temperature, the time and temperature taken together tend to produce more reasonable results when used as input for developing the boundary conditions. In short, given the large number of tests considered, there is a reasonable assurance that the methodology selected for this Standard will produce a conservative result when used as boundary condition input data but that the predicted temperature-temperature profile may not necessarily be the true time-temperature profile for the assumed fuel load and ventilation conditions.

First, the SFPE standard recognizes that fully developed fire exposures can arise "from a fully-developed fire within an enclosure and exposures from a localized fire involving a concentrated fuel load that is not affected by an enclosure." Second, the SFPE Standard recognizes that fully developed fire exposures provide the most critical information regarding input to thermal response and structural design calculations that will include the SFL as part of structural design load combinations [2]. Third, the standard recognizes that the suggested methods shall be "consistently conservative" on the one hand and reasonably heuristically simplified on the other.

Method 1: Constant Compartment Temperature Method (See Secs. 5.3 and 4.3.2.4.2 of the SFPE Standard [23])
In this case (see Fig. 2.11),

1. The "compartment temperature shall be 1,200°C for all times after ignition but before the burnout time t_b" and "the growth time shall not be included in the burnout time."

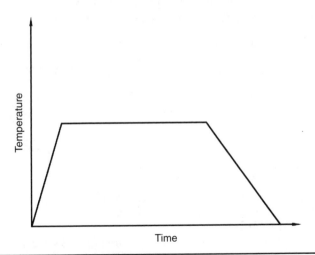

FIGURE 2.11 Temperature-time curve.

2. "The temperature shall decrease to ambient conditions at a constant rate of 7°C/min after the burnout time has been reached." The decay period is 2.86 hours in all severity cases.

3. "The burnout time is the length of time the compartment temperature is equal to 1,200°C. It shall be computed using the following equation:

$$\tau = \frac{EA_f}{90A_0(H_0)^{1/2}} \tag{2.39}$$

where τ = burnout time for the enclosure fire (minutes)
     E = energy load per unit floor area of the enclosure (MJ/m²)
     A_f = floor area of the enclosure over which combustibles are present (m²)
     A_0 = opening area (m²)
     H_0 = opening height (m)

This is a very simple method, but it has an unclear point regarding structural engineering applications. The standard probably implies that the temperature-time load is applied statically to the structural system. However, since the "growth time" is negligible, one can assume that the temperature load should be applied as an impact load. Therefore, the dynamic coefficient in this case (the structural response) is equal to 2. For detailed discussion in this case (fires following an earthquake in a seismologic high zones), see Chap. 7.

Method 2: Tanaka (Refined) Method (See Secs. 5.4 and 4.3.2.4.1 of the SFPE Standard [23])

The temperature-time function for a ventilation-controlled postflashover transient fire is presented by the following equations:

$$T = \beta_{F,1}(2.50 + \beta_{F,1})T_\infty + T_\infty \quad \text{for } \beta_{F,1} \leq 1.0 \tag{2.40}$$

$$T = \beta_{F,1}(4.50 - \beta_{F,1})T_\infty + T_\infty \quad \text{for } \beta_{F,1} > 1.0 \tag{2.41}$$

where

$$\beta_{F,1} = \left(\frac{A_0\sqrt{H_0}}{A}\right)^{1/3} \left(\frac{t}{k\rho c}\right)^{1/6} \tag{2.42}$$

with T = temperature (K)
     T_∞ = 300 K
     A_0 = area of openings (m²)
     A = total surface area of room, excluding opening (m²)
     H_0 = height of opening (m)
     t = time (s)
     k = thermal conductivity of enclosure lining (kW/m K)
     c = specific heat of enclosure lining (kJ/kg K)
     ρ = density of enclosure lining (kg/m³)

The Tanaka calculations use Kawagoe and Sekine's method [24] of predicting the mass burning rate as follows:

$$\dot{m} = 0.1 A_0 \sqrt{H_0} \ (kg/s) \tag{2.43}$$

where A_0 = area of opening (m²)
H_0 = height of opening (m)

Let's now analyze Eqs. (2.40) and (2.41). Substituting $x = \beta_{F,1}$, let's rewrite these equations:

$$T = x(2.50 + x) \qquad \text{if } x < 1 \tag{2.44}$$

and

$$T = x(4.50 - x) \qquad \text{if } x > 1 \tag{2.45}$$

Equations (2.44) and (2.45) represent two quadratic parabolas. The first one has a positive curvature (second derivative is positive), and the second one has the negative curvature (second derivative is negative). This means that if $x > 1$, then the first parabola represents the fire growth period, the point $x = 1$ is the flashover point, and the second parabola represents the burnout time (fully developed fire) and the decay period. The curve has a breaking point at $x = 1$ (the derivative on the left side is not equal to the derivative on the right side), which contradicts the major physical concept of nonsteady combustion (the total curve should be presented by an analytical function) [14].

The total duration of fire (growth period time plus burnout time plus decay) in this case can be calculated from $x = \beta_{F,1} = 4.5$ and Eq. (2.40). The maximum temperature in this case is somewhat higher than in method 1 ($T = 1,200°C = $ const.). However, the total energy released is probably comparable because the total duration of fire is different.

Localized Fire Exposure

The thermal boundary conditions from localized fire exposures may arise from a concentrated fuel load, and they are presented in terms of an incident heat flux at a specific location for an exposure duration. The localized fire could have the following configurations (see Sec. 6.2.2 of the SFPE standard [23]):

1. Unconfined fire exposures, fires beneath a ceiling (see Sec. 6.5.2)

2. Fires adjacent to a flat wall with or without a ceiling and with or without a gap between the wall and burning fuel package (see Sec. 6.5.3)

3. Fires adjacent to a corner with or without a ceiling and with or without a gap between the wall and burning fuel package (see Sec. 6.5.4)

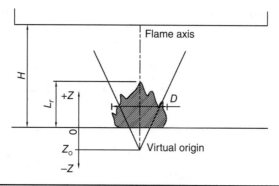

Figure 2.12   Schematic diagram of a localized fire.

The effective diameter of the fuel package should be calculated based on Eq. (6.5.1) of the SFPE standard [23]:

$$D_{eff} = \sqrt{\frac{4LW}{\pi}} \qquad (2.46)$$

where D_{eff} = effective fuel package diameter (m)
$\quad\quad\ L$ = length of the fuel package (m)
$\quad\quad\ W$ = width of the fuel package (m)

Figure 2.12 shows a schematic diagram of a localized fire.

The flame temperature changes with height. It is roughly constant in the continuous flame region and represents the mean flame temperature. The temperature decreases sharply above the flames as an increasing amount of ambient air is entrained into the plume. The SFPE standard provides a design formula to calculate the maximum flame height and the maximum incident heat-flux data for each case of a localized fire configuration just mentioned.

Case 1: Unconfined Fire Exposures, Fires Beneath a Ceiling

The maximum flame height above the reference point can be calculated as

$$F_h = H - 1.02D_{eff} + 0.23(\dot{q}''A_f)^{0.4} \qquad (2.47)$$

where F_h is the maximum flame height above the reference point (m), H is the fuel package height above the reference point (m), D_{eff} is the effective fire diameter determined from Eq. (6.5.1) (m) [23], $\dot{q}''$ is the heat release rate per unit surface area of the fuel package material (kW/m²), A_f is the burning surface area of the fuel package (m²).

1. If $H \leq F_h$, then heat flux will be 120 kW/m², and the fire duration can be calculated from:

$$\tau = \frac{M(\Delta H_c)}{\left(\dfrac{T_H + 1.02D_{\text{eff}}}{0.23}\right)^{5/2}} \qquad (2.48)$$

where τ is the fire duration (s), M is the mass of combustible material available for combustion (kg), ΔH_c is the effective heat of combustion (kJ/kg), T_H is the height of the exposed object or surface for which the boundary condition is computed (m), and D_{eff} is the effective fire diameter determined from Eq. (6.5.1) (m) [23].

2. If $H > F_h > 0.5H$, then heat flux will be 20 kW/m², and the fire duration can be calculated from:

$$\tau = \frac{M(\Delta H_c)}{\left[\dfrac{(T_H/2) + 1.02D_{\text{eff}}}{0.23}\right]^{5/2}} \qquad (2.49)$$

where τ is the fire duration (s), M is the mass of combustible material available for combustion (kg), ΔH_c is the effective heat of combustion (kJ/kg), T_H is the height of the exposed object or surface for which the boundary condition is computed (m), and D_{eff} is the effective fire diameter determined from Eq. (6.5.1) (m) [23].

3. If $H > 2F_h$, the thermal boundary condition cannot be determined using the methodologies presented in Eq. (6.5.2) [23].

Case 2: Fire Adjacent to Wall, With or Without Ceiling Effects

The thermal boundary condition for a flat vertical structural element surface adjacent to the fire can be determined as follows:

1. The boundary condition heat flux is 120 kW/m²

2. The fire duration can be determined from

$$\tau = \frac{M(\Delta H_c)}{A_f \dot{q}''} \qquad (2.50)$$

where τ is the fire duration (s), M is the mass of combustible material available for combustion (kg), A_f is the burning surface area of the fuel package (m²), and $\dot{q}''$ is the heat release rate per unit surface area of the fuel package material (kW/m²).

Case 3: Fire Adjacent to Corner, With or Without Ceiling Effects

The maximum flame height can be determined from

$$F_h = H + 0.03D_{\text{eff}}\left(\frac{\dot{q}'' A_f}{D_{\text{eff}}^{5/2}}\right)^{1/2} \qquad (2.51)$$

where F_h is the maximum flame height above the reference point (m), H is the fuel package height (m), D_{eff} is the effective fire diameter determined from Eq. (1) (m) [23], $\dot{q}''$ is the heat release rate per unit surface area of the fuel package material (kW/m²), and A_f is the burning surface area of the fuel package (m²).

1. If $F_h > T_{H'}$ then the heat flux will be 120 kW/m², and the fire duration can be calculated from

$$\tau = \frac{M(\Delta H_c)}{\dfrac{(T_H - H)^2}{(0.03D_{eff})^2} D_{eff}^{5/2}} \qquad (2.52)$$

where τ is the fire duration (s), M is the mass of combustible material available for combustion (kg), ΔH_c is the effective heat of combustion (kJ/kg), T_H is the height of the exposed object or surface for which the boundary condition is computed (m), H is the fuel package height (m), and D_{eff} is the effective fire diameter (m).

2. If $T_H > F_h > 0.5T_{H'}$ then the heat flux will be 20 kW/m², and the fire duration can be calculated from

$$\tau = \frac{M(\Delta H_c)}{\dfrac{(0.5T_H - H)^2}{(0.03D_{eff})^2} D_{eff}^{5/2}} \qquad (2.53)$$

where τ is the fire duration (s), M is the mass of combustible material available for combustion (kg), ΔH_c is the effective heat of combustion (kJ/kg), T_H is the height of the exposed object or surface for which the boundary condition is computed (m), H is the fuel package height (m), and D_{eff} is the effective fire diameter (m).

3. If $F_h < 0.5T_{H'}$ then the thermal boundary condition cannot be determined using the methodologies presented in Sec. 6.5.4 from [23].

Localized Fires Approach: BSEN 1991-1-2 (2002)

BSEN 1991-1-2 provides a simple approach for determining the thermal action of localized fires in Annex C. Depending on the height of the fire flame relative to the ceiling of the compartment, a *localized fire* can be defined as either a small fire or a big flame. For a small fire, a design formula has been provided to calculate the temperature in the plume at heights along the vertical flame axis. For a big fire, some simple steps have been developed to give the heat flux received by the fire exposed surfaces at the level of the ceiling.

In a localized fire, as shown in Fig. 2.12, the highest temperature is at the vertical flame axis. The temperature decreases sharply above the flames because an increasing amount of ambient air is entrained into the plume. BSEN 1991-1-2 provides a design formula to calculate the temperature in the plume of a small, localized fire. The maximum flame height F_h of the fire is given by

$$F_h = -1.02D + 0.0148Q^{2/5} \tag{2.54}$$

where D is the diameter of the fire (m), and Q is the rate of heat release by the fire (W).

If the fire does not impinge on the compartment ceiling when $F_h < H$, the temperature $T(z)$ in the plume along the symmetric vertical flame axis is given by

$$T(z) = 20 + 0.25Q_c^{2/3}(z - z_0)^{-5/3} \leq 900 \tag{2.55}$$

with

$$z_0 = -1.02D + 5.24(10^{-3})Q^{2/5} \tag{2.56}$$

where Q_c is the convective part of the rate of heat release (W), with $Q_c = 0.8\,Q$ by default, z is the height along the flame axis (m), and z_0 is the virtual origin of the axis (m).

The virtual origin z_0 depends on the diameter of the fire D and the rate of heat release Q. This empirical equation has been derived from experimental data. The value of z_0 may be negative and located beneath the fuel source, indicating that the area of the fuel source is large compared with the energy being released over that area (see Fig. 2.12). For fire sources where the fuel releases high energy over a small area, z_0 may be positive and located above the fuel source.

When a localized fire becomes large enough with $L_f \geq H$ (see Fig. 2.13), the fire flame will impinge on the ceiling of the compartment. The ceiling surface will cause the flame to turn and move horizontally beneath the ceiling. Figure 2.13 is a schematic diagram of a localized fire impacting on a ceiling, with the ceiling jet flowing beneath an unconfined ceiling. As the ceiling jet moves horizontally outward, it loses heat to the cooler ambient air being entrained into the flow, as well as the heat transfer to the ceiling. Generally, the maximum temperature occurs relatively close to the ceiling.

BSEN 1991-1-2 provides design formulas only for determining the heat flux received by the surface area at the ceiling level but not for calculating the ceiling jet temperatures. Simple approaches for determining ceiling jet temperatures will be discussed briefly later.

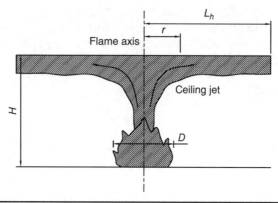

FIGURE 2.13   Localized fire impacting on the ceiling of a compartment.

Considering a localized fire impacting a ceiling of a compartment as shown in Fig. 2.13 ($L_f \geq H$), the horizontal flame length L_h is given by

$$L_h = [2.9H(Q_H^*)^{0.33}] - H \tag{2.57}$$

where L_h is the horizontal flame length as given by Eq. (2.57) [m], H is the distance between the fire source and the ceiling (m) [23], Q_H^* is a nondimensional rate of heat release, and Q is the rate of heat release of the fire (W).

The heat flux $\dot{h}$(W/m²) received by the fire exposed unit surface area at a ceiling level at a distance r from the flame axis is given by

$$\dot{h} = \begin{cases} 100,000y & \text{for } y < 0.3 \\ 136,300 \text{ to } 121,000y & \text{for } 0.3 < y < 1.0 \\ 15,000y^{-3.7} & \text{for } y > 1.0 \end{cases} \tag{2.58}$$

with

$$y = \frac{r + H + z'}{L_h + H + z'} \tag{2.59}$$

where r is the horizontal distance from the vertical flame axis to the point along the ceiling where the thermal flux is calculated (m), z' is the vertical position of the virtual heat source as given by Eq. (2.58) (m), and D is the diameter of the fire (m).

The virtual position of the virtual heat source z' is given by

$$z' = \begin{cases} 2.4D(Q_D^{*2/5} - Q_D^{*2/3}) & \text{for } Q_D^* < 1.0 \\ 2.4D(1.0 - Q_D^{*2/5}) & \text{for } Q_D^* \geq 1.0 \end{cases} \tag{2.60}$$

with

$$Q_D^* = \frac{Q}{1.11 \times 10^6 \cdot D^{2.5}} \tag{2.61}$$

The net heat flux $\dot{h}_{net}$ received by the fire-exposed unit surface at the level of the ceiling is given by

$$\dot{h}_{net} = \dot{h} - \alpha_c(T_m - 20) - \Phi\varepsilon_m\varepsilon_f\sigma[(T_m + 273)^4 - (293)^4] \qquad (2.62)$$

where α_c is the coefficient of heat transfer by convection (W/m² K), ε_f is the emissivity of the fire, ε_m is the surface emissivity of the member, Φ is the configuration factor, T_m is the surface temperature of the member (°C), σ is the Boltzmann constant (= 5.67×10^{-8} W/m² K⁴), and $\dot{h}$ is the heat flux received by the fire-exposed unit surface area at the level of the ceiling, as given by Eq. (2.56).

The following empirical equations are based on experimental data collected for fuels such as wood and plastic pallets, cardboard boxes, plastic products in cardboard boxes, and liquids with heat release rates ranging from 668 kW to 98 MW under ceiling heights from 4.6 to 15.5 m. The maximum temperature T (°C) of a ceiling jet is given by

$$T - T_\infty = \begin{cases} \dfrac{16.9Q^{2/3}}{H^{5/3}} & \text{for } r/H \le 0.18 \\ \dfrac{5.38(Q/r)^{2/3}}{H} & \text{for } r/H \le 0.18 \end{cases} \qquad (2.63)$$

Discussions and Recommendations

All approaches in major standards (Eurocode, SFPE standard, and Lie's formulas) are quite different; therefore, the results are also not identical. It is even difficult to compare them and choose the most conservative approach for a number of reasons:

1. All the results (temperature-time relationships) are not presented in dimensionless forms and different sets of parameters have been used for different approaches. For example: Lie's investigation takes into consideration only two construction types of a compartment (made out of light and heavy materials); SFPE recommends the type of temperature-time function very different from Eurocode and Lie (see Fig. 2.14).

2. It is a common argument (see, e.g., SFPE standard) that the main goal of a parametric method is to provide the maximum gas temperature and the duration of a fire (but not the curve itself). In this case, it is possible to compare the different results of different approaches. However, one can argue that the total energy released (the area under the temperature-time curve up to the maximum temperature) is also very

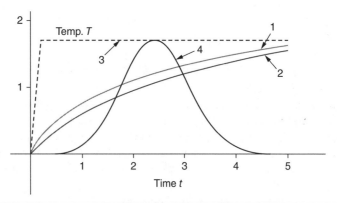

FIGURE 2.14   Temperature-time curves comparison (1–Eurocode; 2–Lie; 3–SFPE standard case 1; and 4–SFPE standard case 2).

important for structural engineering analyses (possible impact or dynamic action on a structural system).

3. The fully developed fire (the most important case in structural analysis and design) is defined as a stage of fire that follows after the flashover point (the maximum of the heat release rate, the second derivative of a temperature-time function is zero). However, since the temperature-time functions in all studies are not presented by dimensionless variables and parameters, one cannot differentiate and analyze Eqs. (2.28) and (2.32), for example.

4. Any structural design load such as wind or seismic load has a very important component: structural system response. For example, it is almost impossible to choose a priori what the most critical fire scenario is: higher maximum temperature with the shorter duration or smaller maximum temperature with larger duration of fire? The answer to this question depends on the reaction of the particular structural system. The high values of HRR could have a bigger effect on long-span structures and tall buildings than on a short-span structural elements and mid-rise buildings. Unfortunately, the parametric methods do not include any structural response coefficients (they have thermal response coefficients, e.g., thermal inertia).

5. The results of parametric methods should be presented in a simple form that can be used by a structural engineer on a day-to-day basis. They also should be coordinated with a proper structural design load combination system (i.e., dead load, live load, wind load, etc.).

References

1. Arbed, S. A., *Natural Fire Safety Concept*. ECSC, Luxembourg, 2001.
2. ASCE. *Minimum Design Loads for Buildings and Other Structures*, ASCE 7-05. ASCE, New York, 2005.
3. Buchanan, Andrew H. *Structural Design for Fire Safety*, Wiley, Hoboken, NJ, p. 91, 2001.
4. Center for Enabling New Technologies (CEN). Actions on Structures, Parts 1–2— Actions on Structures Exposed to Fire, EN 1991-1-2, EUROPEAN STANDARD NORME (EUROCODE 1). CEN Central Secretariat, Brussels, 2002.
5. Kawagoe, Kunio, and Sekine, Takashi. *Estimation of Fire Temperature-Time Curve in Rooms*. Building Research Institute, Ministry of Construction, Japanese Government, Tokyo, June 1963.
6. British Standards Institution (BSI). *Fire Tests on Building Materials and Structures*, BS 476, Parts 1 to 23. BSI, London, 1987.
7. British Standards Institution (BSI). "Structural Steelwork for Use in Building," Part 8, *Code of Practice for Fire Design*, BS 5950-8. BSI, London, 1990.
8. Society of Fire Protection Engineers (SFPE). *Engineering Guide: Fire Exposures to Structural Elements*. SFPE, Bethesda, MD, 2004.
9. Magnusson, S. E., and Thelandersson S., *Temperature-Time Curves of Complete Process of Fire Development: Theoretical Study of Wood Fuel Fires in Enclosed Spaces*. Civil Engineering and Building Construction Series 65, Acta Polytechnica Scandinavica, Boras, Sweden, 1970.
10. National Institute of Standards and Technology (NIST). "Fire Resistance Test for Performance-Based Fire Design of Buildings," GCR 07-910. Final Report. NIST, Gaithersburg, MD, June 2007.
11. National Institute of Standards and Technology (NIST). *Fire Dynamics Simulator*, Version 5 (Special Publication 1018-5). *Technical Reference Guide, Vol. 1: Mathematical Model*. NIST, Gaithersburg, MD, 2008.
12. Babrauskas V., "Temperatures in Flames and Fires." Fire Science and Technology, Inc., written April 28, 1997; Issaquah, WA, revised February 25, 2006.
13. Evans, L. C., "An Introduction to Mathematical Optimal Control Theory," Version 0.2. Department of Mathematics, University of California, Berkeley, 1983.
14. Frank-Kamenestkii, D. A., *Diffusion and Heat Transfer in Chemical Kinetics*. Plenum Press, New York, 1969.
15. Ingberg, S. H., "Tests of the Severity of Building Fires." *NFPA Quarterly* 22(1): 43–61, 1928.
16. Law, M., "Review of Formulae for T-Equivalent," in Y. Hasemi (ed.), *Fire Safety Science: Proceedings of the Fifth International Symposium, March 3–7, 1997, Melbourne, Australia*. International Association for Fire Safety Science, Boston, MA, 1997, pp. 985–996.
17. Ingberg, S. H., "Fire Resistance Requirements in Building Codes." *NFPA Quarterly* 23(2):153–162, 1929.
18. Pettersson, O., et al., *Fire Engineering Design of Steel Structures (Publication 50)*. Swedish Institute of Steel Construction, 1976.
19. Magnusson, S. E., and Thelandersson, S., *Temperature-Time Curves of Complete Process of Fire Development in Enclosed Spaces*. Acts Polytechnica Scandinavia, Boras, Sweden, 1970.
20. Harmathy, T. Z., "On the Equivalent Fire Exposure." *Ontario Fire and Materials* 11(2):95–104, 1987.
21. Harmathy, T. Z., and Mehaffey, J. R., "Normalized Heat Load: A Key Parameter in Fire Safety Design." *Ontario Fire and Materials* 6(1): 27–31, 1982.
22. European Committee for Standardization (EC1). *Eurocode 1:Actions on Structures*, ENV 1991, Part 1–2: General Actions—Actions on Structures Exposed to Fire. ECI, 2002.
23. Society of Fire Protection Engineers (SFPE). *Eng. Guide Standard on Calculating Fire Exposure to Structures*. SFPE, Bethesda, MD, 2011.

24. Kawagoe, Kunio, and Sekine, Takashi, *Estimation of Fire Temperature-Time Curve in Rooms*. Building Research Institute, Ministry of Construction, Japanese Government, Tokyo, June 1963.
25. Kawagoe, Kunio, *Estimation of Fire Temperature-Time Curve in Rooms*. Third Report, Building Research Institute, Ministry of Construction, Japanese Government, Tokyo, October 1967.
26. Lienhard, J. H., IV and Lienhard, J. H., V, *Heat Transfer Textbook*, 3rd ed. Phlogiston Press, Cambridge, MA, 2008.
27. Drysdale, Dougal, *An Introduction to Fire Dynamics*, Bookcraft, London, 1985.
28. Pettersson, O., et al., *Fire Engineering Design of Steel Structures (Publication 50)*. Swedish Institute of Steel Construction, 1976.
29. Babrauskas V., and Williamson, R., *Post-flashover Compartment Fires: Basis of a Theoretical Model*. Heydon & Son, London, 1978.
30. Lie, T. T., "Characteristic Temperature Curves for Various Fire Severities." *Fire Technology* 10(4):315–326, 1974.

CHAPTER 3

Structural Fire Load and Computer Modeling

3.1 Introduction

The structural fire load (SFL, temperature-time relationship of hot gases) can be obtained analytically by using computer models of fire development. The simplest model is a one-zone model for postflashover fires, in which the conditions within the compartment are assumed to be uniformly distributed in space and represented by a single temperature-time function. The parametric models discussed in Chap. 2 are the one-zone models. In general, zone models are simple computer models that divide the fire compartments under consideration into separate zones, where the conditions in each zone are assumed to be uniformly distributed in space. Analytical zone models for predicting fire behavior has been evolving since the 1960s. Since then, the completeness of the models has gone through major development to multizones and multicompartments for modeling localized and preflashover fires. The zone models also model the fire compartments in more detail than the parametric and time-equivalence methods. The geometry of compartments, as well as the dimensions and locations of openings, can be modeled easily.

Zone modeling is the most common type of physically based fire model. It is a deterministic model. It solves the conservation equations for distinct and relatively large regions. Its main characteristic is that it models the room into two distinct regions: one hot upper layer and a cooler layer below. The model estimates the conditions for each layer as a function of time only. There are many zone modeling packages now available on a market. Below is a summary of current zone models available from the Society of Fire Protection Engineers (SFPE) [1]:

CFAST

Name	CFAST (Consolidated Model of Fire Growth and Smoke Transport)
Summary	Upgrade of FAST program that incorporates numerical solution techniques originally implemented in CCFM program
Determines	A multiroom model that predicts fire development within a structure resulting from a user-specified fire
Inputs	Geometric data, thermophysical properties, fire mass loss, generation rates of combustion
Outputs	Temperature, thickness, and species concentration of hot and cool layers, surface temperatures, mass flow rates, and heat transfers
Advantages	Can accommodate up to 30 compartments with multiple openings between compartments and to the outside; also includes mechanical ventilation, ceiling jet algorithm, capability of multiple fires, heat transfer to targets, detection and suppression systems, and flame spread model

FIRST

Name	FIRST (First Software, First Computer Systems Ltd.)
Summary	Predicts the development of a fire and resulting conditions within a room given a user-specific fire or user-specified ignition
Determines	Predicts the heating and possible ignition of up to three targets
Inputs	Geometric data, thermophysical properties, generation rate of soot and other species; input fire as mass loss rate or fuel properties
Outputs	Temperature, thickness, species concentration of hot and cool layers, surface temperatures, heat transfers, and mass flow rates
Advantages	Can be used for a variety of building types

ASET

Name	ASET (Available Safe Egress Time)
Summary	Calculates the temperature and position of the hot upper layer in a single room with closed doors and windows

Determines	Time to the onset of hazardous conditions
Inputs	Heat-loss fractions, height of fuel above ground, criteria for hazardous detection, ceiling height, floor area, heat release rate (HHR), species generation rate
Outputs	Temperature, layer thickness, species concentration (function of time)
Advantages	Can examine multiple cases in a single run
Limitations	Specific room only

COMPBRN III

Name	COMPBRN III (Computer Code for Modeling Compartment Fires)
Summary	For nuclear power industry, used with probabilistic analysis; assumes a relatively small fire in a large space or involving large fuel loads early during preflashover fire growth stage
Determines	Temperature profile of an element and the ignition point
Inputs	Geometric data, thermophysical properties, generation rate of soot and other species; input fire as mass loss rate or fuel properties
Outputs	HHR, temperature, depth of hot layer, mass burning rate, surface temperatures, heat flux
Advantages	Emphasis on the thermal response of elements and model simplicity
Limitations	Designed for nuclear-specific fires

COMPF2

Name	COMPF2 (Computer Program for Calculating Post Flashover Fire-Temperatures by Vytenis Babrauskas)
Summary	Calculates characteristics of a postflashover fire in a single building compartment based on fire-induced ventilation through a single door or window
Determines	Gas temperature, heat-flow terms, flow variables
Advantages	Used for performance design calculations, analysis of experimental data; can be used for a range of fuel types
Limitations	Very specific on compartment and fire type

BRANZFIRE

Name	BRANZFIRE (is a zone including flame spread options)
Summary	Model for predicting the fire environment in an enclosure resulting from a room corner fire involving walls and ceilings
Determines	Predicts the ignition, flame spread, heat released by wall and ceiling material; considers flame spread properties
Outputs	Layer height, species concentration, gas temperatures, visibility, wall temperatures, HRR
Limitations	Very specific in terms of fire location and building geometry

JET

Name	JET (Data Warehouse for FACT 2)
Summary	Two-zone single-compartment model where the compartment is enclosed by a combination of draft curtains and walls
Inputs	Fire characterized by time-dependent HHR, irradiative fraction, and fire diameter; thermal properties of ceiling
Outputs	Ceiling jet temperatures and velocity, link temperature, activation times
Limitations	Building-specific

EPETOOL

Name	EPETOOL [Early Power Estimator (EPE) tool]
Summary	Estimates the potential fire hazard in buildings based on relatively simple engineering equations
Determines	Addresses problems related to fire development in buildings and the resulting conditions and response of fire protection systems
Inputs	Geometric data, material of enclosure, fire description, parameters of detector systems
Outputs	Temperature and volume of hot layer, smoke flow, response of detectors, effects of available oxygen on combustion
Advantages	For both pre- and post-flashover fires

LAVENT

Name	LAVENT (Engineering Laboratory Software and Tools, NIST, U.S. Department of Commerce)

Summary	Simulates the environment and the response of sprinkler elements in compartment fires with draft curtains and ceiling vents
Determines	The heating of the fusible links, including the ceiling jet and hot layer
Inputs	Geometric data, thermophysical properties, fire elevation, HHR, fire diameter, vent details, activation temperature, and ambient temperature
Outputs	Temperature, height of hot layer, temperature of links, ceiling jet temperatures and velocities
Limitations	Maximum of 5 ceiling vents and 10 fusible links

One more computer zone model should be added to the list. This is the ozone model that was developed at the University of Liege, Belgium. Ozone, Version 2.2.0 [2], had been developed originally as part of a European Coal and Steel Community project entitled, *Natural Fire Safety Concept*. It created considerable interest in Europe, and it was suggested that it could replace the parametric temperature-time relationship in *Eurocode 1*. The parametric method in this case is done using the Probabilistic Fire Simulator, Version 2.1, developed at VTT Technical Research Centre of Finland [3]. Ozone belongs to the family of zone models. Ozone permits modeling of only one compartment. The main hypothesis in this case is that the compartment is divided in two zones in which temperature distribution is uniform in time. Transition from a two-zones model to a one-zone model is related to the notion of a fire growth curve, as shown in Fig. 3.1.

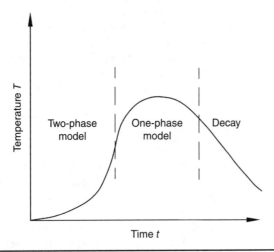

Figure 3.1 Typical fire growth curve.

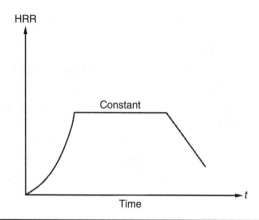

FIGURE 3.2   Heat release rate input.

Before flashover, during fire growth, tests have shown that it is more suitable to model two zones (one hot layer and one cool layer). However, after the flashover point, only one zone is needed. Transition from a two-zone model to a one-zone model is performed automatically by Ozone on the basis of criteria describing the occurrence of a fully developed fire. Before starting a simulation, Ozone sets up a theoretical heat release rate (HRR) curve with the model inputs (Fig. 3.2).

This software was tested against experimental data from fire tests conducted by CORUS Research, Development and Technology, Sweden Technology Centre. In general, the correlation between the fire test and predicted results was found to be weak. There were concerns about the theoretical background of the first model. One major concern was the use of a "design" HRR curve based on a t^2 growth phase, a constant release phase, and a linear descending branch after 70 percent of the fire load has been consumed. Further developments in the software have been made that solved the problems associated with using the software. However, dissimilarities between measured and predicted temperatures still may exist.

Of many two-zone models just listed, let's focus on the CFAST model [4].

3.2   Two-Zone Model: CFAST

The Consolidated Model of Fire Growth and Smoke Transport (CFAST) is a two-zone fire model used to calculate the evolving distribution of smoke, fire gases, and temperature throughout building compartments during a fire. The level of complexity (i.e., combustion behavior and fire development) increases from simple fire models (e.g., time-equivalence and parametric methods) to zone/field models. In CFAST, each compartment is divided into two gas layers. The gas temperature of each zone is assumed to be uniformly distributed in space and is represented by a temperature-time relationship. The advanced two-zone fire models normally are theoretical computer models that simulate the heat and mass-transfer process associated with a compartment fire. They can predict compartment gas temperatures in much more detail. Smoke movement and fire spread also may be taken into account. Zone models are simple computer models. They are valid for localized or preflashover fires. The input parameters for each of these models are quite different, with the advanced models requiring very detailed input data and the simple models requiring little input. The theoretical background of zone models is the conservation of mass and energy in fire compartments. These models take into account the HRR of combustible materials, fire plume, mass flow, and smoke movement and gas temperatures. They rely on some assumptions concerning the physics of fire behavior and smoke movement suggested by experimental observations of real fires in compartments. The upper layer represents the accumulation of smoke and pyrolysis beneath the ceiling. There is horizontal interface between the upper and lower layers. The air entrained by the fire plume from the lower layer into the upper layer is taken into account. The geometry of compartments, as well as the dimensions and locations of openings, can be modeled easily. The zone models require expertise in defining the correct input data and assessing the feasibility of the calculated results. A schematic diagram of a two-zone model is shown on Fig. 3.3. Similar to one-zone models, the two-zone models are based on solving the ordinary differential equations for the conservation of mass and energy in the compartment, but at a higher degree of complexity. The conservation of mass and energy need to be considered for individual zones, as well as the exchange of mass and energy between the different zones. The main interests are the evolution of the gas temperature and the thickness of the upper layer.

In real enclosure fires, a preflashover fire may develop into a postflashover fire under certain circumstances. Annex D of EN 1362-2 [5] lists two situations where a two-zone fire model may develop into a one-zone fire model. They are

- If the gas temperature of the upper layer is higher than 500°C
- If the upper layer is growing to cover 80 percent of the compartment height

Differential Equations for a Two-Layer Model

The differential equations used in CFAST [4] "... are derived using the conservation of mass, the conservation of energy (equivalently the first law of thermodynamics), the ideal gas law. These equations predict as functions of time quantities such as pressure,

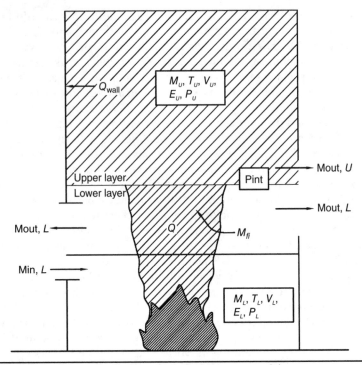

Figure 3.3 Schematic diagram of a typical two-zone model.

layer height and temperatures given the accumulation of mass and enthalpy in the two layers. The assumption of a zone model is that properties such as temperature can be approximated throughout a control volume by an average value." A compartment is divided into two control volumes. Each zone (control volume) is defined by 11 variables—mass, internal energy, density, temperature, and volume—denoted respectively by m_i, E_i, ρ_i, T_i, and V_i, where $i = L$ for the lower layer and $i = U$ for the upper layer. The compartment as a whole has the same (equalized) pressure P. The corresponding differential equations for each layer and equalized pressure are shown in Table 3.1.

These 11 variables are interconnected by the following seven constraints (counting twice, once for each layer):

$$\rho_i = \frac{m_i}{V_i} \qquad \text{(density)} \tag{3.1}$$

$$E_i = c_v m_i T_i \qquad \text{(internal energy)} \tag{3.2}$$

$$P = R\rho_i T_i \qquad \text{(Ideal gas law)} \tag{3.3}$$

$$V = V_L + V_U \qquad \text{(Total volume)} \tag{3.4}$$

Equation Type	Differential Equation
Ith layer mass	$\dfrac{dm_i}{dt} = \dot{m}_i$
Pressure	$\dfrac{dP}{dt} = \dfrac{\gamma - 1}{V}(\dot{h}_L + \dot{h}_U)$
Ith layer energy	$\dfrac{dE_i}{dt} = \dfrac{1}{\gamma}\left(\dot{h}_i + V_i\dfrac{dP}{dt}\right)$
Ith layer volume	$\dfrac{dV_i}{dt} = \dfrac{1}{\gamma P}\left[(\gamma - 1)\dot{h}_i - V_i\dfrac{dP}{dt}\right]$
Ith layer density	$\dfrac{d\rho_i}{dt} = -\dfrac{1}{c_pT_iV_i}\left[(\dot{h}_i - c_p\dot{m}_iT_i) - \dfrac{V_i}{\gamma - 1}\dfrac{dP}{dt}\right]$
Ith layer temperature	$\dfrac{dT_i}{dt} = \dfrac{1}{c_p\rho_iV_i}\left[(\dot{h}_i - c_p\dot{m}_iT_i) + V_i\dfrac{dP}{dt}\right]$

TABLE 3.1   Zone Model Equations

The first law of thermodynamics can be written now in differential form as

$$\frac{dE_i}{dt} + P\frac{dV_i}{dt} = \dot{h}_i \qquad \text{(enthalpy)} \tag{3.5}$$

Based on dimensional analysis, there are only four fundamental independent variables in this case: temperature, mass, length (volume), and time. Therefore, only four (out of 11) differential equations are needed to solve the problem. Obviously, one can have 330 combinations [C(11, 4) = 11!/4! (7!) = 330] of differential equations in this case. The final set of equations used in CFAST is as follows:

$$\frac{dP}{dt} = \frac{\gamma - 1}{V}(\dot{h}_L + \dot{h}_U) \tag{3.6}$$

$$\frac{dV_U}{dt} = \frac{1}{\gamma P}\left[(\gamma - 1)\dot{h}_U - V_U\frac{dP}{dt}\right] \tag{3.7}$$

$$\frac{dT_U}{dt} = \frac{1}{c_p\rho_U V_U}\left[(\dot{h}_U - c_p\dot{m}_U T_U) + V_U\frac{dP}{dt}\right] \tag{3.8}$$

$$\frac{dT_L}{dt} = \frac{1}{c_p\rho_L V_L}\left[(\dot{h}_L - c_p\dot{m}_L T_L) + V_L\frac{dP}{dt}\right] \tag{3.9}$$

The limitations of two-zone model application are stated in CFAST publication [4] as follows:

The zone model concept best applies for an enclosure in which the width and length are not too different. If the horizontal dimensions of the room

Group	Acceptable	Special Consideration Required	Corridor Flow Algorithm
$(L/W)_{max}$	$L/W < 3$	$3 < L/W < 5$	$L/W > 5$
$(L/H)_{max}$	$L/H < 3$	$3 < L/H < 6$	$L/H > 6$
$(W/H)_{max}$	$W/H > 0.4$	$0.2 < W/H < 0.4$	$W/H < 0.2$

TABLE 3.2   Recommended Compartment Dimension Limits

differ too much (i.e., the room looks like a corridor), the flow pattern in the room may become asymmetrical. If the enclosure is too shallow, the temperature may have significant radial differences. The width of the plume may at some height become equal to the width of the room and the model assumptions may fail in a tall and narrow enclosure. Therefore, the user should recognize approximate limits on the ratio of the length (L), width (W), and height (H) of the compartment.

The ratio (L/W, W/H, and L/H) limits listed in Table 3.2 are recommended.

In general, CFAST uses a simple definition of a combustion reaction in the air that includes major products of combustion for hydrocarbon fuels:

$$C_mH_nO_pN_qCl_t + m_1(O_2 + 3.76N_2) \rightarrow$$

$$m_2CO_2 + m_3CO + m_4H_2O + m_5HCl + m_6HCN + 3.76N_2 \quad (3.10)$$

where the coefficients m_1, m_2, and so on represent appropriate molar ratios for a stoichiometric balance of the equation. The only inputs required are the pyrolysis rate and the heat of combustion. Stoichiometry is used to ensure conservation of mass and elements in the reaction. The species that are calculated are oxygen, carbon dioxide, carbon monoxide, water, total unburned hydrocarbons, and soot. Gaseous nitrogen is included, but it acts only as a diluent. The unburned hydrocarbons are tracked in this model. Further combustion of CO to CO_2 is not explicitly included in the model. CFAST includes a calculation of average flame height based on Heskestad [6].

These calculations are valid for a wide range of hydrocarbon and gaseous fuels. The formula is

$$H = -1.02D + 0.235\left(\frac{Q_f}{1,000}\right)^{2/5} \quad (3.11)$$

where H is the average flame height, D is the diameter of the fire, and Q_f is the fire size. (Note that Q_f is in kilowatts.) Some fraction χ_f of Q_f will exit the fire as radiation. The remainder, χ_c, then will be deposited in the layers as convective energy. Within CFAST, the radiative fraction defaults to 0.30 [7]; that is, 30 percent of the fire's energy is released via radiation. The typical range for the radiative fraction is from 0.05 to 0.4.

Flow through vents is a dominant component of any fire model because it is sensitive to small changes in pressure. CFAST models two types of vent flow: vertical flow through horizontal vents (such as ceiling holes or hatches) and horizontal flow through vertical vents (such as doors or windows). Horizontal vent flow is determined using the pressure difference across a vent. Flow at a given elevation may be determined using Bernoulli's law by computing the pressure difference at that elevation. Atmospheric pressure is about 101,000 Pa (2.12 ksf). Fires produce pressure changes from 1 to 1,000 Pa (0.0209 to 20.9 psf). The pressure variables are solved to a higher accuracy than other solution variables because the fire-produced pressure is very small compared with atmospheric pressure. This fact will be used in Chap. 5, where in order to simplify the Navier-Stokes equations the pressure gradients produced by fire are neglected.

Flow through normal vents such as windows and doors is governed by the pressure difference across a vent. Since a momentum equation for the zone boundaries is not solved in the CFAST program, the integrated form of Euler's equation, namely, Bernoulli's solution for the velocity equation, is used to calculate the approximate average velocity of the mass and energy passing through a vent. The general formula in this case is

$$V = C \left(\frac{2 \Delta P}{\rho} \right)^{1/2} \tag{3.12}$$

where $C = 0.7$ is the flow coefficient (0.7), ρ is the gas density on the source side, and ΔP is the pressure across the interface.

The overall mass flow is calculated by the following formula [8]:

$$\dot{m} = C f(\gamma, \varepsilon) \left(\frac{\Delta P}{\rho} \right)^{1/2} A_v \tag{3.13}$$

where
$$C = 0.68 + 0.17\varepsilon \tag{3.14}$$

and
$$\varepsilon = \frac{\Delta P}{P} \tag{3.15}$$

The function f is a weak function of both variables γ and ε [8].

The energy flux then is determined by the size and temperature of the compartment from which the mass is flowing:

$$\dot{q}_g = c_p \dot{m}_u T_u + c_p \dot{m}_l T_l \tag{3.16}$$

The Fire Dynamics Simulator (FDS; see discussion below) is able to compute the density, velocity, temperature, pressure, and species concentration of the gases in a fire compartment. The FDS model uses the laws of conservation of energy, mass, and momentum to track the movement of fire gases. The ability of the FDS model to predict the temperature and velocity of fire gases accurately has been evaluated previously by conducting experiments, both laboratory-scale and full-scale, and measuring quantities of interest. For example, if the velocity of a gas flow is measured at 0.5 m/s (with possible error of ±0.05 m/s), the FDS model gas flow velocity predictions also were in the range of 0.45 to 0.55 m/s. Because, on the one hand, these velocities are relatively small (they are 10 to 20 times smaller than the case of gas or vapor explosions in a compartment) and, on the other hand, there is a strong desire to simplify the final expression for the structural fire load (SFL), the corresponding component in the equation of conservation of energy will be omitted (to be on the conservative side) for the first approximation (see Chap. 5).

3.3   Field Model

Major structural fires in high-rise buildings require a complete and thorough investigation. The analysis of failure modes is important in determining how the fire occurred and spread through the building structure. In such cases, research, experiments, theory, and calculations are needed to confirm or deny a suggested fire scenario. Assessment of fire hazard potential should involve more than just enclosure surface area and total fuel load in terms of mass of equivalent ordinary combustibles per unit area (in kg/m^2 or lbm/ft^2). The arrangement of combustibles and their chemical composition, physical state, ease of ignition, rates of fire growth, and so on are all factors to be evaluated, in addition to room geometry and size, ventilation capability, and fire protection facilities. These issues are very important because different combustible materials, such as low-density fiberboard ceilings and plastic floor coverings, constitute significant factors in fire development, permitting a fire to spread to objects far from the origin of the fire. In such cases, field computer modeling [9,10] is the most appropriate method of fire development investigation. Unlike the two-zone method, the field method divides the fire compartment into thousands of zones (volumes), and the fundamental conservation equations governing fluid dynamics, heat transfer, and combustion are written for each small volume. In field modeling, computational demands are very large, and correct simulation ultimately depends on the empirical specification of such things as ignition, burning rates, fire spread, ventilation limitation, and so on. The differential equations are solved numerically by dividing the physical space where the fire is to be simulated into a large number of rectangular volumes. Within each "simple" volume, the gas velocity, temperature, and so on are assumed to be uniform; changing only with time. From computational point of view, field models use the *finite-volume method* (FVM). This method obviously has a very close relationship with the *finite-difference method* (FDM) and *finite-element method* (FEM). For detailed information about this correlation, see Chung [11]. At present, field models are being developed and applied to the simulation of structural fires, smoke development, egress computations, identified detector response, fire endurance of structural elements and systems, and so on. The rapidly growing number of new and improved (existing) field models has prompted the establishment of a website for an updated international survey of computer models. Similar to the two-zone models, Table 3.3 is a current list of computer field models [10], obviously limited to simulations of structural fires only.

Similar to the two-zone model review process, let's review just the FDS model [13]. The conservation equations for mass, momentum, and energy for a Newtonian fluid are used in this case.

Differential Equations for a Field Model

Conservation of mass:

$$\frac{\partial \rho}{\partial t} + \nabla \cdot \rho \vec{u} = \dot{m}_b'''$$

(3.17)

Conservation of momentum (Newton's second law):

$$\frac{\partial}{\partial t}(\rho \vec{u}) + \nabla \cdot \rho \vec{u}\vec{u} + \nabla p = \rho \vec{g} + \vec{f}_b + \nabla \tau_{ij}$$

(3.18)

Transport of sensible enthalpy:

$$\frac{\partial}{\partial t}(\rho h_s) + \nabla \cdot \rho h_s \vec{u} = \frac{Dp}{Dt} + \dot{q}''' - \dot{q}_b''' - \dot{q}'' + \varepsilon$$

(3.19)

Equation of state for a perfect gas:

$$p = \frac{\rho R T}{\overline{W}}$$

(3.20)

This is a set of partial differential equations (six equations for six unknowns), all functions of three spatial coordinates (independent variables) and time: the density ρ, the three components of a velocity vector $u = [u; v; w]$, the temperature T, and the pressure p.

The sensible enthalpy h_s is a function of the temperature:

$$h_s(T) = \int_{T_0}^{T} c_p(T') dT'$$

(3.21)

Model	Country	Identifying Reference	Description
CFX	UK	[12]	General-purpose CFD (computation fluid dynamics) software, applicable to fire and explosions
FDS	US	[13]	Low-Mach-number CFD code specific to fire-related flows
FIRE	Australia	[14]	CFD model with water sprays and coupled with solid/liquid-phase fuel to predict burning rate and extinguishment
FLUENT	US	[15]	General purpose CFD software
JASMINE	UK	[16]	Field model for predicting consequences of fire to evaluate design issues (based on PHOENICS)
KAMELEON FireEx	Norway	[17]	CFD model for fire linked to a finite-element code for thermal response of structures
KOBRA-3D	Germany	[18]	CFD for smoke spread and heat transfer in complex geometries
MEFE	Portugal	[19]	CFD model for one or two compartments, includes time response of thermocouples
PHOENICS	UK	[20]	Multipurpose CFD code
RMFIRE	Canada	[21]	Two-dimensional field model for the transient calculation of smoke movement in room fires
SMARTFIRE	UK	[22]	Fire field model
SOFIE	UK/ Sweden	[23]	Fire field model
SOLVENT	US	[24]	CFD model for smoke and heat transport in a tunnel
SPLASH	UK	[25]	Field model describing interaction of sprinkler sprays with fire gases
STAR-CD	UK	[26]	General purpose CFD software
UNDSAFE	US/ Japan	[27]	Fire field model for use in open spaces or in enclosures
ALOFT-FT	US	[28]	Smoke movement from large outdoor fires
FIRES-T3	US	[29]	Finite-element heat transfer for 1D, 2D, or 3D conduction
HSLAB	Sweden	[30]	Transient temperature development in a heated slab composed of one or several materials
LENAS	France	[31]	Mechanical behavior of steel structures exposed to fire

TABLE 3.3   Identified Field Models

Note: If c_p = const., then $h_s = c_p(T - T_0)$, and Eq. (3.19) has the temperature T as an unknown function of three spatial coordinates and time.

The term $\dot{q}'''$ is the HRR per unit volume from a chemical reaction, and it can be expressed as a function of temperature based on Arrhenius's law and the type of a chemical reaction.

The term $\dot{q}''$ represents the conductive and radiative heat fluxes:

$$\dot{q}'' = -k\nabla T - \sum_{\alpha} h_{s,\alpha}\rho D_{\alpha}\nabla Y_{\alpha} + \dot{q}_r'' \qquad (3.22)$$

where k is the thermal conductivity, and $\dot{q}_r''$ is the irradiative heat flux that can be presented as a function of temperature based on the Boltzmann law.

The term ε in the sensible enthalpy equation is known as the *dissipation rate*. It is the rate at which kinetic energy is transferred to thermal energy owing to the viscosity of the fluid. This term is usually neglected because it is very small relative to the HRR of the fire.

In the sensible enthalpy equation (3.19), the material derivative

$$\frac{DP}{Dt} = \frac{\partial P}{\partial t} + \vec{u} \cdot \nabla P \qquad (3.23)$$

For most compartment fire applications, pressure P changes very little with height or time and therefore can be neglected. The main feature of the computational fluid dynamics (CFD) model is the regime of flow with the low Mach number (low-speed solver). This assumption allows elimination of the compressibility effects that are connected with the acoustic wave's propagation.

The approximation has been made that the total pressure from Eq. (3.2) was broken up into two components: a "background" component and a perturbation. In other words, the pressure within any "elementary" small zone volume was presented as a linear combination of its background component and the low speed flow-induced perturbation. That is,

$$p(\vec{x},t) = p_m(z,t) + \bar{p}(\vec{x},t) \qquad (3.24)$$

The background pressure $p_m(z,t)$ is a function of the vertical special coordinate z (only) and time. For most compartment fire applications, $p_m(z,t)$ changes very little with height or time. The equation of state (3.20) for any given "elementary" small zone volume (m) can be approximated now as

$$\bar{p}_m = \frac{\rho TR}{\bar{W}} \qquad (3.25)$$

To summarize the effect of the main assumption in the FDS model (low-speed flow) from a computational point of view, the National

Institute of Standards and Technology (NIST) report [9] states the following: "The low Mach number assumption serves two purposes. First, the filtering of acoustic waves means that the time step in the numerical algorithm is bound only by the flow speed as opposed to the speed of sound, and second, the modified state equation leads to a reduction in the number of dependent variables in the system of equations by one."

Combustion Model

The second major part of the FDS model is the mixture-fraction combustion model that is defined by the ratio of a subset of species to the total mass present in the volume. The mixture fraction is a function $Z(x, t)$ of space and time, and it varies between 0 and 1. If it can be assumed that the chemical reaction of fuel and oxygen occurs rapidly and completely, then the combustion process is referred to as "mixing-controlled." In many instances, such as large, open-floor volumes, atriums and so on, "mixed is burned" is a reasonable assumption. However, for some fire scenarios in small, underventilated compartments, where it cannot be assumed that fuel and oxygen react completely on mixing, the "mixed is burned" model cannot be accepted. In this case, instead of solving a single transport equation for the mixture fraction Z, multiple transport equations are solved for components of the mixture fraction Z_a. Similar to a two-zone model, a single-step instantaneous chemical reaction of fuel and oxygen is written as follows:

$$C_xH_yO_zN_aM_b + v_{O_2}O_2 \rightarrow$$

$$v_{CO_2}CO_2 + v_{H_2O}H_2O + v_{CO}CO + v_sS + v_{N_2}N_2 + v_MM \qquad (3.26)$$

where v_{CO}, v_s, and so on are stoichiometric coefficients, S is soot (mixture of carbon and hydrogen), and M is the average molecular weight of addition product species.

The mixture fraction Z satisfies the conservation equation:

$$\rho \frac{DZ}{Dt} = \nabla \cdot \rho D \nabla Z \qquad (3.27)$$

It is assumed that combustion occurs so rapidly that the fuel and oxygen cannot coexist, and both vanish simultaneously at the flame surface. In previous versions of the FDS model, a one-step instantaneous reaction of fuel and oxygen was assumed. However, starting in version 5, a more generalized formulation has been implemented: a single-step reaction but with local extinction and a two-step reaction with extinction. There is a possibility to simplify the modeling of combustion process based on chemical kinetics theory. For example, as stated in the NIST publication [9]: "Thus, possible to implement a relatively simple set of one or more chemical reactions to model the combustion." Consider the reaction of oxygen and a hydrocarbon fuel:

$$C_xH_y + v_{O_2}O_2 \rightarrow v_{CO_2}CO_2 + v_{H_2O}H_2O \qquad (3.28)$$

If this were modeled as a single-step reaction, the reaction rate would be given by the expression

$$\frac{d[C_xH_y]}{dt} = -B[C_xH_y]^a[O_2]^b e^{-E/RT} \qquad (3.29)$$

Suggested values of B, E, a, and b for various hydrocarbon fuels are given in [32] and [33]. It should be understood that the implementation of any of these one-step reaction schemes is still very much a research exercise because it is not universally accepted that combustion phenomena can be represented by such a simple mechanism. Improved predictions of the HRR may be possible by considering a multistep set of reactions. This simplified approach, however, will be used in Chap. 4 for our approximate analysis and development of SFL. The two-step model of a chemical reaction is needed in fuel-rich fire scenarios (underventilated fires), where soot and CO are produced at higher rates. Knowledge about soot and CO concentrations in a compartment is very important from many other aspects of fire protection and safety as a whole, but it is a secondary issue with respect to SFL only. Chemical kinetics deals with the experimental determination of reaction rates from which rate laws and rate constants are derived. The rate equation is a differential equation (3.29), and it can be integrated to obtain an integrated rate equation that links concentrations of reactants or products with time. Parameters a and b are the rate constants of a chemical reaction. Relatively simple rate laws exist for first- and second-order reactions and can be derived for the consecutive reactions also. The Arrhenius equation has been used in the theory and practice of nonsteady combustion and explosion very successfully for many years [34,35], and it will be used in Chap. 4.

It is worthwhile also to underline here that Eq. (3.27) is very similar to a regular equation of thermoconductivity. The main difference is that the thermodiffusivity coefficient is substituted by the material diffusivity coefficient. This analogy will be used in Chap. 4, where the similarity theory is used.

3.4 Summary

Any simulation of a real fire scenario involves specifying material properties for the walls, floor, ceiling, and furnishings. Describing these materials in any field model is the most challenging task. Thermal properties such as conductivity, specific heat, density, and thickness can be found in various handbooks [36,37], or from bench-scale measurements. The burning behavior of materials at different heat fluxes is more difficult to describe and the properties more difficult to obtain. NIST has recognized this and has identified the necessary capabilities of a standard fire resistance test to support performance-based structural fire engineering (PBSFE) [38]. These recommendations are intended to be applied to the entire range of fire-resistive assemblies. The key issues in any CDS type of applications are the verification and validation of the results. The difficulties connected with any computer field modeling of a fire scenario are described in the NIST publication [9], and they are as follows:

> The difficulties revolve about three issues: First, there are an enormous number of possible fire scenarios to consider due to their accidental nature. Second, the physical insight and computing power required to perform all the necessary calculations for most fire scenarios are limited. Any fundamentally based study of fires must consider at least some aspects of bluff body aerodynamics, multi-phase flow, turbulent mixing and combustion, radiative transport, and conjugate heat transfer; all of which are active research areas in their own right. Finally, the 'fuel' in most fires was never intended as such. Thus, the mathematical models and the data needed to characterize the degradation of the condensed phase materials that supply the fuel may not be available. Indeed, the mathematical modeling of the physical and chemical transformations of real materials as they burn is still in its infancy.

> In order to make progress, the questions that are asked have to be greatly simplified. To begin with, instead of seeking a methodology that can be applied to all fire problems, we begin by looking at a few scenarios that seem to be most amenable to analysis. Hopefully, the methods developed to study these 'simple' problems can be generalized over time so that more complex scenarios can be analyzed. Second, we must learn to live with idealized descriptions of fires and approximate solutions to our idealized equations.

Simplification and approximation of a field model are the main goals of Chap. 4.

References

1. Society of Fire Protection Engineers (SFPE). *Handbook of Fire Protection Engineering*, 4th ed. SFPE, Bethesda, MD, 2008.
2. Cadorin, J. F., Pintea, D., and Franssen, J. M., *The Design Fire Tool Ozone* Version 2.0: *Theoretical Description and Validation on Experimental Fire Tests*. University of Liege, Belgium, December 1, 2008.
3. Hostikka, S., Keski-Rahkonen, O., and Korhonen, T., *Probabilistic Fire Simulator: Theory and User's Manual*. VTT Technical Research Centre of Finland, Vuorimiehentie 5, P.O. Box 2000, FIN.02044 VTT, Finland, JULKAISIJA. UTGIVARE. PUBLISHER, 2003.

4. CFAST: *Consolidated Model of Fire Growth and Smoke Transport*, Version 6. NIST Special Publication 1026. NIST, Gaithersburg, MD, 2008.

5. British Standards Institution (BSI). *Fire Resistance Tests–Part 2: Alternative and Additional Procedures*, EN 1363–2. BSI, Brussels, 1999.

6. Heskestad, G., "Fire Plumes, Flame Height, and Air Entrainment" in *SFPE Handbook of Fire Protection Engineering*, 3rd ed., National Fire Protection Association, Quincy, MA, 2002.

7. Drysdale, D., *An Introduction to Fire Dynamics*, Wiley, New York, 1985.

8. Cooper, L. Y., "Calculation of the Flow Through a Horizontal Ceiling/Floor Vent" (NISTIR 89-4052). NIST, Gaithersburg, MD, 1989.

9. National Institute of Standards and Technology, *Fire Dynamics Simulator, Version 5, Technical Reference Guide*, Vol. 1: *Mathematical Model* (NIST Special Publication 1018-5). NIST, Gaithersburg, MD, 2008.

10. Olenick, Stephen M., and Carpenter, Douglas J., "An Updated International Survey of Computer Models for Fire and Smoke," *SFPE Journal of Fire Protection Engineering*, 13(2):87–110, 2003.

11. Chung, T. J., *Computational Fluid Dynamics*. Cambridge University Press, Cambridge, UK, 2002.

12. *CFX-5 User Manual*. AEA Technology, Harwell, UK, 2000.

13. McGrattan, K. B., and Forney, G. P., *Fire Dynamics Simulator: User's Manual* (NISTIR 6469). NIST, Gaithersburg, MD, 2000.

14. Novozhilov, V., Harvie, D. J. E., Green, A. R., and Kent, J. H., "A Computational Fluid Dynamic Model of Fire Burning Rate and Extinction by Water Sprinkler," *Combustion Science and Technology* 123 (1–6):227–245, 1997.

15. *Fluent/UNS and Rampant 4.2 User's Guide*, 1st ed., TNO-MEPPO, Box 3427300, AH Apeldoorn, Netherlands, 1997.

16. Cox, G., and Kumar, S., "Field Modelling of Fire in Forced Ventilated Enclosures." *Combustion Science and Technology*, 52(7):1986 (info. from web site).

17. *Kameleon FireEx 99 User Manual*, Report TRF5119. SINTEF Energy Research, Trondheim, Norway, 1998.

18. Schneider, V., *WinKobra 4.6: User's Guide*. Intefrierte Sischerheits-Technik GmbH, Frankfurt, Germany, 1995.

19. Viegas, J. C. G., "Seguranca Contra Incendios Em Edificios: Modelacao Matematica De Incendios E Validacao Experimental" (Fire Safety in Buildings: Mathematical Modelling of Fire and Experimental Validation). Ph.D. thesis, Instituto Superior Tecnico, Lisbon, Portugal, 1999.

20. *PHOENICS User's Guide*; available at www.cham.co.uk/phoenics/d_polis/d_docs/tr326/tr326top.htm.

21. Hadjisophocleous, G. V., and Yakan, A., "Computer Modeling of Compartment Fires" (Internal Report No. 613). Institute for Research in Construction, National Research Council of Canada, Ottawa, ON, 1991.

22. *SMARTFIRE V2.0 User Guide and Technical Manual* (Doc. Rev. 1.0), The University of Greenwich, UK SE10 9LS; smartfire@gre.ac.uk, July 1998.

23. Rubini, P. A., "SOFIE: Simulation of Fires in Enclosures," in *Proceedings of the 5th International Symposium on Fire Safety Science*, International Association for Fire Safety Science, pp. 1009–1020, Melbourne, Australia, March 1997.

24. http://www.tunnelfire.com.

25. Gardiner, A. J., "The Mathematical Modeling of the Interaction Between Sprinkler Sprays and the Thermally Buoyant Layers of Gases from Fires." Ph.D. thesis, South Bank Polytechnic, 1998 (now under development by FRS).

26. Star-CD V3.100A User Guide, Computational Dynamics Ltd., http://www.cd.co.uk. London, UK, 2007.

27. Yang, K. T., and Chang, L. C., *UNDSAFE-1: A Computer Code for Buoyant Flow in an Enclosure* (NBS GCR 77-84), National Bureau of Standards (now National Institute of Standards and Technology), Gaithersburg, MD, 1977.

28. McGrattan, K. B., Baum, H. R., Walton, W. D., and Trelles, J. J., *Smoke Plume Trajectory from In Situ Burning of Crude Oil in Alaska: Field Experiments and Modeling of Complex Terrain* (NISTIR 5958). NIST, Gaithersburg, MD, 1997.

29. Bresler, B., Iding, R., and Nizamuddin, Z., *FIRES-T3: A Computer Program for the Fire Response of Structure-Thermal* (UCB FRG 77–15). University of California, Berkeley, 1996. (Also NIST GCR 95-682, NIST, 1996.)

30. *A User's Guide for HSLAB: HSLAB—A Program for One-Dimensional Heat Flow Problems* (FOA Report C20827). National Defense Research Institute, Stockholm, Sweden, 1990.

31. Personal communication; Dr. Zhao Bin, Centre Technique Industriel de la Construction Métallique, France; binzhao@cticm.com.

32. Puri I. K., and Seshadri K., "Extinction of Diffusion Flames Burning Diluted Methane and Diluted Propane in Diluted Air." *Combustion and Flame* 65:137–150, 1986.

33. Westbrook, C. K., and Dryer, F. L., "Simplified Reaction Mechanisms for the Oxidation of Hydrocarbon Fuels in Flames." *Combustion Science and Technology* 27:31–43, 1981.

34. Frank-Kamenestkii, D. A., *Diffusion and Heat Transfer in Chemical Kinetics.* Plenum Press, New York, 1969.

35. Zeldovich, Ya. B., Barenblatt, G. I., Librovich, V. B., and Makhviladze, G. M., *The Mathematical Theory of Combustion and Explosions.* Consultants Bureau, New York, 1985.

36. Lewis, B., and Von Elbe, G., *Combustion, Flames and Explosions of Gases.* Academic Press, New York, 1987.

37. Lienhard, John H., IV, and Lienhard, John H., V, *Heat Transfer Textbook*, 3rd ed. Phlogiston Press, Cambridge, MA, 2008.

38. *Fire Resistance Testing for Performance-Based Fire Design of Buildings* (NIST GCR 07-910). NIST, Gaithersburg, MD, June 2007.

CHAPTER 4

Differential Equations and Assumptions

Notation

q	Heat flux rate per unit area.
k	Thermal conductivity, which obviously must have the dimensions W/m K or J/m s K
T	Temperature
d	Thickness in the direction of heat flow
ρ	Air density
c	Specific heat capacity
K	Number of collisions that result in a reaction per second
A	Total number of collisions
E	Activation energy
R	Ideal gas constant
P	Losses of heat owing to thermal radiation
e	Emissivity factor
σ	Boltzmann constant ($\sigma = 5.6703 \times 10^{-8}$ W/m^2 K^4)
T_o	Ambient temperature
A_v	Area of openings

c_p	Average specific heat at constant pressure
t	Time
$\vec{v}(u;v;w)$	Velocity vector
Q	Heat rate (heat effect of chemical reaction)
V	Compartment volume
C_i	Mass fractions (concentrations) of the individual gaseous species
v_i and v_1	Stoichiometric coefficients
M	Molecular weight
i and k	Gas component numbers
C_{mi}	Concentrations of mass fractions
C	Mass fraction (concentration) of a component of one-step chemical reaction
D	Diffusion coefficient (m^2/s)
k_1	Portion of a chemical reaction velocity that is a function of temperature only
$m = m_A + m_B + \cdots$	Order of a chemical reaction
p	Pressure
S_{ij}	Stress tensor
$\vec{f}$	Vector representing body forces (per unit volume) acting on the fluid
∇	Delta operator
∇^4	Two-dimensional biharmonic operator
ν	Kinematic viscosity; $\nu = \mu/\rho$
θ	Dimensionless temperature
τ	Dimensionless time
h	Height of the compartment (m)
a	Thermal diffusivity (m^2/s)
$\vec{v}(u;v;w)$	Velocity vector
Time	$t = \dfrac{h^2}{a}\tau$ (s)
Temperature	$T = \dfrac{RT_*^2}{E}\theta + T_*$ (K), where $T_* = 600$ K is the baseline temperature

Coordinates	$\bar{x} = x/h$ and $\bar{z} = z/h$, where x and z are dimensionless coordinates
Velocities	$\bar{u} = \dfrac{v}{h}u$ (m/s) and $\bar{w} = \dfrac{v}{h}w$ (m/s), the horizontal and vertical components of velocity
v	Kinematic viscosity (m²/s)
$\text{Pr} = v/a$	Prandtl number
$\text{Fr} = \dfrac{gh^3}{va}$	Froude number
g	Gravitational acceleration
$\bar{\Psi}$ (m²/s)	Stream function; $\bar{\Psi} = v\Psi$, where Ψ is the dimensionless stream function
$\text{Le} = a/D = \text{Sc}/\text{Pr}$	Lewis number
$\text{Sc} = v/D$	Schmidt number
$\beta = \dfrac{RT_*}{E}$	Dimensionless parameter
$\gamma = \dfrac{c_p RT_*^2}{QE}$	Dimensionless parameter
$P = \dfrac{e\sigma K_v (\beta T_*)^3 h}{\lambda}$	Thermal radiation dimensionless coefficient
$\sigma = 5.67(10^{-8})$ (W/m² K⁴)	Boltzman constant
$K_v = A_o h/V$	Dimensionless opening factor
A_o	Total area of vertical and horizontal openings
$\delta = \left(\dfrac{E}{RT_*^2}\right) Qz \left[\exp\left(-\dfrac{E}{RT_*}\right)\right]$	Frank-Kamenetskii's parameter
$C = [1 - P(t)/P_o]$	Concentration of the burned fuel product in the fire compartment
$\bar{W} = \dfrac{v}{h}W$	Vertical component of gas velocity
$\bar{U} = \dfrac{v}{h}U$	Horizontal component of gas velocity
$b = L/h$	Length L (width) and height h of fire compartment, accordingly
W, U	Dimensionless velocities

4.1   Definitions and Classifications

Fire is a very complex chemical process that is influenced by many factors that affect its growth, spread, and development. Structural fire load (SFL) establishment requires a basic understanding of the chemical and physical nature of fire. This includes information describing sources of heat energy, composition and characteristics of fuels, and environmental conditions necessary to sustain the combustion process. It is important first to define certain terms and conditions that will be used in this book because some definitions differ from one source to another. For example, the definition of flashover in fire compartment (room) development has qualitative and quantitative differences in different reference sources. Flashover is defined by different authors in the literature are as follows:

1. "The transition from the fire growth period to the fully developed stage in the enclosure fire development."

2. "A dramatic event in a room fire that rapidly leads to full room involvement; an event that can occur at a smoke temperature of 500 to 600°C."

3. "The transition from a localized fire to the general conflagration within the compartment when all fuel surfaces are burning."

The latest definition of *flashover* given by National Fire Protection Association (NFPA) 921-2004 [4] is as follows: "a transitional phase in the development of a compartment fire in which surfaces exposed to thermal radiation reach ignition temperature more or less simultaneously and fire spreads rapidly throughout the space, resulting in full room involvement or total involvement of the compartment or enclosed area." It is important to underline here that the latest NFPA definition has a reference to thermal radiation as a main source of energy that creates the flashover conditions and as a consequence leads to the fully developed compartment fire. The design values of an SFL in most practical cases are based on fully developed stage in the enclosure fire development; therefore, simplification can be made in the conservation of energy equation with respect to other sources of energy, such as conduction and convection [see p. 82, formulas (4.9) and (4.10)]. For all practical purposes, in structural engineering analyses and design, the SFL is temperature-time function that is obtained from conservation of energy, mass, and momentum equations. As has been indicated in Chap. 2, this function is presented by a double-convex curve; therefore, the "mathematical" definition of a flashover point could be stated as follows: the point where the curvature sign changes is (second derivative is 0 or first derivation is maximum at this point). The first derivative of temperature with respect to time multiplied by specific heat and density is the *heat release rate* (HRR). Therefore, the definition of a flashover point could be stated also as a point on the temperature-time curve where HRR reaches its maximum.

HRR should not be confused with heat of combustion. Heat of combustion simply represents how much energy the fuel would release if it were consumed completely. HRR pertains to the speed at which combustibles will burn. Therefore, it can be computed as follows: HRR = mass burning rate (mass/time) multiplied by the heat of combustion (energy/mass). The measure of HRR is kilowatts (kW). Below are some other definitions that are needed in this chapter.

The *diffusion flame process* (fire) consists of three basic elements: fuel, oxygen, and heat. There are six elements of the life cycle, and they are as follows: input heat, fuel, oxygen, proportioning, mixing, and ignition continuity. All these elements are essential for both the initiation and continuation of the diffusion flame combustion process. For combustion to take place, solid or liquid materials must be heated sufficiently to produce vapors. The lowest temperature at which a solid or liquid material produces sufficient vapors to burn under laboratory conditions is known as the *flashpoint*. A few degrees above the flashpoint is the *flame point*, the temperature at which the fuel will continue to produce sufficient vapors to sustain a continuous flame. If the source of the heat is an open flame or spark, it is referred to as *piloted ignition*. In general terms, *combustible* means capable of burning, generally in air under normal conditions of ambient temperature and pressure, whereas *flammable* is defined as capable of burning with a flame.

The primary source of oxygen normally is the atmosphere, which contains approximately 20.8 percent oxygen. A concentration of at least 15 to 16 percent is needed for the continuation of flaming *combustion*, whereas charring or smoldering (pyrolysis) can occur with as little as 8 percent. Pyrolysis is defined as the transformation of a compound into one or more other substances by heat alone.

Mixing and proportioning are reactions that must be continuous in order for fire to continue to propagate. The fuel vapors and oxygen must be mixed in the correct proportions. Such a mixture of fuel vapors and oxygen is said to be within the *explosive limits* or *flammable limits*. Explosive or flammable limits are expressed as the concentration (percentage) of fuel vapors in air. A mixture that contains fuel vapors in an amount less than necessary for ignition to occur is *too lean*, whereas a mixture that has too high a concentration of fuel vapors is *too rich*. The lowest concentration that will burn is known as the *lower explosive limit* (LEL), whereas the highest level is known as the *upper explosive limit* (UEL). For example, the explosive or flammable limits for propane are 2.15 (LEL) to 9.6 (UEL). This means that any mixture of propane and air between 2.15 and 9.6 percent will ignite if exposed to an open flame, spark, or other heat source equal to or greater than its ignition temperature, which is between 920 and 1,120°F (493.3 and 604.4°C). Another important characteristic of gases is *vapor density*—the weight of a volume of a given gas to an equal volume of dry air, where air is given a value of 1.0. A vapor density of less than 1.0 means that the gas is lighter than air and will tend to rise in a relatively calm atmosphere, whereas a vapor density of more than 1.0 means that the gas is heavier than air and will tend to sink to ground/floor level. NFPA 325M [5] contains an extensive listing of the flashpoints, ignition temperatures, flammable limits, vapor densities, and specific gravities of various materials.

Ignition continuity is the thermal feedback from the fire to the fuel. Heat is transferred by conduction, convection, radiation, and direct flame contact. *Conduction* is the transfer of heat by direct contact through a solid body. Fire can move from one compartment to another via heat conduction. *Convection* is the transfer of heat caused by changes in density of liquids and gases. It is the most common method of heat transfer; when liquids or gases are heated, they become less dense and will expand and rise. *Convection* is the transfer of heat through the motion of heated matter, that is, through the motion of smoke, hot air, heated gases produced by the fire, and flying embers. When it is confined (as in a compartment), convected heat moves in predictable patterns. The fire produces lighter-than-air gases that rise toward high parts of the compartment. Heated air, which is lighter than cool air, also rises, as does the smoke produced by combustion. As these heated combustion products rise, cool air takes their place; the cool air is heated in turn and then also rises to the highest point it can reach. As the hot air and gases rise from the fire, they begin to cool; as they do, they drop down to be reheated and rise again. This is the *convection cycle*.

Transfer of heat by radiation is less commonly understood or appreciated than conduction or convection. *Radiation* is the transfer of heat by infrared radiation (heat waves, e.g., the sun), which generally is not visible to the naked eye. Heat radiation is the transfer of heat from a source across an intervening space; no material substance is involved. The heat travels outward from the fire in the same manner as light, that is, in straight lines. When it contacts a body, it is absorbed, reflected, or transmitted. Absorbed heat increases the temperature of the absorbing body. Heat radiates in all directions unless it is blocked. Radiant heat extends fire by heating combustible substances in its path, causing them to produce vapor, and then igniting the vapor. As hot gases from the flame rise into contact with additional fuel, the heat is transferred to the fuel by convection and radiation until the additional fuel begins to vaporize. The flames then will ignite these additional vapors. *Heat* is defined as a form of energy characterized by the vibration of molecules and capable of initiating and supporting chemical changes. *Fire gases* include carbon monoxide, hydrogen cyanide, ammonia, hydrogen chloride, and acrolein. *Flame* is the luminous portion of burning gases or vapors. *Smoke* is the airborne particulate products of incomplete combustion, suspended in gases, vapors, or solid or liquid aerosols. *Soot,* black particles of carbon, is contained in smoke.

A fire in a room or defined space generally will progress through three predictable developmental stages. The first stage of fire development is the *incipient stage* (*growth*). This begins at the moment of ignition, and at this time, the flames are localized. At this stage, the fire is fuel-regulated; that is, fire propagation is regulated not by the available oxygen but by the configuration, mass, and geometry of the fuel itself. The oxygen content is within the normal range, and normal ambient temperatures still exist. A plume of hot fire gases will begin to rise to the upper portions of the room. As convection causes the plume to raise, it will draw additional oxygen into the bottom of the flames. Fire gases such as sulfur dioxide, carbon monoxide, and others will begin to accumulate in the room. If there is any solid fuel above the flame, both convection and direct flame contact will cause upward and outward fire spread, producing the characteristic V-pattern charring on vertical surfaces. Second is the *free-burning stage* (*development*). In this stage, more fuel is being consumed, and the fire is intensifying. Flames have spread upward and outward from the initial point of origin by convection, conduction, and direct flame impingement. A hot, dense layer of smoke and fire gases is collecting at the upper levels of the room and is beginning to radiate heat downward. This upper layer of smoke and fire gases contains not only soot but also toxic gases such as carbon monoxide, hydrogen cyanide, hydrogen chloride, arcolein, and others. The fire continues to grow in intensity, and the layer of soot and fire gases drops lower and lower. The soot and combustible gases continue to accumulate until one (or more) of the fuels reaches its ignition temperature. *Rollover* occurs when ignition of the upper layer results in fire extending across the compartment at its upper levels. This rollover causes the overhead temperature to increase at an even greater rate and also increases the heat being radiated downward into the compartment. Secondary fires can and do result from the heat being generated. The fire is still fuel-regulated at this time.

When the upper layer reaches a temperature of approximately 1,100°F (593.3°C), sufficient heat is generated to cause simultaneous ignition of all fuels in the room. This is called *flashover*. Once flashover has occurred, survival for more than a few seconds is impossible. Temperatures in the space will reach 2,000°F (1,093.3°C) or more at the overhead level down to over 1,000°F (593.3°C) at the floor. At the point of flashover, the fire is still fuel-regulated, but if the fire stays confined to the compartment of origin, it quickly becomes oxygen-regulated. If the fire conditions provide free communication with the atmosphere outside the compartment, the unlimited supply of oxygen causes the fire to remain in the fuel-regulated phase. As a general rule, once flashover has occurred, full involvement of the compartment follows quickly. Flashover results in intense burning of the entire compartment and its contents. The length of time necessary for a fire to go from the incipient stage to flashover depends on the fuel package, the compartment geometry, and ventilation. If the fire has been contained to a compartment or space and the oxygen level drops below 15 to 16 percent, *open-flaming combustion* will stop even if unburned fuel is still present. At this point, *glowing combustion* will take place; this is known as the *smoldering stage* (*decay*). High temperature and considerable quantities of soot and combustible fire gases have accumulated, and at this point the fire is oxygen-regulated. The temperatures may exceed the ignition temperatures of the accumulated gases. If a source of oxygen is introduced to the area, the accumulated soot and fire gases may ignite with explosive force. This smoke explosion is known as a *backdraft*. The pressures generated by a backdraft are enough to cause significant structural damage and endanger the lives of firefighting personnel and bystanders. Backdrafts can take place in any enclosed space. The behavior of a fire in a corridor is affected by the same conditions as a room fire. The physical configuration of a corridor can cause the fire to spread rapidly because the corridor will function as a horizontal chimney or flue. Rapid fire spread in a corridor can occur with normal materials providing the fuel load.

4.2   Conservation of Energy and Mass Equations

The processes of heat and mass transfer owing to conduction or convection are similar to each other. However, the heat transfer owing to radiation does not follow the analogy in the mass-transfer process. Since heat transfer and mass transfer are two similar processes, there is no need to analyze each of them separately. The differential equations that describe the heat transfer in a fire compartment are similar to the corresponding equations that describe the mass-transfer process. The heat flux resulting from thermal conduction is proportional to the magnitude gradient of the temperature and opposite to it in sign (based on Fourier's law):

$$\vec{q} = -k\nabla T \tag{4.1}$$

In a one-dimensional heat-conduction problem, it is convenient to write Fourier's law in simple scalar form:

$$q = k\frac{\Delta T}{d} \tag{4.2}$$

Since both quantities in Eq. (4.2) are positive, one must remember that q always flows from high to low temperatures.

Now consider the one-dimensional element illustrated in Fig. 4.1.

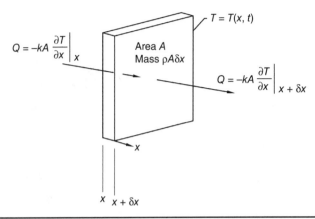

Figure 4.1 Schematic diagram.

From Fourier's law applied at each side of the element,

$$Q_{net} = -kA \left\{ \frac{\left. \dfrac{\partial T}{\partial x} \right|_{at \; x+\delta x} - \left. \dfrac{\partial T}{\partial x} \right|_{at \; x}}{\delta x} \right\} \delta x \equiv \frac{\partial^2 T}{\partial x^2} \qquad (4.3)$$

On the other hand, from the first law of thermodynamics,

$$-Q_{net} = \frac{dU}{dt} = \rho c A \frac{d(T - T_{ref})}{dt} \delta x = \rho c A \frac{dT}{dt} \delta x \qquad (4.4)$$

Equations (4.3) and (4.4) can be combining to give

$$k \frac{\partial^2 T}{\partial x^2} = \rho c \frac{\partial T}{\partial t} \qquad (4.5)$$

Let's now extend the analysis and allow the volume containing the gas, illustrated in Fig. 4.2, to move with a field velocity $\bar{u}(u, v, w)$.

In the case of natural convection, the following approximations are made (all these approximations and assumptions are very similar to the assumptions made in field and zone models):

- The pressure and derivatives of it in the flow are small and do not affect thermodynamic properties. The effect of dp on enthalpy, internal energy, and density shall be neglected. This approximation is reasonable for gas flows moving at speeds of less than about a third the speed of sound. In the case of fire, the flame propagation is in a range of not more than 10 m/s.

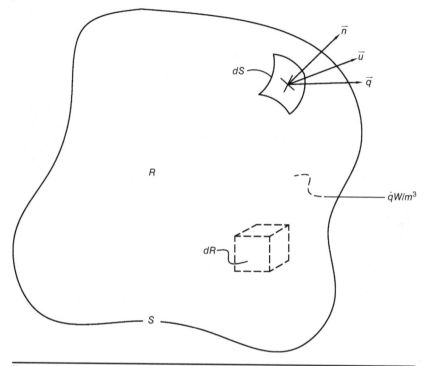

Figure 4.2   Control volumes in a heat-flow field.

- The flow can be assumed as uncompressible because the density changes are small. For such flows, $\nabla \cdot \vec{u} = 0$.
- All physical constants such as c, k, v and so on should be assumed to be constant in combustion process, and the corresponding values should be taken at the maximum gas temperature [6,7].
- Potential and kinetic energies are negligible in comparison with thermal energy changes.
- The viscous stresses do not dissipate enough energy to warm the fluid significantly.
- The speed of chemical reaction (the fuel-burning process) has a major dependence on temperature and therefore follows the Arrhenius equation [7]:

$$K = Ae^{-(E/RT)} \tag{4.6}$$

The units of the pre-exponential factor A are identical to those of the rate constant and will vary depending on the order of the reaction. In the first-order chemical reaction, the dimensional unit is per second (s^{-1}). In other words, K is the number of collisions that result in a reaction per second, A is the total number of collisions (leading to a reaction or not and, obviously, is a function of the reactant gas concentration) per second, and $e^{-E/RT}$ is the probability that any given collision will result in a reaction. The *activation energy* is the energy barrier that the reactants must surmount in order to react. Therefore, the activation energy is viewed as an energetic threshold for a fruitful reaction. Given the small temperature range in which kinetic studies are carried out, it is reasonable to approximate the activation energy as being independent of temperature. The Arrhenius equation can be converted to a different form by taking the natural log of both sides of Eq. (4.6):

$$\ln k = \ln A - \frac{E}{RT} \qquad (4.7)$$

This form of the Arrhenius equation is a straight-line equation ($y = mx + b$), where $y = \ln k$ and $x = 1/T$. The slope is $-E/R$, and the intercept is $\ln A$. Equation (4.7) can be defined by just using rate constants at two temperatures; it would be more realistic and reliable, however, to use at least three rate constants (preferably more) at three different temperatures and determine A and E using a statistical method.

This relationship allows us to make some meaningful interpretations. We can determine the frequency of collisions and the fraction of collisions that have the proper orientation for the reaction to occur. In other words, we can directly predict the energy of activation required for activation. The most comprehensive source of information regarding activation energy, heat of combustion process, speed of chemical reaction, velocity of flame propagation, and so on is contained in Lewis and Von Elbe [8].

- The combustion model is presented by a chemical reaction with stoichiometric coefficients, where individual gas species react according to the specified Arrhenius equation. Based on the ideal gas theory, all three coefficients—thermodiffusivity, material diffusivity, and kinematical viscosity—are approximately equal ($a = D = v$); therefore, both processes, heat transfer and mass transfer, are similar, and the differential equations are similar.

- Losses of heat owing to thermal radiation (losses owing to conduction and convection are assumed to be much smaller) are based on a given ventilation area (openings) and the Boltzmann law. The net radiation loss rate takes the form

$$P = e\sigma A_v (T^4 - T_o^4) \qquad (4.8)$$

Differential Equations

Based on all these assumptions, the differential equations for heat and mass transfer can be written as follows [6]:

$$c_p \rho \frac{\partial T}{\partial t} = \text{div}(\lambda \text{grad} T - c_p \rho \vec{v} \nabla T) + Qze^{-E/RT} - \frac{e\sigma A_v (T^4 - T_o^4)}{V} \quad (4.9)$$

$$\frac{\partial C_i}{\partial t} = D_i \Delta C_i - \text{div} \vec{v} C_i - \frac{v_i}{v_1} Qze^{-E/RT} \quad (4.10)$$

The mass fractions are defined as

$$C_{mi} = \frac{M_i C_i}{\sum_k M_k C_k} = \frac{M_i C_i}{\rho} \quad (4.11)$$

where i and k are gas component numbers; and M_k are molecular weights.

For a binary mixture of gas species;

$$C_{m1} + C_{m2} = 1 \quad (4.12)$$

Fick's law for a multi–mass fraction mixture diffusion process can be written as

$$g = -D\rho \text{grad} C_{mi} + \vec{v} C_{mi} \quad (4.13)$$

However, if the density of the mixture is assumed to be constant or the diffusivity coefficients for the gas components are approximately equal, then one can assume that the diffusion process is independent for each component, and therefore, Fick's law can be written as

$$g = -D\text{grad} C + C \quad (4.14)$$

where C is the mass fraction (concentration) of a component of a one-step chemical reaction (reactant or product of a chemical reaction). This assumption simplifies considerably the number of partial differential equations (4.10). Instead, there will be only one equation (4.10).

All chemical reactions can be divided in two groups: simple and complex. Simple reactions are reactions where velocity is a function of the mass fraction (concentration) of reactant components only and is not dependent on the mass fractions of the products of the chemical reaction. As stated in Laider, Meiser, and Sanctuary [9];

$$W = k_1 C_A^{m_A} C_B^{m_B} \quad (4.15)$$

where k_1 is the portion of a chemical reaction velocity that is a function of temperature only, and $m = m_A + m_B + \cdots$ is the order of the chemical reaction. Again, to simplify the computational process, it will be assumed that the burning process of fuel in a compartment during a postflashover fire stage can be presented as a first-order chemical reaction. The assumption that the fire shell be presented by a second-order chemical reaction has relatively small effect on the main parameters of the temperature-time curve—the SFL (see above, formula 4.15). Many of combustion processes can be described as first-order chemical reactions, except for autocatalytic reactions: They are chemical reactions in which at least one of the products is also a reactant. The rate of the equations for autocatalytic reactions is fundamentally nonlinear.

4.3 Conservation of Momentum

The Navier-Stokes equations describe the motion of fluid and gas substances that can flow. These equations arise from applying Newton's second law to fluid motion, together with the assumption that the fluid stress is the sum of a diffusing viscous term plus a pressure term. These equations are most useful because they describe the physics of a large number of phenomena of academic and economic interest. They may be used to model weather, ocean currents, flow around an airfoil (wing), and so on. The Navier-Stokes equations dictate not position but rather velocity. A solution of the Navier-Stokes equations is called a *velocity field* or *flow field*, which is a description of the velocity of the fluid or gas at a given point in space and time. Once the velocity field is solved for, other quantities of interest (e.g., flow rate or drag force) may be found. This is different from what one normally sees in structural engineering, where solutions typically are trajectories of deflection of a structural element. The Navier-Stokes equations are nonlinear differential equations in almost every real situation. In some cases, such as one-dimensional flow and Stokes flow (or creeping flow), the equations can be simplified to linear equations. The nonlinearity makes most problems difficult or impossible to solve and is the main contributor to the turbulence that the equations model.

The nonlinearity is due to convective acceleration, which is an acceleration associated with a change in velocity over position. Hence any convective flow, whether turbulent or not, will involve nonlinearity. An example of convective but laminar (non turbulent) flow is the passage of a viscous fluid through a small, converging opening. Such flows, whether exactly solvable or not, often can be thoroughly studied and understood. The derivation of the Navier-Stokes equations begins with an application of the conservation of momentum being written for an arbitrary control volume.

The most general form of the Navier-Stokes equation ends up being

$$\rho\left(\frac{\partial \vec{v}}{\partial t} + \vec{v} \bullet \nabla \vec{v}\right) = -\nabla p + \nabla S_{ij} + \vec{f} \tag{4.16}$$

This is a statement of the conservation of momentum in a fluid, and it is an application of Newton's second law to a continuum. A very significant feature of the Navier-Stokes equations is the presence of convective acceleration: the effect of time-independent acceleration of a fluid with respect to space, represented by the quantity $\vec{v}\nabla\vec{v}$, where $\nabla\vec{v}$ is the tensor derivative of the velocity vector, equal in Cartesian coordinates to the component-by-component gradient. This may be expressed in x, y, and z coordinates as

$$u\frac{\partial u}{\partial x} + v\frac{\partial u}{\partial y} + w\frac{\partial u}{\partial z} \qquad \text{Projection on } x \text{ coordinates}$$

$$u\frac{\partial v}{\partial x} + v\frac{\partial v}{\partial y} + w\frac{\partial v}{\partial z} \qquad \text{Projection on } y \text{ coordinates}$$

$$u\frac{\partial w}{\partial x} + v\frac{\partial w}{\partial y} + w\frac{\partial w}{\partial z} \qquad \text{Projection on } z \text{ coordinates}$$

The effect of stress in the fluid is represented by ∇p and ∇S_{ij} terms; these are gradients of surface forces, analogous to stresses in structural engineering analysis. ∇p is called the *pressure gradient* and arises from normal stresses. ∇S_{ij} conventionally describes *viscous forces*, for incompressible (Newtonian) flow, and it has only a shear stresses with the quantity of $\nu\rho\nabla^2\vec{v}$. The vector $\vec{f}$ represents body forces. Typically, the vector represents only gravity forces, but it may include other fields (e.g., centrifugal force). Often, these forces may be represented as the gradient of some scalar quantity. Gravity in the z coordinate direction, for example, is the gradient of $-\rho g z$.

Regardless of the flow assumptions, a statement of the conservation of mass generally is necessary. This is achieved through the mass continuity equation, given in its most general form as

$$\frac{\partial \rho}{\partial t} + \nabla \cdot (\rho \vec{v}) = 0 \tag{4.17}$$

The Navier-Stokes equations are strictly a statement of the conservation of momentum. In order to fully describe fluid flow, more information is needed: This may include boundary conditions, the conservation of mass, the conservation of energy, and an equation of state.

Incompressible Flow of Newtonian Fluids

The vast majority of work on the Navier-Stokes equations is done under an incompressible flow assumption for Newtonian fluids. The incompressible flow assumption typically holds well even when dealing with a "compressible" fluid such as air at room temperature (even when flowing up to about Mach number 0.3). Taking the incompressible flow assumption into account and assuming constant viscosity, the Navier-Stokes equations will read (in vector form) [10]

$$\rho\left(\frac{\partial \vec{v}}{\partial t} + \vec{v} \cdot \nabla \vec{v}\right) = -\nabla p + \nu\rho\nabla^2\vec{v} + \vec{f} \qquad (4.18)$$

Note that only the convective terms are nonlinear for incompressible Newtonian flow. The convective acceleration is an acceleration caused by a (possibly steady) change in velocity over position, for example, the speeding up of fluid (gas) entering a window opening. Although individual fluid particles are being accelerated and thus are under unsteady motion, the flow field (a velocity distribution) will not necessarily be time-dependent. Another important observation is that the viscosity is represented by the vector Laplacian of the velocity field. This implies that Newtonian viscosity is diffusion of momentum; this works in much the same way as the diffusion of heat seen in the heat equation (which also involves the Laplacian). Under the incompressible assumption, density is a constant, and it follows that the mass continuity equation will simplify to

$$\nabla\vec{v} = 0 \qquad (4.19)$$

This is more specifically a statement of the conservation of volume.

While the Cartesian equations seem to follow directly from the preceding vector equation, the vector form of the Navier-Stokes equation involves some tensor calculus, which means that writing it in other coordinate systems is not as simple as doing so for scalar equations (such as the heat equation).

Cartesian Coordinates

Writing the vector equations explicitly, that is

$$\rho\left(\frac{\partial u}{\partial t} + u\frac{\partial u}{\partial x} + v\frac{\partial u}{\partial y} + w\frac{\partial u}{\partial z}\right) = -\frac{\partial p}{\partial x} + \nu\rho\left(\frac{\partial^2 u}{\partial x^2} + \frac{\partial^2 u}{\partial y^2} + \frac{\partial^2 u}{\partial z^2}\right) + \rho g_x \quad (4.20)$$

$$\rho\left(\frac{\partial v}{\partial t} + u\frac{\partial v}{\partial x} + v\frac{\partial v}{\partial y} + w\frac{\partial v}{\partial z}\right) = -\frac{\partial p}{\partial y} + \nu\rho\left(\frac{\partial^2 v}{\partial x^2} + \frac{\partial^2 v}{\partial y^2} + \frac{\partial^2 v}{\partial z^2}\right) + \rho g_y \quad (4.21)$$

$$\rho\left(\frac{\partial w}{\partial t} + u\frac{\partial w}{\partial x} + v\frac{\partial w}{\partial y} + w\frac{\partial w}{\partial z}\right) = -\frac{\partial p}{\partial z} + \nu\rho\left(\frac{\partial^2 w}{\partial x^2} + \frac{\partial^2 w}{\partial y^2} + \frac{\partial^2 w}{\partial z^2}\right) + \rho g_z \quad (4.22)$$

the continuity equation reads

$$\frac{\partial u}{\partial x}+\frac{\partial v}{\partial y}+\frac{\partial w}{\partial z}=0 \qquad (4.23)$$

The velocity components (the dependent variables to be solved for) typically are named u, v, and w. This system of four equations comprises the most commonly used and studied form. Although comparatively more compact than other representations, this is a nonlinear system of partial differential equations for which solutions are difficult to obtain.

Cylindrical coordinates are chosen to take advantage of symmetry so that a velocity component can disappear. A very common case is axisymmetric flow, where there is no tangential velocity ($u_\varphi = 0$) and the remaining quantities are independent of φ:

$$\rho\left(\frac{\partial u_r}{\partial t}+u_r\frac{\partial u_r}{\partial r}+w\frac{\partial u_r}{\partial z}\right)=-\frac{\partial p}{\partial r}+\mu\left[\frac{1}{r}\frac{\partial}{\partial r}\left(r\frac{\partial u_r}{\partial r}\right)+\frac{\partial^2 u_r}{\partial z^2}-\frac{u_r}{r^2}\right]+\rho g_r \qquad (4.24)$$

$$\rho\left(\frac{\partial w}{\partial t}+u_r\frac{\partial w}{\partial r}+w\frac{\partial w}{\partial z}\right)=-\frac{\partial p}{\partial z}+\mu\left[\frac{1}{r}\frac{\partial}{\partial r}\left(r\frac{\partial w}{\partial r}\right)+\frac{\partial^2 w}{\partial z^2}\right]+\rho g_z \qquad (4.25)$$

$$\frac{1}{r}\frac{\partial}{\partial r}(ru_r)+\frac{\partial w}{\partial z}=0 \qquad (4.26)$$

Cylindrical coordinates are shown in Fig. 4.3.

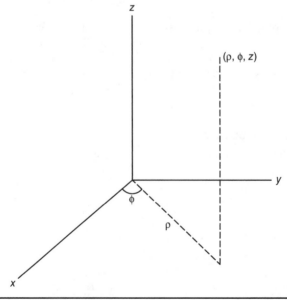

Figure 4.3   Cylindrical coordinates system.

Consider a cylindrical coordinates system (r, φ, and z), with the z axis the line around which the incompressible flow is axisymmetrical, φ the azimuthal angle, and r the distance to the z axis. Then the flow velocity components u_r and w can be expressed in terms of the Stokes stream function ψ by [11]

$$u_r = -\frac{1}{r}\frac{\partial \Psi}{\partial z} \tag{4.27}$$

$$w = +\frac{1}{r}\frac{\partial \Psi}{\partial r} \tag{4.28}$$

The azimuthal velocity component u_φ does not depend on the stream function. Owing to the axisymmetry, all three velocity components (u_r, u_φ, and w) depend only on r and z and not on the azimuth φ.

In cylindrical coordinates, the divergence of the velocity field $\vec{u}$ becomes zero, that is,

$$\nabla \cdot \vec{u} = \frac{1}{r}\frac{\partial}{\partial r}\left(-\frac{\partial \Psi}{\partial z}\right) + \frac{\partial}{\partial z}\left(\frac{1}{r}\frac{\partial \Psi}{\partial r}\right) = 0 \tag{4.29}$$

Let's rewrite Eqs. (4.20), (4.21), and (4.22) for 2D Cartesian flow (assuming $v = 0$ and no dependence of the y coordinate):

$$\rho\left(\frac{\partial u}{\partial t} + u\frac{\partial u}{\partial x} + w\frac{\partial u}{\partial z}\right) = -\frac{\partial p}{\partial x} + \mu\left(\frac{\partial^2 u}{\partial x^2} + \frac{\partial^2 u}{\partial z^2}\right) + \rho g_x \tag{4.30}$$

$$\rho\left(\frac{\partial w}{\partial t} + u\frac{\partial w}{\partial x} + w\frac{\partial w}{\partial z}\right) = -\frac{\partial p}{\partial z} + \mu\left(\frac{\partial^2 w}{\partial x^2} + \frac{\partial^2 w}{\partial z^2}\right) + \rho g_z \tag{4.31}$$

Stream Function

Let's now define the stream function as follows:

$$u = \frac{\partial \Psi}{\partial z} \quad \text{and} \quad w = -\frac{\partial \Psi}{\partial x} \tag{4.32}$$

Differentiating Eq. (4.30) with respect to z and Eq. (4.31) with respect to x and subtracting the resulting equations will eliminate pressure and any potential force. The mass continuity equation (4.23) is unconditionally satisfied (given that the stream function is continuous), and Eqs. (4.30) and (4.31) are degrading into one equation:

$$\frac{\partial}{\partial t}(\nabla^2\Psi) + \frac{\partial \Psi}{\partial z}\frac{\partial}{\partial x}(\nabla^2\Psi) - \frac{\partial \Psi}{\partial x}\frac{\partial}{\partial z}(\nabla^2\Psi) = \nu\nabla^4\Psi \tag{4.33}$$

4.4 Dimensionless Parameters and Equations

Scale analysis is a powerful tool used in combustion theory for the simplification of differential equations with many parameters. It also allows one to identify some small parameters that approximate the magnitude of individual terms in the equations and their impact on the solution as a whole. Let's introduce the dimensionless parameters and variables in conservation of energy, mass, and momentum equations [7].

Stream function has a dimension: $\bar{\Psi}$ (m^2/s); therefore,

$$\bar{\Psi} = \nu\Psi$$

where Ψ is the dimensionless stream function, and Eq. (4.33) can be rewritten in dimensionless form as

$$\frac{\partial}{\partial\tau}(\nabla^2\Psi) + \Pr\frac{\partial\Psi}{\partial z}\frac{\partial}{\partial x}(\nabla^2\Psi) - \Pr\frac{\partial\Psi}{\partial x}\frac{\partial}{\partial z}(\nabla^2\Psi) = \Pr\nabla^4\Psi \qquad (4.34)$$

The definitions of dimensionless velocities u and w are similar to Eq. (4.32) in this case and can be written as

$$u = \frac{\partial\Psi}{\partial z} \qquad \text{and} \qquad w = -\frac{\partial\Psi}{\partial x} \qquad (4.35)$$

Let's now rewrite Eqs. (4.9), (4.10), (4.20), (4.21), and (4.22) in dimensionless form (assuming for Newtonian fluids that pressure $p = $ const. and air density $\rho = $ const.):

$$\frac{\partial\theta}{\partial\tau} + \Pr\left(u\frac{\partial\theta}{\partial x} + w\frac{\partial\theta}{\partial z}\right) = \nabla^2\theta + \delta(1-C)^k e^{\frac{\theta}{1+\beta\theta}} - P\theta^4 \qquad (4.36)$$

$$\frac{\partial C}{\partial\tau} + \Pr\left(u\frac{\partial C}{\partial x} + w\frac{\partial C}{\partial z}\right) = \Le\nabla^2 C + \gamma\delta(1-C)^k e^{\frac{\theta}{1+\beta\theta}} \qquad (4.37)$$

$$\frac{\partial u}{\partial\tau} + \Pr\left(u\frac{\partial u}{\partial x} + w\frac{\partial u}{\partial z}\right) = \frac{4}{3}\Pr\nabla^2 u \qquad (4.38)$$

$$\frac{\partial w}{\partial\tau} + \Pr\left(u\frac{\partial w}{\partial x} + w\frac{\partial w}{\partial z}\right) = \frac{4}{3}\Pr\nabla^2 w + \Fr \qquad (4.39)$$

$$\frac{\partial u}{\partial x} + \frac{\partial w}{\partial z} = 0 \qquad (4.40)$$

Initial conditions are as follows:

$$\tau = 0, \qquad C(0, x, z) = u(0, x, z) = w(0, x, z) = 0, \qquad \theta = \theta_o \qquad (4.41)$$

Boundary conditions are as follows:

$$x = 0, \qquad x = 1, \qquad z = 0, \qquad z = 1, \qquad \theta = 0, \qquad \frac{\partial C}{\partial n} = 0 \qquad (4.42)$$

Then

$$\gamma = \frac{c_p R T_*^2}{QE}$$

is a dimensionless parameter that characterizes the amount of fuel burned in the compartment before the temperature had reached the baseline point of $T_* = 300°C$ ($0 < \lambda < 1$). If this parameter is small, then the fire will have a flashover point, and if it is large, the fire will proceed in a steady-state motion until the decay stage.

Parameter δ is calculated based on [7]

$$\delta = 12.1(\ln \theta_*)^{0.6} \tag{4.43}$$

Then $C = [1 - P(t)/P_o]$

is the concentration of the burned fuel product in the compartment.

The Prandtl number Pr in Eqs. (4.38) and (4.39) is the ratio of the kinematic viscosity and the thermal diffusivity. Pr is most insensitive to temperature in gases made up of the simplest molecules. However, if the molecular structure is very complex (e.g., in case of long-chain hydrocarbons), Pr might reach values on the order of 10^5. The convective heat-transfer coefficient can be characterized by the so-called Nusselt number, which is a function of the gas properties, temperature, and velocity. In the case of natural convection, the Nusselt number is related to the Rayleigh number (or Froude number in our case), which is recognized by the CFAST software [27]. For small Rayleigh numbers Ra (up to 10^6), this relationship is very weak; therefore, the Prandtl number Pr = 1 can be assumed in Eqs. (4.1) and (4.2). However, if $10^8 < Ra < 10^{10}$, CFAST uses the following relationship between Nusselt numbers and Rayleigh numbers [13]:

$$Nu = 0.12 Ra^{1/3} \tag{4.44}$$

On the other hand, the Prandtl numbers are estimated on the basis of the scaling laws and experimental data as follows [7]:

$$Nu = 1.615 \left(Re Pr \frac{h}{L} \right)^{1/3} \tag{4.45}$$

Finally, the estimated Prandtl numbers are as shown in Table 4.1 after equating Eqs. (4.19) and (4.20).

Fr	Fr < 10^6	Fr = 10^7	Fr = 10^8	Fr = 10^9	Fr = 10^{10}
Pr	1.0	7.0	10.0	27.0	120.0

TABLE 4.1   Estimated Prandtl Numbers

4.5   Real Fire Modeling with Convection

Let's note that the solutions of Eqs. (4.38) and (4.39) do not include the effect of convection in real fire compartment modeling processes. It is possible to neglect convection if the size of fire compartment is small (e.g., the size of a standard fire test camera). However, in the case of a real fire in a building (particularly in a large open space), the convection process plays a substantial role in establishing the velocities field in the fire compartment, as well as providing the limitations on the size of the compartment when the assumption about steady gas flow is not valid anymore and the turbulent effect has to be taken into consideration. In order to achieve this goal, let's find the approximate solution of Eqs. (4.38) and (4.39) by using the Galerkin method.

The *discontinuous Galerkin* (DG) *method* is often referred to as a hybrid or mixed method because it combines features of both finite-element and finite-volume methods. The solution is represented within each element as a polynomial approximation (as in FEM), whereas the interelement convection terms are resolved with upwind numerical flux formulas (as in FVM). While the DG method was developed in the early 1970s, it was not used for the computational fluid dynamics (CFD) simulations until the early 1990s. The solution of the Navier-Stokes equations with the DG method was presented in publications [15,16]. As the method gained more attention in the CFD research community, further advances came fairly rapidly. Researchers are now using the DG method to perform simulations of a wide variety of flow regimes. The method has been adapted for use with compressible and incompressible, steady and unsteady, and laminar and turbulent conditions.

The DG method is derived from the finite-element method, which is itself a variational method. In contrast to most other finite-element techniques, the DG method specifies that there be no explicit continuity requirements for the solution representation at the element boundaries. The solutions state is represented by a collection of piecewise discontinuous functions. Formally, this is accomplished by setting the basis (or shape) functions within each element. In order to simplify the end results, only one element will represent the whole fire compartment (similar to a one-zone model).

To obtain the governing equations for the DG method, we begin by defining the basis functions:

$$U = U(t) \sum_m \cos \frac{m\pi x}{b} \sum_n \sin n\pi z \qquad (4.46)$$

$$W = W(t) \sum_i \sin \frac{i\pi x}{b} \sum_j \sin j\pi z \qquad (4.47)$$

Let's now introduce trial solutions of Eqs. (4.46) and (4.47) to Eqs. (4.38) and (4.39) ($m = 1, n = 1, i = 1, l = 1$):

$$U = U(t)\cos\frac{\pi x}{b}\sin\pi z \qquad (4.48)$$

$$W = W(t)\sin\frac{\pi x}{b}\sin\pi z \qquad (4.49)$$

To obtain the solution, let's multiply Eqs. (4.38) and (4.39) by the basis function Eqs. (4.48) and (4.49) and integrate over the interval $[0; 1]$:

$$\int_0^1\int_0^1 U(t,x,z)[L_1(u;w)]\,dx\,dz = 0 \qquad (4.50)$$

$$\int_0^1\int_0^1 W(t,x,z)[L_2(u;w)]\,dx\,dz = 0 \qquad (4.51)$$

where $L_1(u; w)$ and $L_2(u; w)$ are the differential operators of Eqs. (4.38) and (4.39), respectively. Let's assume for now that $b = L/h = 1$ and $\text{Pr} = 1$, L is the horizontal dimension of fire compartment, h is the height of the compartment, dimensionless parameter Froude number $\text{Fr} = gh^3/va$ is a characteristic velocity of flow, g is the acceleration of gravity, a is the thermal diffusivity, v is the kinematic viscosity. The results are as follows:

$$\dot{U}(t) = 1.14U^2 - 26.3U \qquad (4.52)$$

$$\dot{W}(t) = -1.14UW - 26.3W - 0.406\text{Fr} \qquad (4.53)$$

Equations (4.52) and (4.53) are quasi-linear ordinary differential equations (ODEs) with dimensionless parameters and dimensionless variables. The solutions are presented in Chap. 5, along with the approximate solutions of heat and mass conservation equations.

References

1. Karlsson, Bjorn, and Quintiere, James G., *Enclosure Fire Dynamics.* CRC Press, Boca Raton, FL, 2000.
2. Quintiere, James G., *Principles of Fire Behavior.* Delmar Publishers, Albany, NY, 1998.
3. Drysdale, Dougal, *An Introduction to Fire Dynamics*, 2nd ed. Wiley, West Sussex, England, 1999.
4. National Fire Protection Association (NFPA). Guide for Fire and Explosion Investigations, NFPA 921–2001. NFPA, Quincy, MA, 2004.
5. National Fire Protection Association (NFPA). *Manual on Fire Hazard Properties of Flammable Liquids, Gases, and Volatile Solids*, NFPA 325M. NFPA, Quincy, MA, 1991.

6. Zeldovich, Ya. B., Barenblatt, G. I., Librovich V. B., and Makhviladze, G. M., *The Mathematical Theory of Combustion and Explosions*. Consultants Bureau, New York, 1985.

7. Frank-Kamenestkii, D. A., *Diffusion and Heat Transfer in Chemical Kinetics*. Plenum Press, New York, 1969.

8. Lewis, B., and Von Elbe, G., *Combustion, Flames, and Explosions of Gases*. Academic Press, New York, 1987.

9. Laider, J., Meiser, John H., and Meiser, Bryan C., *Physical Chemistry*, 4th ed. Houghton-Mifflin, Boston, 2003.

10. Acheson, D. J., *Elementary Fluid Dynamics* (Oxford Applied Mathematics and Computing Science Series). Oxford University Press, Oxford, UK, 1990.

11. Batchelor, G. K., *An Introduction to Fluid Dynamics*. Cambridge University Press, Cambridge, UK, 1967.

12. Lienhard, John H., IV, and Lienhard, John H., V, *Heat Transfer Textbook*, 3rd ed. Phlogiston Press, Cambridge, MA, 2008.

13. National Fire Protection Association (NFPA). CFAST: Consolidated Model of Fire Growth and Smoke Transport (Version 6), NIST Special Publication 1026. NIST, Gaithersburg, MD, 2008.

14. Atreya, A., "Convection Heat Transfer," in *The SFPE Handbook of Fire Protection Engineering*, 3rd ed. National Fire Protection Association, Quincy, MA, 2002.

15. Arnold, D. N., Brezzi, F., Cockburn, B., and Marini, L. D., "Unified Analysis of Discontinuous Galerkin Methods for Elliptic Problems," *Siam J. Numer. Anal.* 39(5):1749–1779, 2002.

16. Hesthaven, J. S., and Warburton, T., *Nodal Discontinuous Galerkin Methods: Algorithms, Analysis, and Applications* (Springer Texts in Applied Mathematics 54). Springer Verlag, New York, 2008.

CHAPTER 5

Simplifications of Differential Equations

Notation

k	Thermal conductivity, which has the dimensions W/m K or J/m s K
T	Temperature
d	Thickness in the direction if heat flow
ρ	Air density
c	Specific heat capacity
K	Number of collisions that result in a reaction per second
A	Total number of collisions
E	Activation energy
R	Ideal gas constant
P	Losses of heat owing to thermal radiation
e	Emissivity factor
σ	Boltzmann constant ($\sigma = 5.6703 \times 10^{-8}$ W/m^2 K^4)
T_o	Ambient temperature
A_v	Area of openings
c_p	Average specific heat at constant pressure

t	Time
$\bar{v}(u;v;w)$	Velocity vector
D	Diffusion coefficient (m²/s)
p	Pressure
ν	Kinematic viscosity; $\nu = \mu/\rho$
θ	Dimensionless temperature
τ	Dimensionless time
h	Height of the compartment (m)
a	Thermal diffusivity (m²/s)
Time	$t = \dfrac{h^2}{a}\tau$ (s)
Temperature	$T = \dfrac{RT_*^2}{E}\theta + T_*$ (K), where $T_* = 600$ K is the baseline temperature
Coordinates	$\bar{x} = x/h$ and $\bar{z} = z/h$, where x and z are dimensionless coordinates
Velocities	$\bar{u} = \dfrac{\nu}{h}u$ (m/s) and $\bar{w} = \dfrac{\nu}{h}w$ (m/s) are the horizontal and vertical components of velocity, where u and w are dimensionless velocities
ν	Kinematic viscosity (m²/s)
$\mathrm{Pr} = \nu/a$	Prandtl number
$\mathrm{Fr} = \dfrac{gh^3}{\nu a}$	Froude number
g	Gravitational acceleration
$\mathrm{Le} = a/D = \mathrm{Sc}/\mathrm{Pr}$	Lewis number
$\mathrm{Sc} = \nu/D$	Schmidt number
$\beta = \dfrac{RT_*}{E}$	Dimensionless parameter
$\gamma = \dfrac{c_p RT_*^2}{QE}$	Dimensionless parameter

$$P = \frac{e\sigma K_v (\beta T_*)^3 h}{\lambda}$$

Thermal radiation dimensionless coefficient

$\sigma = 5.67 \times 10^{-8}$ (W/m^2 K^4)

Boltzman constant

$K_v = A_o h / V$

Dimensionless opening factor

A_o

Total area of vertical and horizontal openings

$$\delta = \left(\frac{E}{RT_*^2}\right) Qz \left[\exp\left(-\frac{E}{RT_*}\right)\right]$$

Frank-Kamenetskii's parameter

$C = [1 - P(t)/P_o]$

Concentration of the burned fuel product in the fire compartment

$$\bar{W} = \frac{v}{h} W$$

Vertical component of gas velocity

$$\bar{U} = \frac{v}{h} U$$

Horizontal component of gas velocity

$b = L/h$

Length L (width) and height h of fire compartment, accordingly

W, U

Dimensionless velocities

5.1 International Code Requirements Review

The aim of structural fire engineering design is to ensure that structures do not collapse when subjected to high temperatures in a fire. Design of structures for fire still relies on single-element behavior in the fire-resistance test. The future of structural fire engineering design has to be evaluated in terms of the whole performance-based design of structures for fire. This should include natural fire exposures, heat-transfer calculations, and whole structural system behavior, recognizing the interaction of all elements of the structure in the region of the fire and any cooler elements outside the boundary of the compartment. Prescriptive fire grading and design methods based on heating single elements in a fire-resistance test oversimplify the whole fire design process. The real problem can be addressed by performance-based design methods, where possible fire scenarios are investigated and fire temperatures are calculated based on the compartment size, shape, ventilation, assumed fire load, and thermal properties of the fuel itself. The temperatures achieved by the connected structure then can be determined by heat-transfer analysis. Traditionally, steel and reinforced-concrete fire design has been based on fire resistance testing, although fire resistance by calculation also has been implemented for many years.

Analysis of a small number of room fire tests revealed that fire load was an important factor in determining fire severity. It has been suggested that fire severity could be related to the fire load of a room and expressed as an area under the temperature-time curve. The severities of two fires were equal if the area under the temperature-time curves were equal (above a baseline of 300°C). Thus, any fire temperature-time history could be compared with the standard curve. This approach obviously has limited applicability with respect to structural design. The structural engineer obviously is interested in knowing not only the temperature-time relationship but also the second derivative of such a function, which creates the acceleration and therefore the dynamic forces that are acting on the structural system on top of static forces owing to temperature elongations. The real fire test normally is presented by the double-curvature temperature-time function, whereas the standard test is presented by a single-curvature function—and this results in a significant difference for structural design. In addition, real fire computer simulations [1] of the temperature-time curves have "small" oscillations along the curve that create additional dynamic forces. The area under the temperature-time curve obviously doesn't provide the answers to all these questions.

The Eurocodes are a collection of the most recent methodologies for structural fire load design. *Eurocode 3: Design of Steel Structures*, Part 1.2: Structural Fire Design (EC3), and *Eurocode 4: Design of Steel and Composite Structures*, Part 1.2: Structural Fire Design (EC4), were formally approved in 1993 [2]. Each Eurocode is supplemented by a National Application Document (NAD) appropriate to the country. This document details safety factors and other issues specific to that country. The Steel Construction Institute (SCI) of the United Kingdom has published a guide comparing EC3 and EC4 with BS 5950 to aid the transition for designers in the United Kingdom. All Eurocodes are presented in a limit-state format where partial safety factors are used to modify loads and material strengths. EC3 and EC4 are very similar to BS 5950, Part 8, although some of the terminology differs. EC3 and EC4, Parts 1.2, and BS 5950, Part 8, are concerned only with calculating the fire resistance of steel or composite sections. Three levels of calculation are described in EC3 and EC4: tabular methods, simple calculation models, and advanced calculation models. Tabular methods are look-up tables for direct design based on parameters such as loading, geometry, and reinforcement. They relate to most common designs. Simple calculations are based on principles such as plastic analysis taking into account reductions in material strength with temperature. These are more accurate than the tabular methods. Advanced calculation methods relate to computer analyses and are not used in general design.

Building codes worldwide are moving from prescriptive to performance-based approaches. Performance-based codes establish fire safety objectives and leave the means for achieving those objectives to the designer. One of the main advantages of performance-based designs is that the most recent models and fire research can be used by practicing engineers, inevitably leading to innovative and cost-effective designs. Prescriptive codes are easy to use, and building officials can quickly determine whether a design follows code requirements. However, the application of prescriptive methods to many modern-design buildings is very questionable. This is especially true of modern steel-framed buildings. The fire-resistance ratings in building codes were not made for these types of structures. By assuming a worst-case but realistic natural fire scenario and calculating the heat transfer to the steel, the load-carrying capacity of the steel members can be checked at high temperatures, and requirements for fire protection, if any, can be judged in a rational manner.

Performance-based design has been documented in the literature extensively over the past 10 years [3,4]. It has been reported that by 1996 there were 14 countries (Australia, Canada, Finland, France, England, Wales, Japan, The Netherlands, New Zealand, Norway, Poland, Spain, Sweden, and the United States) and 3 organizations [ICC, ISO, and International Council for Research and Innovation in Building and Construction (CIB)] actively developing or using performance-based design codes for fire safety. Performance-based fire safety engineering design is now implemented and accepted in many countries. The design methodology has key advantages over prescriptive-based design. Structural behavior in fire depends on a number of variables. These include material degradation at elevated temperature and restraint stiffness of the structure around the fire compartment.

The energy- and mass-balance equations for the fire compartment can be used to determine the actual thermal exposure and fire duration. This is known as the *natural fire method*. This method allows the combustion characteristics of the fire load, the ventilation effects, and the thermal properties of the compartment enclosure to be considered. It is the most rigorous means of determining fire duration. The rapid growth of computing power and the corresponding maturing of computational fluid dynamics (CFD) have led to the development of CFD-based field models applied to fire research problems. The use of CFD models has allowed the description of fires in complex geometries and the incorporation of a wide variety of physical phenomena. The differential equations are solved numerically by dividing the physical space where the fire is to be simulated into a large number of rectangular cells. Within each cell, the gas velocity, temperature, and so on are assumed to be uniform, changing only with time. The accuracy with which the fire dynamics can be simulated depends on the number of cells that can be incorporated into the simulation structural analysis; therefore, the simplifications and approximations of the structural fire load are absolutely essential.

5.2   Structural Fire Load Design

The analytical approach in the structural fire engineering field typically comprises thermal and subsequent structural analyses of a building. When designing structures for structural fire load, the first step is to calculate the temperature field within the floor area (or part of it) and then the ultimate strength capacity, based on the temperatures assessed. This is possible by using the simplified (but conservative) design method or the more sophisticated global analysis and design in accordance with the structural code requirements [ACI 318 or American Institute of Steel Construction (AISC)]. The simplification (where it is possible for the determination of structural fire load only) is the key element of the methodology proposed here. The overall system of conservation of energy, mass, and momentum equations that are analyzed here is similar to the Fire Dynamics Simulator (FDS) model [1]. However, the limitations and simplifications are different because they are concentrated on a narrowly focused problem: structural fire load (SFL). For example, the large eddy simulation technique, the mixture fraction combustion model, pyrolysis, sprinkler and smoke detector locations and activations, and so on are not needed (or can be simplified) in our case. The FDS model solves the conservation equations of mass, momentum, and energy by using the finite-difference method, and the solution is updated in time on a three-dimensional (3D) rectilinear grid. However, the thermal radiation is computed using a finite-volume technique. The method proposed in this book uses the spatial averaging of variables; therefore, it is similar to the two-zone method in this respect. Consequently, this method has an intermediate position between the FDS and two-zone methods.

Heat can travel throughout a burning building by conduction, convection, or radiation. Since the existence of heat within a substance is caused by molecular action, the greater the molecular activity, the more intense is the heat. *Conduction* is heat transfer by means of molecular agitation within a material without any motion of the material as a whole. The energy in this case will be transferred from the higher-speed particles to the slower ones with a net transfer of energy to the slower ones. *Convection* is heat transfer by mass motion of a fluid such as air when heated is caused to move away from the source of heat, carrying energy with it. Convection above a hot surface occurs because hot air expands and rises upward, causing convection currents that transport energy (see Fig. 5.1). Configurations of these

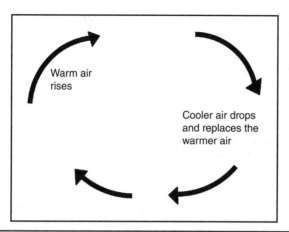

FIGURE **5.1**   Convection currents in a fire compartment.

currents depend in part on the geometry of the fire compartment and the opening factor.

Radiated heat is one of the major sources of fire spread. This method of heat transmission is known as *radiation of heat waves*. Heat and light waves are similar in nature, but they differ in length per cycle. Heat waves are longer than light waves, and they are sometimes called *infrared rays*. Radiated heat will travel through space in all directions. *Flame* is the visible, luminous body of a burning gas. When a burning gas is mixed with the proper amount of oxygen, the flame becomes hotter and less luminous. This loss of luminosity is due to a more complete combustion of the carbon. For these reasons, flame is considered to be a product of combustion. Heat, smoke, and gas, however, can develop in certain types of smoldering fires without evidence of flame.

In order to simplify the heat- and mass-balance equations, the following is assumed in this book:

1. The heat transfer owing to conduction can be neglected, similar to the assumption in a two-zone model—the spatial averaging of temperature.

2. The increase in energy flux (in addition to heat of combustion release) is due to natural convection.

3. The loss of energy through the openings is due to radiation only (the conductive heat loss to the walls and the convective heat loss through the openings are neglected owing to their much weaker dependence on temperature).

The overall SFL design process can be separated into the activities illustrated in Fig. 5.2. In addition, the following flowchart emphasizes that the ultimate strength and overall stability of a structure very much depends on the assessed design fire scenario:

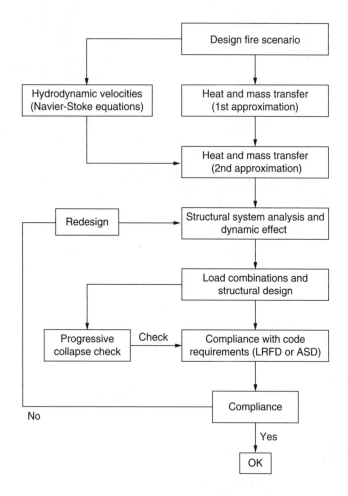

Consider now nonlinear singularly perturbed parabolic system representing the energy and mass conservation law [5]:

$$\frac{\partial \theta}{\partial \tau} + \Pr\left(U\frac{\partial \theta}{\partial x} + W\frac{\partial \theta}{\partial z}\right) = \delta(1-C)^k \exp\left(\frac{\theta}{1+\beta\theta}\right) - P\theta^4 \tag{5.1}$$

$$\frac{\partial C}{\partial \tau} + \Pr\left(U\frac{\partial C}{\partial x} + W\frac{\partial C}{\partial z}\right) = \gamma\delta(1-C)^k \exp\left(\frac{\theta}{1+\beta\theta}\right) \tag{5.2}$$

where U and W are the horizontal and vertical velocities, respectively, that should be obtained from the Navier-Stoke equations [Eqs. (4.38) and (4.39)].

Thermal input data are as follows: heat-transfer coefficients, thermal conductivity and thermal capacity of the air-gas mixture, heat effect (chemical reaction heat effect), activation energy, and so on. The thermal analysis comprises a determination of the temperature field versus time in the structural components under design consideration.

Based on the results obtained in the thermal analysis and on the structural input data, the reduced stiffness can be calculated in the structural analyses. Structural input data encompass mechanical properties (strength, modulus of elasticity, and stress-strain relation) as a function of temperature and structural boundary conditions. If the calculated ultimate strength capacity of structural members does not exceed the SFL effect, redesign actions must be taken (see the preceding flowchart).

Let's now consider the spatial averaging of temperature and combustion rate (the unsteady process of chemical reaction). Equations (5.1) and (5.2) are simplified further [7]:

$$\frac{\partial \theta}{\partial \tau} = \delta(1-C)^k \exp\left(\frac{\theta}{1+\beta\theta}\right) - P\theta^4 \qquad (5.3)$$

$$\frac{\partial C}{\partial \tau} = \gamma\delta(1-C)^k \exp\left(\frac{\theta}{1+\beta\theta}\right) \qquad (5.4)$$

Equations (5.3) and (5.4) describe the unsteady combustion process at any temperature level. However, as stated in Frank-Kamenestkii [8], the parameter δ is calculated for the temperature close to the flashover point. In our case of SFL, the postflashover stage of the fire (the fully developed fire) is the most important one. As was stated in Chap. 4, the gas temperature ranges in this case from approximately 1,100°C at the overhead level down to over 600°C at the floor; therefore, the average temperature in the space is 850°C (or 1,150 K). The dimensionless temperature and parameter δ now can be calculated based on Eq. (4.43) as follows:

$$\theta = \frac{E(T-T_*)}{RT_*^2} = \frac{T-T_*}{\beta T_*} = \frac{1{,}150-600}{0.1(600)} = 9.17$$

$$\delta = 12.1(\ln 9.17)^{0.6} = 20 \qquad (5.5)$$

It is assumed here that the fire starts at the floor level of a compartment and occupies (in the 2D case) 20 percent of the linear dimension and 35 percent of the compartment height. The average temperature in this area is assumed also to be 850°C. The equivalent rise in temperature in the whole volume, then, is $\Delta T = (0.2)(0.3)850 = 60°C$. The initial average temperature in the fire compartment (above the baseline temperature of 300°C) is $T = 300 + 60 = 360°C$ (or 660 K). The initial dimensionless average temperature in the fire compartment is

$$\theta_1 = \theta(\tau = 0) = \frac{T - T_*}{\beta T_*} = \frac{660 - 600}{0.1(600)} = 1.0 \qquad (5.5a)$$

Equation (5.5a) represents the initial condition for differential Eq. (5.3). The second equation (5.4) has an obvious initial condition, that is, $C(\tau = 0) = 0$, so there is no product of chemical reaction at the beginning of fire.

The fire engineering begins with the development of a design fire exposure to the structure. This normally takes the form of a time-temperature curve based on the fire load, ventilation, and thermal properties of the bounding surfaces (i.e., walls, floor, and ceiling). Design fire loads depend on the occupancy and other fire protection features of the building. The analysis involves defining the design fire exposure and the thermal response of the structural system. In Annex E of *Eurocode 1* the fuel-load densities per floor area for different occupancies are presented, and they are illustrated in Table 5.1. In some other European documents, the fuel load is presented as a density per total enclosed area of a compartment. The transformation of floor area to enclosed area must be done first in these cases and is presented below. For example, if the floor area is chosen to be 20 or 50 m² the room size can be $l \times b \times h = 5 \times 4 \times 2.5$ or $10 \times 5 \times 2.5$, and the enclosing area is 85 and 175 m² respectively. For a dwelling unit, the 80 percent fractile fuel load for the enclosed area can be calculated as follows: $q = 948(50)/175 = 270$.

The corresponding values are given in columns 3 and 4 of the Table 5.1.

	Fire Load Densities $q_{f,k}$ (MJ/m² Floor Area)			
Occupancy	Average	80% Fractile	80% Fractile	
Floor area	—	—	20 m²	50 m²
Dwelling	780	948	225	270
Hospital (room)	230	280	66	80
Hotel (room)	310	377	89	108
Library	1500	1824	120	146
Office	420	511	82	99
Classroom of a school	285	347	—	104
Shopping center	600	730	—	35
Theater (cinema)	300	365	—	—
Transport (public space)	100	122	—	—

TABLE 5.1 Fire Load Densities (of Floor Area)

Category	Fuel Load L (MJ/m²)	Maximum Temperature T_{max} (K)	Maximum Dimensionless Temperature θ_{max}	Parameter γ from Eq. (5.4)
Ultrafast	$500 < L < 700$	$1020 < T_{max} < 1300$	$7.0 < \theta_{max} < 11.67$	$0 < \gamma < 0.05$
Fast	$300 < L < 500$	$880 < T_{max} < 1020$	$4.67 < \theta_{max} < 7.0$	$0.05 < \gamma < 0.175$
Medium	$100 < L < 300$	$820 < T_{max} < 880$	$3.67 < \theta_{max} < 4.67$	$0.175 < \gamma < 0.275$
Slow	$50 < L < 100$	$715 < T_{max} < 820$	$1.92 < \theta_{max} < 3.67$	$0.275 < \gamma < 1.0$

Note: If fuel load $L > 700$ MJ/m² select $\gamma = 0$.

TABLE 5.2   Fire Severity

Based on the Society of Five Protection Engineers (SFPE) guide [13] and Swedish fire curves [9,10] for postflashover realistic fire exposure, we can standardize fires as shown in Table 5.2.

The direct solution of Eqs. (5.3) and (5.4) is the normal way of solving the problem (obtaining the temperature-time function in a fire compartment). However, in the case of a developed fire in a large building volume, mathematical modeling of the physical and chemical transformations of real materials is known to have only with a small degree of confidence. At the same time, based on many full fire test results data, one can expect that curtain parameters, such as the maximum temperature, type of temperature-time function, and so on, are well known. On the other hand, some other parameters [e.g., parameter γ from Eq. (5.4)] are known with some degree of approximation. From a physical point of view, this parameter characterizes the ratio of heat loss (e.g., from considerable quantities of soot) during the development stage of a fire (incipient and free-burning) divided by total energy released (heat rate) [11], that is,

$$\gamma = \frac{c_p R T_*^2}{QE}$$

If, for example, the heat rate of a chemical reaction is large and/or the losses are small, then parameter γ is small. Therefore, parameter γ has a bounded variation between 0 and 1.

It is also important to underline here that for any given value of parameter γ from the interval [0; 1], only one solution of Eqs. (5.3) and (5.4) exists, and the temperature-time function in this case has only one maximum value. It can be seen by observation (see Figs. 5.2–5.5) that this maximum temperature value increases when parameter γ decreases from 1 to 0. On the other hand, the maximum gas temperature in a real fire compartment and the fuel load define the category of fire severity (see Table 5.2); therefore, there is a correlation between the fire severity category and the value of parameter γ. In order to establish this correlation, the *mathematical optimum control theory* will be used here. For the mathematical background of this theory, see Evans [12]. The idea and application of this theory in our case are presented below.

In the case of a fully developed fire in a large building volume, the physical and chemical transformations of real materials occur in a very small flame zone under very high temperatures (much higher than the average gas temperature in a fire compartment); therefore, it is very difficult (if not impossible) to obtain these data (i.e., specific heat, thermal conductivity parameter, thermal diffusivity, etc.) under the regular laboratory conditions. The fire engineering community is fully aware of this fact, and corresponding tasks and recommendations regarding possible improvements in this area of expertise are provided in the SFPE guide [13]. A National Institute of Standards and Technology (NIST) special publication (*Mathematical Model of FDS*) [14] calls them "uncertain parameters." In this book, we will call them simply "unknown parameters." Therefore, in our case, any solution of differential equations (5.3) and (5.4) is a function of two independent variables: t (time) and γ from an interval [0; 1]. Now, in order to select the solution that is needed, an additional condition has to be imposed on any given system of differential equations. Each fire severity category (see Table 5.2) is defined by a corresponding maximum (averaged up in space) gas temperature T_{max} in a given fire compartment and the fuel load. These values are used here as an additional condition required by the optimum control method. The mathematical model of a real fire in a compartment now can be formulated as follows:

1. For each fixed number of γ from the interval [0; 1], find the discrete number of solutions of differential equations (5.3) and (5.4)—temperature-time curves—collection of functions.

2. Find the maximum values of temperature from this collection of functions.

3. Select the temperature-time curve from this collection of functions and the corresponding parameter γ if the difference between the maximum temperature from Table 5.2 (for each fire severity case) and the maximum value from item 2 is less than 1.0 percent.

4. Obviously, all solutions of differential equations have to be obtained in dimensionless forms (temperature θ and time τ) and then should be transferred into real temperatures and time variables (see "Notation" and Examples 6.3 and 6.4).

5. The optimum control method not only allows one to connect the old prescriptive method and the new approximate performance-based design method but also provides a partial verification of the results that are obtained by this method.

The solutions of Eqs. (5.3) and (5.4) (using the simple mathematical software Polymath) are presented below in tabular and analytical forms (the analytical formula in this case is the regression curve based on the tabular solution data). The reason for presenting the results in both forms is that the tabular solution allows the user to analyze some other regimes of fire development, such as fire growth period, decay, flashover period, and so on.

Now we can set up a typical computational problem from mathematical modeling of a real fire in a compartment for each category of fire exposure from Table 5.2. Each case of fire exposure is presented below by the envelope of solutions of differential equations (conservatively), and the data selected in each case are as follows:

Case 1: 1022 K < T_{max} < 1305 K, Ultrafast Fire

Data $T_* = 600$ K; $\delta = 20$; $K_v = 0.05$; $\beta = 0.1$; $P = 0.233$; $0 < \tau < 0.2$

Select $0 < \gamma < 0.05$

Differential Equations   Equations (5.3) and (5.4) are rewritten as an input for Polymath software:

1. $d(y0)/d(t) = 20*(1-y2)*exp(y0/(1+.1*y0)) -.233*y0^4$
2. $d(y2)/d(t) = 1.0*(1-y2)*exp(y0/(1+.1*y0))$
3. $d(y1)/d(t) = 0*(1-y1)^1.0*exp(y/(1+.1*y))$
4. $d(y)/d(t) = (1)* 20*(1-y1)^1.0*exp(y/(1+.1*y)) -.233*y^4$

where y is the dimensionless temperature θ with the corresponding parameter $\gamma = 0$, y0 is the dimensionless temperature θ with the corresponding parameter $\gamma = 0.05$, y1 is the concentration of the burned fuel product C in the fire compartment with the corresponding parameter $\gamma = 0$, and y2 is the concentration of the burned fuel product C in the fire compartment with the corresponding parameter $\gamma = 0.05$.

	Variable	Initial Value	Minimal Value	Maximal Value	Final Value
1	t	0	0	0.2	0.2
2	y	1	1	11.74485	11.74485
3	y0	1	1	7.049816	2.601609
4	y1	0	0	0	0
5	y2	0	0	0.9727725	0.9727725

Calculated Values of DEQ Variables

The differential equations [Eqs. (5.3) and (5.4)] cannot be integrated in closed form. A numerical integration of these equations is derived, and the results are presented in tabulated form in the appendix at the end of this chapter (see Table 5A.1) and in graph form in Fig. 5.2.

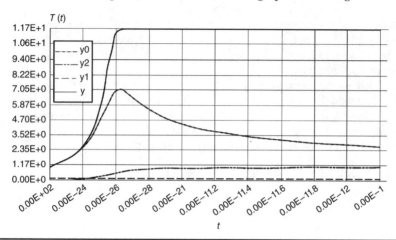

Figure 5.2 Dimensionless time-temperature curves.

Based on the tabulated data shown above (see also Table 5A.1), the final approximation of the dimensionless temperature-time curve can be presented as follows:

$$\theta = A\exp[-(\tau - a)^2/2\sigma^2] \tag{5.5}$$

with a second derivative of

$$\theta = -A/\sigma^2[\exp(-\tau/a-1)^2/2(\sigma/a)^2][1-(\tau/a-1)^2/(\sigma/a)^2] \tag{5.6}$$

where A, a, and σ are variables used in the Polymath software for nonlinear approximation of the dimensionless temperature θ. The second derivative of the temperature-time function is needed to account for possible dynamic effect of the SFL. Detailed analyses in these cases are presented and discussed in Chap. 7.

The nonlinear approximation is provided by the following model:

Model $y = A*(\exp(-(1*t-B)^2/(2*(C^2))))$

Variable	Initial Guess	Value
A	11	11.98158
B	1	0.0971132
C	1	0.0575742

$$A = 11.98; \qquad a = 0.097; \qquad \sigma = 0.0576 \tag{5.7}$$

Case 2: 882 K < T_{max} < 1022 K, Fast Fire

Data $T_* = 600$ K; $\delta = 20$; $K_v = 0.05$; $\beta = 0.1$; $P = 0.157$; $0 < \tau < 0.2$

Select $0.05 < \gamma < 0.175$

Differential Equations Equations (5.3) and (5.4) are rewritten as an input for the Polymath software:

$d(y0)/d(t) = 20*(1-y2)*exp(y0/(1+.1*y0)) -.157*y0^4$

$d(y2)/d(t) = 3.5*(1-y2)*exp(y0/(1+.1*y0))$

$d(y1)/d(t) = 1.0*(1-y1)^1.0*exp(y/(1+.1*y))$

$d(y)/d(t) = (1)* 20*(1-y1)^1.0*exp(y/(1+.1*y)) -.157*y^4$

where y is the dimensionless temperature θ with the corresponding parameter $\gamma = 0.05$, y0 is the dimensionless temperature θ with the corresponding parameter $\gamma = 0.175$, y1 is the concentration of the product of the first-order chemical reaction with $\gamma = 0.05$, y2 is the concentration of the product of the first-order chemical reaction with $\gamma = 0.175$.

	Variable	Initial Value	Minimal Value	Maximal Value	Final Value
1	t	0	0	0.2	0.2
2	y	1	1	8.201185	2.744677
3	y0	1	1	4.707775	2.389874
4	y1	0	0	0.9867073	0.9867073
5	y2	0	0	0.9997236	0.9997236

The tabulated solutions of Eqs. (5.3) and (5.4) are presented in Table 5A.2, and the graphs are shown in Fig. 5.3.

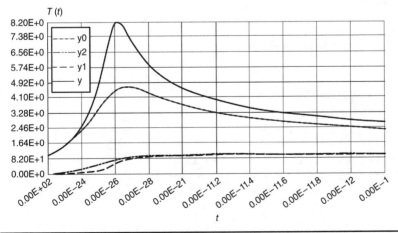

FIGURE 5.3 Dimensionless time-temperature curves.

Based on the tabulated data shown above, the final approximation of the dimensionless temperature-time curve can be presented as follows:

$$\boxed{\theta = A\exp[-(\tau - a)^2/2\sigma^2]}$$ (5.8)

with a second derivative of

$$\boxed{\theta = -A/\sigma^2[\exp(-(\tau/a - 1)^2/2(\sigma/a)^2][1-(\tau/a-1)^2/(\sigma/a)^2]}$$ (5.9)

where A, a, and σ are variables used in the Polymath software for nonlinear approximation of dimensionless temperature θ.

The nonlinear approximation is provided by the following model:

Model   y1 = A*(exp (−(t−B) ^2/ (2*(C^2))))

Variable	Initial Guess	Value
A	8	6.950116
B	1	0.0646352
C	1	0.0382137

$$A = 6.95; \quad a = 0.0646; \quad \sigma = 0.0382$$ (5.10)

Case 3: 822 K < T_{max} < 882 K, Medium Fire

Data $T_* = 600$ K; $\delta = 20$; $K_v = 0.05$; $\beta = 0.1$; $P = 0.157$; $0 < \tau < 0.2$

Select $0.175 < \gamma < 0.275$

Differential Equations   Equations (5.3) and (5.4) are rewritten as an input for the Polymath software:

1. d(y0)/d(t) = 20*(1−y2)*exp(y0/(1+.1*y0)) −.157*y0^4
2. d(y2)/d(t) = 5.5*(1−y2)*exp(y0/(1+.1*y0))
3. d(y1)/d(t) = 3.5*(1−y1)^1.0*exp(y/(1+.1*y))
4. d(y)/d(t) = (1)* 20*(1−y1)^1.0*exp(y/(1+.1*y)) −.157*y^4

where y is the dimensionless temperature θ with the corresponding parameter $\gamma = 0.175$, y0 is the dimensionless temperature θ with the corresponding parameter $\gamma = 0.275$, y1 is the concentration of the product of the first-order chemical reaction with $\gamma = 0.175$, and y2 is the concentration of the product of the first-order chemical reaction with $\gamma = 0.275$.

	Variable	Initial Value	Minimal Value	Maximal Value	Final Value
1	t	0	0	0.2	0.2
2	y	1	1	4.70778	2.389874
3	y0	1	1	3.707141	2.286563
4	y1	0	0	0.9997236	0.9997236
5	y2	0	0	0.9999598	0.9999598

Calculated Values of DEQ Variables

The tabulated solutions of Eqs. (5.3) and (5.4) are presented in Table 5A.3, and the graphs are shown in Fig. 5.4.

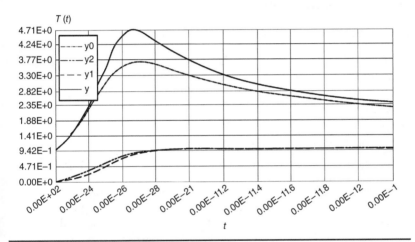

FIGURE 5.4   Dimensionless time-temperature curves.

Based on the tabulated data shown above, the final approximation of the dimensionless temperature-time curve can be presented as follows:

$$\theta = A\exp[-(\tau - a)^2/2\sigma^2] \tag{5.11}$$

with a second derivation of

$$\theta = -A/\sigma^2[\exp(-(\tau/a - 1)^2/2(\sigma/a)^2][1 - (\tau/a - 1)^2/(\sigma/a)^2] \tag{5.12}$$

where A, a, and σ are variables used in the Polymath software for nonlinear approximation of dimensionless temperature θ.

The nonlinear approximation is provided by the following model:

Model $y = A*(\exp(-(1*t-B)^2/(2*(C^2))))$

Variable	Initial Guess	Value
A	8	4.549296
B	1	0.0801842
C	1	0.0597693

$$A = 4.55; \qquad a = 0.0802; \qquad \sigma = 0.0598 \tag{5.13}$$

Case 4: 715 K < T_{max} < 822 K, Slow Fire

Data $T_* = 600$ K; $\delta = 20$; $K_v = 0.05$; $\beta = 0.1$; $P = 0.157$; $0 < \tau < 0.2$

Select $0.275 < \gamma < 1.0$

Differential Equations Equations (5.3) and (5.4) are rewritten as an input for the Polymath software:

1. $d(y0)/d(t) = 20*(1-y2)*\exp(y0/(1+.1*y0))-2.53*0-.157*y0^4$
2. $d(y2)/d(t) = 20*(1-y2)*\exp(y0/(1+.1*y0))$
3. $d(y1)/d(t) = 5.5*(1-y1)^1.0*\exp(y/(1+.1*y))$
4. $d(y)/d(t) = (1)* 20*(1-y1)^1.0*\exp(y/(1+.1*y))-2.53*0-.157*y^4$

where y is the dimensionless temperature θ with the corresponding parameter $\gamma = 0.275$, y0 is the dimensionless temperature θ with the corresponding parameter $\gamma = 1.0$, y1 is the concentration of the product of the first-order chemical reaction with $\gamma = 0.275$, and y2 is the concentration of the product of the first-order chemical reaction with $\gamma = 1.0$.

	Variable	Initial Value	Minimal Value	Maximal Value	Final Value
1	t	0	0	0.2	0.2
2	y	1	1	3.706608	2.286563
3	y0	1	1	1.915441	1.68161
4	y1	0	0	0.9999598	0.9999598
5	y2	0	0	1	1

Calculated Values of DEQ Variables

The tabulated solutions of Eqs. (5.3) and (5.4) are presented in Table 5A.4 and the graphs are shown in Fig. 5.5.

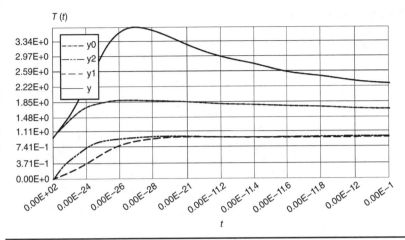

FIGURE 5.5   Dimensionless time-temperature curves.

Based on the tabulated data shown above, the final approximation of the dimensionless temperature-time curve can be presented as follows:

$$\theta = A\exp[-(\tau - a)^2/2\sigma^2]$$ (5.14)

with a second derivation of

$$\theta = -A/\sigma^2[\exp(-\tau/a - 1)^2/2(\sigma/a)^2][1-(\tau/a-1)^2/(\sigma/a)^2]$$ (5.15)

where A, a, and σ are variables used in the Polymath software for nonlinear approximation of dimensionless temperature θ.

The nonlinear approximation is provided by the following model:

Model   y = A*(exp (−(1*t−B) ^2/ (2*(C^2))))

Variable	Initial Guess	Value
A	8	3.7263
B	1	0.0893914
C	1	0.075

$$A = 3.73; \quad a = 0.0893; \quad \sigma = 0.075$$ (5.16)

Case 5: Impact Temperature Action

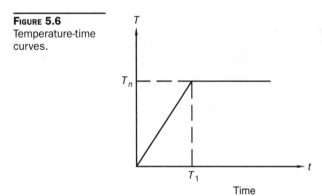

The simple bilinear graph in Fig. 5.6 illustrates the temperature impact load on a structural system (when $t_1 \to 0$). For more information, see the SFPE guide [13].

Case 6: Impact and Fluctuations

The real temperature-time curve has some fluctuations of maximum temperature owing to the hydrodynamic effect of fire propagation. It will be assumed in this case that these fluctuations are small ($\pm10°C$), but they appear with the frequency very close to the natural frequency of the structural system $(\theta = 0.95\omega)$: $T = T_m(1 + 0.01\sin\theta t)$ (see Fig. 5.7).

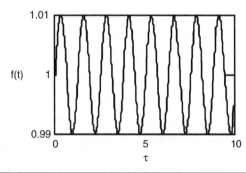

FIGURE 5.7   Fluctuation of T_{max}.

For the dynamic effect on structural systems in these two cases and examples of structural design, see Chap. 7.

5.3   Hydrodynamic Model

The hydrodynamic model [the Navier-Stokes equations (4.38) and (4.39) for the 2D case] is limited (similar to the FDS model) to low-speed, thermally driven flow with an emphasis on heat transport from fires owing to natural convection and radiation processes [14]. Irradiative heat transfer is included in the model via the Boltzmann law for a gray gas. The initial conditions in this case are

$$V(0) = -10{,}000 \quad \text{and} \quad W(0) = 0 \qquad (5.17)$$

The minus sign indicates that the gas is moving out from the fire compartment. The initial dimensionless horizontal velocity is based on the assumption that originally the fire compartment doesn't have any openings; therefore, there is some initial pressure and mass flow afterwards. The absolute value $V = 0.2$ m/s [kinematic viscosity at the referenced temperature $T = 600$ K, assumed $v = 50 \times 10^{-6}$ (m²/s) and height of the compartment $h = 2.5$ m] is very small, and it has some practical influence only when the Froude's parameter (Fr $= gh^3/av$) is small. The quantity av/h^3 is known as a characteristic velocity of flow, g is the acceleration of gravity, a is the thermal diffusivity, and h is the height of the fire compartment. The Froude number can be interpreted as the ratio of the inertial to gravity forces in the flow. For any large Froude's number (Fr $> 10^7$) that represents the real size of a fire compartment in a building, the vertical component W of the velocity governs the heat-convection process. The geometry of the fire compartment (the height and the length: $b = L/h$) obviously has a great influence on the velocities and also indicates the limits, where the turbulent effect cannot be neglected (similar to the CFAST limitations [6]).

Equations (4.17) and (4.18) are quasi-linear differential equations with dimensionless parameters and dimensionless variables. Again, these differential equations cannot be integrated in closed form. A numerical integration in this case is derived, and the results are presented in tabulated form in Tables 5A.5 through 5A.12 in the appendix at the end of this chapter.

Case 1: Fr = 1

	Variable	Initial Value	Minimal Value	Maximal Value	Final Value
1	t	0	0	0.2	0.2
2	u	−10000	−10000	−0.1202038	−0.1202038
3	w	0	0	0.0300632	0.0161793

Calculated Values of DEQ Variables

Differential Equations

1. $d(w)/d(t) = -1.14*u*w-26.3*w+0.406*10^0$
2. $d(u)/d(t) = 1.14*u^2-26.3*u$

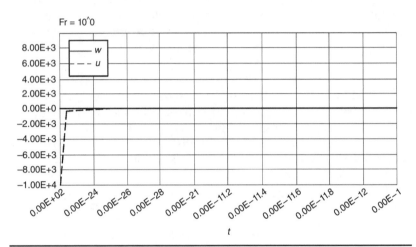

FIGURE 5.8   Dimensionless velocities U and W.

Model $u = a0 + a1*t$

Variable	Value
a0	−445.2943
a1	3292.236

The average dimensionless velocities are

$$U = -445.3 \quad \text{and} \quad W = 0 \tag{5.18}$$

Case 2: Fr = 10⁴

	Variable	Initial Value	Minimal Value	Maximal Value	Final Value
1	t	0	0	0.2	0.2
2	u	−10000	−10000	−0.1202038	−0.1202038
3	w	0	0	300.5611	161.7926

Calculated Values of DEQ Variables

Differential Equations

1. $d(w)/d(t) = -1.14*u*w-26.3*w+0.406*10^4$
2. $d(u)/d(t) = 1.14*u^2-26.3*u$

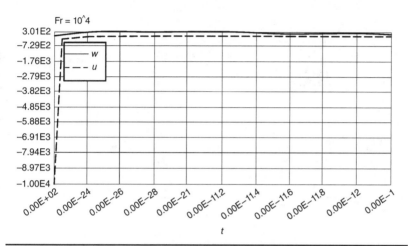

FIGURE 5.9   Dimensionless velocities U and W.

Finally,

$$U = 0 \quad \text{and} \quad W = 300.6 \tag{5.19}$$

Case 3: Fr = 10^5

	Variable	Initial Value	Minimal Value	Maximal Value	Final Value
1	t	0	0	0.2	0.2
2	u	−10000	−10000	−0.1202038	−0.1202038
3	w	0	0	3005.611	1617.926

Calculated Values of DEQ Variables

Differential Equations

1. $d(w)/d(t) = -1.14*u*w-26.3*w+0.406*10^5$
2. $d(u)/d(t) = 1.14*u^2-26.3*u$

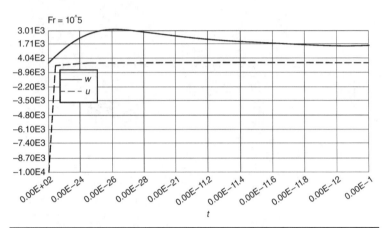

FIGURE 5.10   Dimensionless velocities U and W.

Model $w = a0 + a1*t + a2*t^2 + a3*t^3 + a4*t^4$

Variable	Value
a0	272.9223
a1	1.388E+05
a2	−2.298E+06
a3	1.388E+07
a4	−2.858E+07

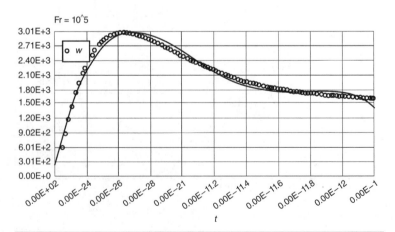

FIGURE 5.11   Dimensionless velocity W.

Finally,

$$W = 273+1.4(10^5)\ t-2.3(10^6)(t^2)+1.4(10^7)t^3-2.86(10^7)t^4 \quad (5.20)$$

Case 4: Fr = 10^6

	Variable	Initial Value	Minimal Value	Maximal Value	Final Value
1	t	0	0	0.2	0.2
2	u	−10000	−10000	−0.1202038	−0.1202038
3	w	0	0	3.006E+04	1.618E+04

Calculated Values of DEQ Variables

Differential Equations

1. $d(w)/d(t) = -1.14*u*w-26.3*w+0.406*10^6$
2. $d(u)/d(t) = 1.14*u^2-26.3*u$

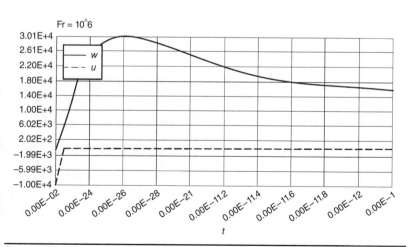

FIGURE 5.12   Dimensionless velocities *U* and *W*.

Model w = a0+a1*t+a2*t^2+a3*t^3+a4*t^4

Variable	Value
a0	2729.523
a1	1.388E+06
a2	−2.298E+07
a3	1.388E+08
a4	−2.858E+08

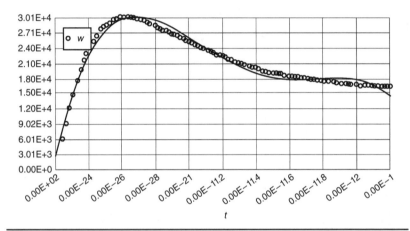

FIGURE 5.13 Dimensionless velocities *U* and *W*.

Finally,

$$W = 2730+1.4(10^6)t-2.3(10^7)t^2+1.4(10^8)t^3-2.86(10^8)t^4 \quad (5.21)$$

Case 5: Fr = 10^7

	Variable	Initial Value	Minimal Value	Maximal Value	Final Value
1	t	0	0	0.2	0.2
2	u	−10000	−10000	−2.345E−15	−2.345E−15
3	w	0	0	4.296E+04	2.205E+04

Calculated Values of DEQ Variables

Differential Equations

1. $d(w)/d(t) = -8*u*w - 184.1*w + 0.406*10^7$
2. $d(u)/d(t) = 8*u^2 - 184.1*u$

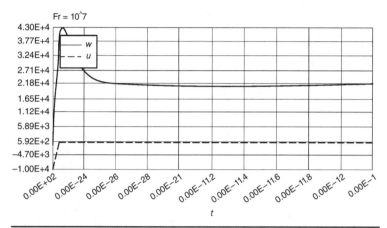

FIGURE 5.14 Dimensionless velocities U and W.

Model $w = a0 + a1*t$

Variable	Value
a0	2.575E+04
a1	-2.646E+04

Finally,

$$U = 0; \quad W = 2.58(10^4) \tag{5.22}$$

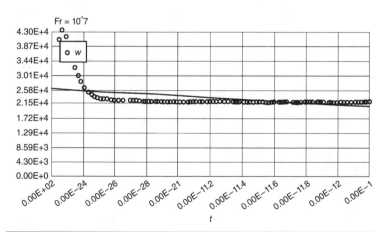

FIGURE 5.15 Dimensionless velocities U and W.

Case 6: Fr = 10⁸

	Variable	Initial Value	Minimal Value	Maximal Value	Final Value
1	t	0	0	0.2	0.2
2	u	−10000	−10000	−3.296E−22	−3.296E−22
3	w	0	0	3.006E+05	1.544E+05

Calculated Values of DEQ Variables

Differential Equations

1. $d(w)/d(t) = -11.4*u*w-263*w+0.406*10^8$
2. $d(u)/d(t) = 11.4*u^2-263*u$

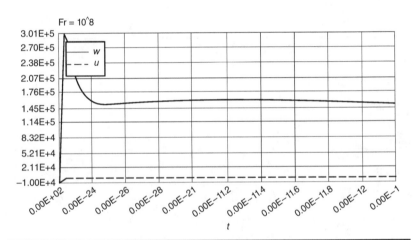

FIGURE 5.16 Dimensionless velocities U and W.

Model $w = a0 + a1*t$

Variable	Value
a0	1.689E+05
a1	−1.044E+05

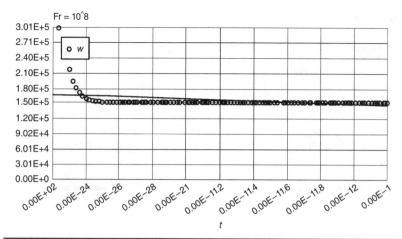

FIGURE 5.17 Dimensionless velocities U and W.

Finally,

$$U = 0; W = 1.69(10^{\wedge}5) \tag{5.23}$$

Case 7: Fr = 10^9

	Variable	Initial Value	Minimal Value	Maximal Value	Final Value
1	t	0	0	0.2	0.2
2	u	−10000	−10000	−3.058E–61	−3.058E-61
3	w	0	0	7.672E+05	5.718E+05

Calculated Values of DEQ Variables

Differential Equations

1. $d(w)/d(t) = -30.78*u*w-710*w+0.406*10^{\wedge}9$
2. $d(u)/d(t) = 30.78*u^{\wedge}2-710*u$

Model $w = a0 + a1*t$

Variable	Value
a0	5.597E+05
a1	9.133E+04

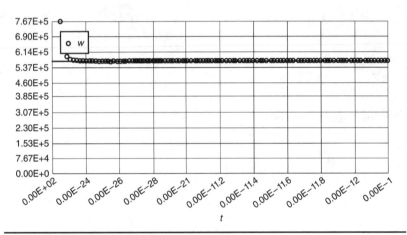

FIGURE 5.18 Dimensionless velocities U and W.

Finally,

$$U = 0; W = 5.6(10^5) \qquad (5.24)$$

Case 8: Fr $= 10^{10}$

	Variable	Initial Value	Minimal Value	Maximal Value	Final Value
1	t	0	0	0.2	0.2
2	u	−10000	−10000	1.058E−13	−5.834E−16
3	w	0	0	1.287E+06	1.286E+06

Calculated Values of DEQ Variables

Differential Equations

1. $d(w)/d(t) = -136.8*u*w-3156*w+0.406*10^{10}$
2. $d(u)/d(t) = 136.8*u^2-3156*u$

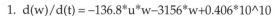
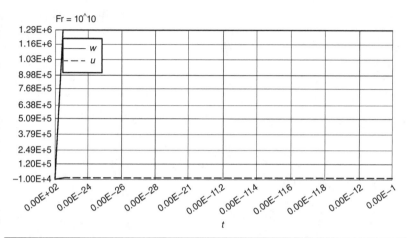

FIGURE 5.19   Dimensionless velocities U and W.

Model $w = a0 + a1*t$

Variable	Value
a0	1.233E+06
a1	3.909E+05

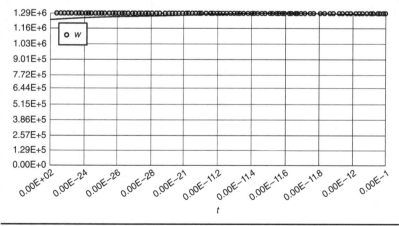

FIGURE 5.20   Dimensionless velocities U and W.

Finally,

$$U = 0; \quad W = 1.23(10^6) \quad (5.24)$$

One can see now (from all the preceding data) that the vertical component of stream flow velocity increases rapidly with an increase of the fire compartment height. The following relationship is established now between the average dimensionless velocities W and log Fr (see Table 5.15 and Fig. 5.20):

Log Fr	0	4	5	6	7	8	9	10
W (10×4)	0	0.04	0.30	3.0	3.0	14.5	53.7	129.0

TABLE 5.15   Velocity W versus Log Fr

Model $W = a0 + a1*t + a2*t^2 + a3*t^3 + a4*t^4$

Variable	Value
a0	0.0021672
a1	−26.91562
a2	14.90899
a3	−2.689609
a4	0.1597144

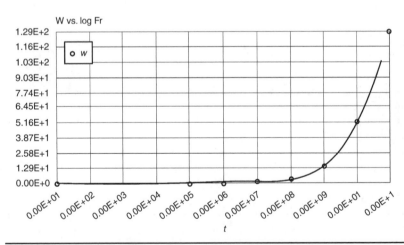

FIGURE 5.21   The average dimensionless velocity W versus log Fr.

Finally,

$$W = 0.0217 - 26.92(\log Fr) + 14.91(\log Fr)^2$$
$$- 2.69(\log Fr)^3 + 0.16(\log Fr)^4 \quad (5.25)$$

The velocities U and W now should be inserted into conservation of energy and mass equations [Eqs. (5.1) and (5.2)]. The horizontal component U is much smaller than the vertical component W and therefore can be neglected. It also should be underlined here that the range of Froude numbers to be considered here are as follows: $10^8 < \text{Fr} < 10^{10}$. Anything above this limit requires the turbulent theory approach to the hydrodynamic problem and is beyond the scope of this book.

Equations (5.1) and (5.2) now will have the addition term: $W(\partial\theta/\partial z)$. The temperature change is a very weak function of vertical coordinate z (as is stated, for example, in two-zone models [6]); therefore, it will be assumed as a constant value. In order to obtain this constant value, let's consider a steady-stream flow process ($W = \text{const.}$; see above). In this case, Eq. (5.1) can be rewritten as

$$W\frac{d\theta}{dz} = \frac{d^2\theta}{dz^2} + \delta\left(\exp\frac{\theta}{1+\beta\theta}\right) - P\theta^4 \qquad (5.26)$$

The numerical solution of Eq. (5.26) is presented below for the following cases.

1. $\text{Fr} = 10^{\wedge}8 \ W = 1.7 \ (10^{\wedge}5)$
2. $\text{Fr} = 10^{\wedge}9 \ W = 5.6 \ (10^{\wedge}5)$
3. $\text{Fr} = 10^{\wedge}10 \ W = 1.3 \ (10^{\wedge}6)$

Case 1: $W = 1.7 \times 10^5$ and $\text{Fr} = 10^8$

	Variable	Initial Value	Minimal Value	Maximal Value	Final Value
1	x	0	0	1	1
2	y	0	−0.0002911	0	−0.000291
3	z	1	0.999709	1	0.999709

Calculated Values of DEQ Variables

Differential Equations

1. $d(z)/d(x) = y$
2. $d(y)/d(x) = 1.0*.157*z^{\wedge}4-(1.0)*20*\exp(z/(1+.1*z))-y*1.7*10^{\wedge}5$

where functions y and z are as follows: $y = d\theta/dz$ and $z = \theta$.

Model   y = a0 + a1*x

Variable	Value
a0	−0.0002793
a1	−1.747E−05

$$\frac{d\theta}{dz} = 2.793 \times 10^{-4}$$

$$\boxed{W\frac{d\theta}{dz} = 47.48}$$ (5.28)

Case 2: W = 5.6 × (10⁵) and Fr = 10⁹

	Variable	Initial Value	Minimal Value	Maximal Value	Final Value
1	x	0	0	1	1
2	y	0	−8.836E−05	0	−8.836E−05
3	z	1	0.9999116	1	0.9999116

Calculated Values of DEQ Variables

Differential Equations

1. d(z)/d(x) = y
2. d(y)/d(x)=1.0*.157*z^4−(1.0)*20*exp(z/(1+.1*z))−y*5.6*10^5

Model   y = a0 + a1*x

Variable	Value
a0	−8.478E−05
a1	−5.348E−06

$$\frac{d\theta}{dz} = 8.478 \times 10^{-5}$$ (5.29)

$$\boxed{W\frac{d\theta}{dz} = 49.28}$$ (5.30)

Case 3: $W = 1.3 \times 10^6$ and $Fr = 10^{10}$

	Variable	Initial Value	Minimal Value	Maximal Value	Final Value
1	x	0	0	1	1
2	y	0	−3.806E−05	0	−3.806E−05
3	z	1	0.9999619	1	0.9999619

Calculated Values of DEQ Variables

Differential Equations

1. $d(z)/d(x) = y$
2. $d(y)/d(x) = 1.0*.157*z^4-(1.0)*20*\exp(z/(1+.1*z))- y*1.3*10^6$

Model $y = a0 + a1*x$

Variable	Value
a0	−3.652E−05
a1	−2.303E−06

$$\frac{d\theta}{dz} = 3.652 \times 10^{-5} \tag{5.31}$$

$$\boxed{W\frac{d\theta}{dz} = 47.45} \tag{5.32}$$

Now substitute values W and $d\theta/dz$ from Eqs. (5.27) through (5.32) into Eqs. (5.1) and (5.2):

$$\frac{d\theta}{d\tau} = \delta(1-C)\left(\exp\frac{\theta}{1+\beta\theta}\right) - P\theta^4 + A \tag{5.33}$$

$$\frac{dC}{d\tau} = \gamma\delta(1-C)\left(\exp\frac{\theta}{1+\beta\theta}\right) - P\theta^4 \tag{5.34}$$

The A value in Eq. (5.33) is as follows: Case 1: $A = 47.48$; case 2: $A = 49.28$; case 3: $A = 47.45$. Since all the A values are very close to each other, let's use only $A = 49.28$. Now, solving Eqs. (5.33) and (5.34) for each severity case (very fast, fast, medium, and slow; tabulated solutions are presented in Tables 5A.14 through 5A.17 in the appendix at the end of this chapter), and comparing the results with the previous solutions (without the effect of W and $d\theta/dz$), we are going to have solutions as follows.

Case 11: 1022 K < T_{max} < 1305 K, Ultrafast Fire

Data $T_* = 600$ K; $\delta = 20$; $K_v = 0.05$; $\beta = 0.1$; $P = 0.157$; $0 < \tau < 0.2$; $0 < \gamma < 0.05$; $A = 49.28$

	Variable	Initial Value	Minimal Value	Maximal Value	Final Value
1	t	0	0	0.2	0.2
2	y	1	1	11.82924	11.82924
3	y0	1	1	7.591702	3.885845
4	y1	0	0	0	0
5	y2	0	0	0.9919975	0.9919975

Calculated Values of DEQ Variables

Differential Equations

1. $d(y0)/d(t) = 20*(1-y2)*\exp(y0/(1+.1*y0))-2.53*0-.233*y0^4+ 49.28$

2. $d(y2)/d(t) = 1.0*(1-y2)*\exp(y0/(1+.1*y0))$

3. $d(y1)/d(t) = 0*(1-y1)^1.0*\exp(y/(1+.1*y))$

4. $d(y)/d(t)=(1)*20*(1-y1)^1.0*\exp(y/(1+.1*y))-2.53*0-.233*y^4+ 49.28$

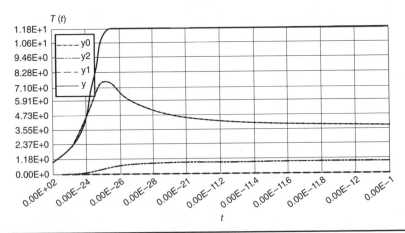

FIGURE 5.22   Dimensionless time-temperature curves.

The maximum temperature increase is 0.5 percent.

Case 21:   882 K < T_{max} < 1022 K; Fast Fire

Data $T_* = 600$ K; $\delta = 20$; $K_v = 0.05$; $\beta = 0.1$; $P = 0.157$; $0 < \tau < 0.2$; $0.05 < \gamma < 0.175$; $A = 49.28$

	Variable	Initial Value	Minimal Value	Maximal Value	Final Value
1	t	0	0	0.2	0.2
2	y	1	1	8.825593	4.250322
3	y0	1	1	5.742239	4.209868
4	y1	0	0	0.9971628	0.9971628
5	y2	0	0	0.9999996	0.9999996

Calculated Values of DEQ Variables

Differential equations

1. $d(y0)/d(t) = 20*(1-y2)*exp(y0/(1+.1*y0))-.157*y0^4+49.28$
2. $d(y2)/d(t) = 3.5*(1-y2)*exp(y0/(1+.1*y0))$
3. $d(y1)/d(t) = 1.0*(1-y1)^{1.0}*exp(y/(1+.1*y))$
4. $d(y)/d(t) = (1)*20*(1-y1)^{1.0}*exp(y/(1+.1*y))-.157*y^4+49.28$

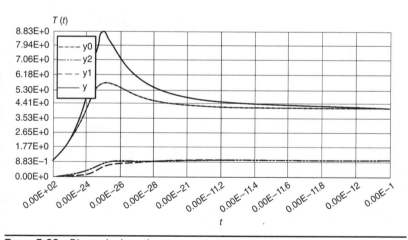

FIGURE 5.23   Dimensionless time-temperature curves.

The maximum temperature increase is 7.7 percent.

Case 31: 822 K < T_{max} < 882 K, Medium Fire

Data $T_* = 600$ K; $\delta = 20$; $K_v = 0.05$; $\beta = 0.1$; $P = 0.157$; $0 < \tau < 0.2$; $0.175 < \gamma < 0.275$; $A = 49.28$

	Variable	Initial Value	Minimal Value	Maximal Value	Final Value
1	t	0	0	0.2	0.2
2	y	1	1	5.739494	4.209868
3	y0	1	1	4.895717	4.209492
4	y1	0	0	0.9999996	0.9999996
5	y2	0	0	1	1

Calculated Values of DEQ Variables

Differential Equations

1. $d(y0)/d(t) = 20*(1–y2)*\exp(y0/(1+.1*y0))–.157*y0^4+49.28$
2. $d(y2)/d(t) = 5.5*(1–y2)*\exp(y0/(1+.1*y0))$
3. $d(y1)/d(t) = 3.5*(1–y1)^{1.0}*\exp(y/(1+.1*y))$
4. $d(y)/d(t)=(1)*20*(1–y1)^{1.0}*\exp(y/(1+.1*y))–.157*y^4+49.28$

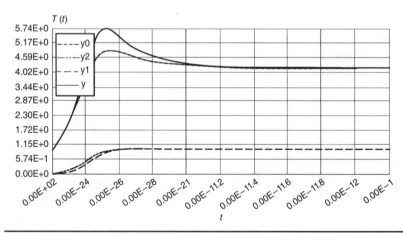

FIGURE 5.24   Dimensionless time-temperature curves.

The maximum temperature increase is 21.9 percent.

Case 41: 715 K < T_{max} < 822 K, Slow Fire

Data $T_* = 600$ K; $\delta = 20$; $K_v = 0.05$; $\beta = 0.1$; $P = 0.157$; $0 < \tau < 0.2$; $0.275 < \gamma < 1.0$; $A = 49.28$

	Variable	Initial Value	Minimal Value	Maximal Value	Final Value
1	t	0	0	0.2	0.2
2	y	1	1	4.896173	4.209492
3	y0	1	1	4.208664	4.208664
4	y1	0	0	1	1
5	y2	0	0	1	1

Calculated Values of DEQ Variables

Differential Equations

1. $d(y0)/d(t) = 20*(1-y2)*\exp(y0/(1+.1*y0)) -.157*y0^4+49.28$
2. $d(y2)/d(t) = 20*(1-y2)*\exp(y0/(1+.1*y0))$
3. $d(y1)/d(t) = 5.5*(1-y1)^{1.0}*\exp(y/(1+.1*y))$
4. $d(y)/d(t) = (1)*20*(1-y1)^{1.0}*\exp(y/(1+.1*y))-.157*y^4+49.28$

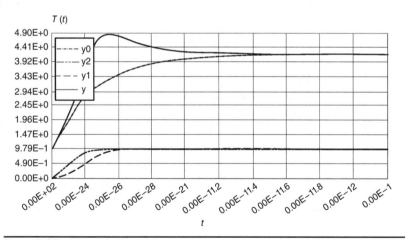

FIGURE 5.25   Dimensionless time-temperature curves.

The maximum temperature increase is 32.1 percent.

5.4   Large-Size Fire Compartment

It has been assumed previously that the fire compartment has approximately equal height, width, and length (the ratios not more then 2). If the largest ratio (width/height or length/height) is more then 2, then Eqs. (4.52) and (4.53) will be modified as follows:

$$\dot{U}(t) = \frac{1.14}{b}U^2 - 13.13\left(1+\frac{1}{b^2}\right)U \tag{5.35}$$

$$\dot{W}(t) = -\frac{1.14}{b}UW - 13.13\left(1+\frac{1}{b^2}\right)W - 0.406\text{Fr} \tag{5.36}$$

where $b = L/h$, and L and h are the length (width) and height of the fire compartment, accordingly.

Solutions to Eqs. (5.35) and (5.36) are similar to those presented earlier for Eqs. (4.52) and (4.53). Since the most important cases from practical point of view in building design are $10^8 < \text{Fr} < 10^{10}$, the following data are provided for $b = 2$ and $b = 4$ (linear interpolation for intermediates values of b is allowed). The tabulated solutions to Eqs. (5.35) and (5.36) are presented in Tables 5A.18 through 5A.23 in the appendix at the end of this chapter.

Case 1: $b = 2$; Fr $= 10^8$; Pr $= 10$

	Variable	Initial Value	Minimal Value	Maximal Value	Final Value
1	t	0	0	0.2	0.2
2	u	−10000	−10000	−1.601E−13	−1.601E−13
3	w	0	0	4.681E+05	2.474E+05

Calculated Values of DEQ Variables

Differential Equations

1. $d(w)/d(t) = -5.7*u*w - 164.1*w + 0.406*10^8$
2. $d(u)/d(t) = 5.7*u^2 - 164.1*u$

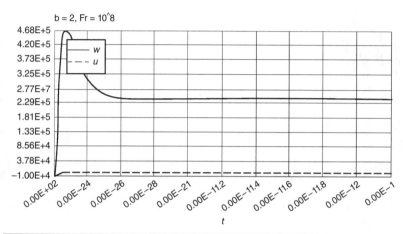

FIGURE 5.26 Dimensionless velocities U and W.

Model $w = a0 + a1*t$

Variable	Value
a0	2.928E+05
a1	-3.235E+05

Finally,

$$U = 0 \quad \text{and} \quad W = 2.93 \times 10^5 \quad (5.37)$$

Case 2: $b = 2$; $Fr = 10^9$; $Pr = 27.0$

	Variable	Initial Value	Minimal Value	Maximal Value	Final Value
1	t	0	0	0.2	0.2
2	u	-10000	-10000	-9.186E-38	-9.186E-38
3	w	0	0	1.557E+06	9.163E+05

Calculated Values of DEQ Variables

Differential Equations

1. $d(w)/d(t) = -15.39*u*w-443.1*w+0.406*10^9$
2. $d(u)/d(t) = 15.39*u^2-443.1*u$

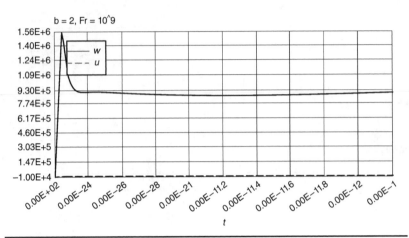

FIGURE 5.27 Dimensionless velocities U and W.

Model $w = a0 + a1*t$

Variable	Value
a0	9.305E+05
a1	−9.979E+04

Finally,

$$U = 0 \quad \text{and} \quad W = 9.3 \times 10^5 \tag{5.38}$$

Case 3: $b = 2$; Fr $= 10^{10}$; Pr $= 120$

	Variable	Initial Value	Minimal Value	Maximal Value	Final Value
1	t	0	0	0.2	0.2
2	u	−10000	−10000	1.914E−15	−4.893E−59
3	w	0	0	2.069E+06	2.061E+06

Calculated Values of DEQ Variables

Differential Equations

1. $d(w)/d(t) = -68.4*u*w - 1970*w + 0.406*10^{10}$
2. $d(u)/d(t) = 68.4*u^2 - 1970*u$

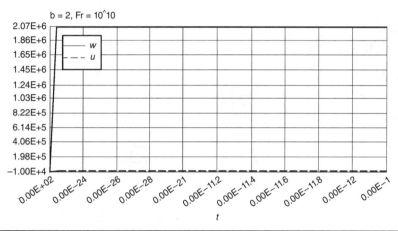

FIGURE 5.28   Dimensionless velocities U and W.

Model $w = a0 + a1*t$

Variable	Value
a0	1.977E+06
a1	6.241E+05

Finally,

$$U = 0 \quad \text{and} \quad W = 1.98 \times 10^6 \tag{5.39}$$

Case 4: $b = 4$; $Fr = 10^8$; $Pr = 10.0$

	Variable	Initial Value	Minimal Value	Maximal Value	Final Value
1	t	0	0	0.2	0.2
2	u	−10000	−10000	−3.722E−11	−3.722E−11
3	w	0	0	5.192E+05	2.91E+05

Calculated Values of DEQ Variables

Differential Equations

1. $d(w)/d(t) = -2.85*u*w - 139.5*w + 0.406*10^8$
2. $d(u)/d(t) = 2.85*u^2 - 139.5*u$

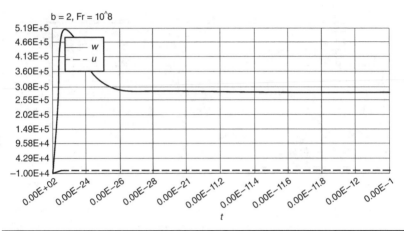

FIGURE **5.29**   Dimensionless velocities U and W.

Model $w = a0 + a1*t$

Variable	Value
a0	3.473E+05
a1	−3.973E+05

Finally,

$$U = 0 \quad \text{and} \quad W = 3.47 \times 10^5 \tag{5.40}$$

Case 5: $b = 4$; Fr $= 10^9$; Pr $= 27.0$

	Variable	Initial Value	Minimal Value	Maximal Value	Final Value
1	t	0	0	0.2	0.2
2	u	−10000	−10000	−9.234E−32	−9.234E−32
3	w	0	0	1.858E+06	1.078E+06

Calculated Values of DEQ Variables

Differential Equations

1. $d(w)/d(t) = -7.695*u*w - 376.7*w + 0.406*10^9$
2. $d(u)/d(t) = 7.695*u^2 - 376.7*u$

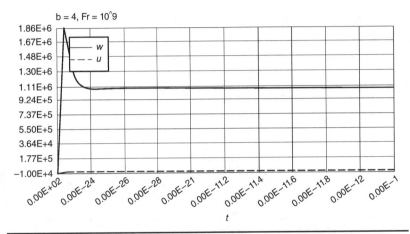

FIGURE 5.30   Dimensionless velocities U and W.

Model $w = a0 + a1*t$

Variable	Value
a0	1.108E+06
a1	−2.159E+05

Finally,

$$U = 0 \quad \text{and} \quad W = 1.11 \times 10^6 \tag{5.41}$$

Case 6: $b = 4$; $Fr = 10^{10}$; $Pr = 120$

	Variable	Initial Value	Minimal Value	Maximal Value	Final Value
1	t	0	0	0.2	0.2
2	u	−10000	−10000	5.056E−16	1.218E−117
3	w	0	0	2.454E+06	2.425E+06

Calculated Values of DEQ Variables

Differential Equations

1. $d(w)/d(t) = -34.2*u*w - 1674*w + 0.406*10^{10}$
2. $d(u)/d(t) = 34.2*u^2 - 1674*u$

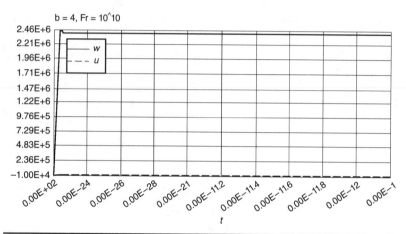

FIGURE 5.31   Dimensionless velocities U and W.

Model $w = a0 + a1*t$

Variable	Value
a0	2.327E+06
a1	7.28E+05

Finally,

$$U = 0 \quad \text{and} \quad W = 2.33 \times 10^6 \quad (5.42)$$

The summary of solutions to Eqs. (5.35) and (5.36) are presented below for the following cases (see Table 5.3).

b	Fr = 10^8	Fr = 10^9	Fr = 10^{10}	Fr = 10^8	Fr = 10^9	Fr = 10^{10}
2	2.928(10^5)	9.305(10^5)	1.977(10^6)			
4				3.473(10^5)	1.108(10^6)	2.327(10^6)

TABLE 5.3   Maximum Dimensionless Vertical Velocities W

Again, the velocities U and W now should be inserted into conservation of energy and mass equations [Eqs. (5.1) and (5.2)]. These equations will have the addition term: $W(\partial\theta/\partial z)$. The temperature change is a very weak function of vertical coordinate z; therefore, it will be assumed as a constant value. In order to obtain this constant value, let's consider a steady-stream flow process (W = const.; see above). Similar to the preceding case (b = 1) the following scenarios should be analyzed.

Case 1a: $W = 2.928 \times 10^5$; Fr $= 10^8$; $b = 2$

	Variable	Initial Value	Minimal Value	Maximal Value	Final Value
1	x	0	0	2	2
2	y	0	–0.000169	0	–0.000169
3	z	1	0.999662	1	0.999662

Calculated Values of DEQ Variables

Differential Equations

1. d(z)/d(x) = y
2. d(y)/d(x)=1.0*.157*z^4–(1.0)*20*exp(z/(1+.1*z))–y*2.928* 10^5

Model   y = a0 + a1*x

Variable	Value
a0	–0.0001621
a1	–5.114E–06

$$\frac{d\theta}{dz} = 1.621 \times 10^{-4} \tag{5.43}$$

$$\boxed{W\frac{d\theta}{dz} = 47.46} \tag{5.44}$$

Case 2a: $W = 9.305 \times 10^5$; Fr $= 10^9$; $b = 2$

	Variable	Initial Value	Minimal Value	Maximal Value	Final Value
1	x	0	0	2	2
2	y	0	–5.318E–05	0	–5.318E–05
3	z	1	0.9998936	1	0.9998936

Calculated Values of DEQ Variables

Differential Equations

1. $d(z)/d(x) = y$
2. $d(y)/d(x) = 1.0*.157*z^4-(1.0)*20*exp(z/(1+.1*z))-y*9.305*10^5$

Model $y = a0 + a1*x$

Variable	Value
a0	−5.102E−05
a1	−1.609E−06

$$\frac{d\theta}{dz} = 5.102 \times 10^{-5} \qquad (5.45)$$

$$\boxed{W\frac{d\theta}{dz} = 47.47} \qquad (5.46)$$

Case 3a: $W = 1.977 \times 10^6$; Fr $= 10^{10}$; $b = 2$

	Variable	Initial Value	Minimal Value	Maximal Value	Final Value
1	x	0	0	2	2
2	y	0	−2.503E−05	0	−2.503E−05
3	z	1	0.9999499	1	0.9999499

Calculated Values of DEQ Variables

Differential Equations

1. $d(z)/d(x) = y$
2. $d(y)/d(x) = 1.0*.157*z^4-(1.0)*20*exp(z/(1+.1*z))-y*1.977*10^6$

Model $y = a0 + a1*x$

Variable	Value
a0	−2.401E−05
a1	−7.574E−07

$$\frac{d\theta}{dz} = 2.401 \times 10^{-5} \qquad (5.47)$$

$$\boxed{W\frac{d\theta}{dz} = 47.47} \qquad (5.48)$$

There is no change in this case; therefore, recalculation of the temperature-time relationship is not required for $b = 2$.

Case 1b: $W = 3.473 \times 10^5$; $Fr = 10^8$; $b = 4$

	Variable	Initial Value	Minimal Value	Maximal Value	Final Value
1	x	0	0	4	4
2	y	0	−0.0001425	0	−0.0001424
3	z	1	0.9994302	1	0.9994302

Calculated Values of DEQ Variables

Differential Equations

1. $d(z)/d(x) = y$
2. $d(y)/d(x) = 1.0*.157*z^4 - (1.0)*20*\exp(z/(1+.1*z)) - y*3.473*10^5$

Model $y = a0 + a1*x$

Variable	Value
a0	−0.0001367
a1	−2.119E–06

$$\frac{d\theta}{dz} = 1.367 \times 10^{-4} \tag{5.49}$$

$$\boxed{W\frac{d\theta}{dz} = 47.47} \tag{5.50}$$

Case 2b: $W = 1.108 \times 10^6$; $Fr = 10^9$; $b = 4$

	Variable	Initial Value	Minimal Value	Maximal Value	Final Value
1	x	0	0	4	4
2	y	0	−4.466E–05	0	−4.465E–05
3	z	1	0.9998214	1	0.9998214

Calculated Values of DEQ Variables

Differential Equations

1. $d(z)/d(x) = y$
2. $d(y)/d(x) = 1.0*.157*z^4 - (1.0)*20*exp(z/(1+.1*z)) - y*1.108*10^6$

Model $y = a0 + a1*x$

Variable	Value
a0	−4.285E−05
a1	−6.744E−07

$$\frac{d\theta}{dz} = 4.285 \times 10^{-5} \tag{5.51}$$

$$\boxed{W\frac{d\theta}{dz} = 47.48} \tag{5.52}$$

Case 3b: $W = 2.327 \times 10^6$; $Fr = 10^{10}$; $b = 4$

	Variable	Initial Value	Minimal Value	Maximal Value	Final Value
1	x	0	0	4	4
2	y	0	−2.127E−05	0	−2.126E−05
3	z	1	0.9999149	1	0.9999149

Calculated Values of DEQ Variables

Differential Equations

1. $d(z)/d(x) = y$
2. $d(y)/d(x) = 1.0*.157*z^4 - (1.0)*20*exp(z/(1+.1*z)) - y*2.327*10^6$

Model $y = a0 + a1*x$

Variable	Value
a0	−2.04E−05
a1	−3.196E−07

$$\frac{d\theta}{dz} = 2.04 \times 10^{-5} \tag{5.53}$$

$$\boxed{W\frac{d\theta}{dz} = 47.47} \tag{5.54}$$

There is no change in this case; therefore, recalculation of the temperature-time relationship is not required for $b = 4$.

Conclusion

In large fire compartments ($1 < b < 4$), the product $W (d\theta/dz)$ remains constant, although the velocities increase with an increase in the linear dimensions of a compartment's footprints. All practical examples and comparisons with the time-equivalent and parametric methods are presented in Chap. 6.

References

1. National Institute of Standards and Technology (NIST). Fire Dynamics Simulator (Version 5) Technical Reference Guide, Vol. 1: Mathematical Model. NIST Special Publication 1018-5. NIST, Gaithersburg, MD, 2008.
2. European Committee for Standardization, Draft for Eurocode I: part 2.7 (CEN TC 250/SC1). CEN, Brussels, April 1993.
3. Babrauskas, V., "Performance-Based Fire Safety Engineering Design: The Role of Fire Models and Fire Tests." Paper presented at Interflam 99, Edinburgh, Scotland, June 1999.
4. Lane, B., "Performance-Based Approach to the Design of Steel Structures in Fire," in *Proceedings of the Society of Fire Protection Engineers (SFPE) and National Institute of Standards and Technology (NIST)*, June 11–15, 2001, San Francisco, CA. Arup Fire, New York, 2001, pp. 415–426.
5. Razdolsky, L., Petrov, A., and Shtessel, E., "Critical Conditions of Local Ignition in a Large Medium with Convective Heat Transfer," in *Physics of Combustion and Explosions*, Academy of Science, Moscow, 1977.
6. National Institute for Standards and Technology (NIST). CFAST: Consolidated Model of Fire Growth and Smoke Transport (Version 6). NIST Special Publication 1026. NIST, Gaithersburg, MD, 2008.
7. Razdolsky, L., "Mathematical Modeling of Fire Dynamics," in *Proceedings of the World Congress on Engineering and Computer Science 2009*. WCE, London, 2009.
8. Frank-Kamenestkii, D. A., *Diffusion and Heat Transfer in Chemical Kinetics*. Plenum Press, New York, 1969.
9. Magnusson, S. E., and Thelandersson, S., "Temperature-Time Curves of Complete Process of Fire Development in Enclosed Spaces," *Acta Polytechnica Scandinavia*, 1970.
10. Society of Fire Protection Engineers, *The SFPE Engineering Guide to Performance-Based Fire Protection Analysis and Design*. SFPE, Quincy, MA, 1998.
11. Razdolsky, L., "Structural Fire Loads in a High-Rise Building Design," in *Proceedings of the SFPE Engineering Technology Conference*. NFPA-Publisher, Scottsdale, 2009.
12. Evans, L. C., *An Introduction to Mathematical Optimal Control Theory*, Version 0.2. Department of Mathematics, University of California, Berkeley, 1983.
13. SFPE, *Engineering Guide: Fire Exposures to Structural Elements*, Society of Fire Protection Engineers, Quincy, MA, 2011.
14. Razdolsky, L., "Structural Fire Loads in a Modern Tall Building Design," in *Proceedings of SFPE Engineering Technology Conference, 2010*, Lund University, Sweden, 2010.
15. Lane, B., "Performance-Based Approach to the Design of Steel Structures in Fire," in *Proceedings of the Society of Fire Protection Engineers (SFPE) and National Institute of Standards and Technology (NIST)*, June 11–15, 2001, San Francisco, CA. Arup Fire, New York, 2001, pp. 415–426.

Appendix 5A

t	Y0	Y2	Y1	Y
0	1.	0	0	1.
0.0046323	1.250966	0.0126359	0	1.252688
0.0062323	1.349469	0.0176143	0	1.352911
0.0094323	1.569195	0.0287704	0	1.578975
0.0110323	1.692407	0.0350629	0	1.70744
0.0126323	1.826133	0.0419279	0	1.848433
0.0142323	1.971858	0.0494567	0	2.004126
0.0174323	2.306642	0.0669785	0	2.371202
0.0190323	2.500224	0.0772799	0	2.590437
0.0206323	2.714956	0.0888787	0	2.840678
0.0222323	2.954166	0.1020431	0	3.1295
0.0254323	3.521305	0.1345053	0	3.867017
0.0269266	3.833766	0.1533144	0	4.313324
0.0282793	4.145828	0.1729136	0	4.795025
0.0306335	4.75519	0.2141772	0	5.867583
0.0326618	5.333863	0.2584788	0	7.094548
0.0345798	5.890116	0.3085869	0	8.496253
0.036806	6.466278	0.3756045	0	10.09218
0.0380691	6.720702	0.4162509	0	10.76418
0.0401642	6.983779	0.4837994	0	11.39501
0.0422384	7.049816	0.5461719	0	11.63165
0.0440082	6.98769	0.5933149	0	11.70302
0.0461123	6.826273	0.6412345	0	11.73221
0.0487499	6.561462	0.6898546	0	11.74205
0.0509712	6.325012	0.7226619	0	11.74406
0.052284	6.187723	0.7392307	0	11.74448
0.0553771	5.882639	0.7717659	0	11.74478
0.0569771	5.737074	0.7857214	0	11.74482
0.0585771	5.600171	0.7981189	0	11.74484
0.0601771	5.471671	0.8091967	0	11.74484
0.0633771	5.238037	0.8281399	0	11.74485
0.0649771	5.131832	0.8362984	0	11.74485
0.0665771	5.031983	0.8437357	0	11.74485
0.0681771	4.937976	0.8505437	0	11.74485

TABLE 5A.1   Tabulated Solutions to Eqs. (5.3) and (5.4)

t	Y0	Y2	Y1	Y
0.0713771	4.765618	0.8625688	0	11.74485
0.0729771	4.686427	0.8679065	0	11.74485
0.0745771	4.611396	0.8728603	0	11.74485
0.0761771	4.540194	0.8774707	0	11.74485
0.0793771	4.408114	0.8857976	0	11.74485
0.0809771	4.34672	0.8895712	0	11.74485
0.0825771	4.28812	0.893117	0	11.74485
0.0841771	4.232113	0.8964555	0	11.74485
0.0873771	4.127166	0.9025807	0	11.74485
0.0889771	4.077909	0.9053976	0	11.74485
0.0905771	4.030609	0.9080682	0	11.74485
0.0921771	3.985139	0.9106038	0	11.74485
0.0953771	3.89924	0.91531	0	11.74485
0.0969771	3.85861	0.9174979	0	11.74485
0.0985771	3.819403	0.9195861	0	11.74485
0.1001771	3.781539	0.9215812	0	11.74485
0.1033771	3.709541	0.9253168	0	11.74485
0.1049771	3.675272	0.927068	0	11.74485
0.1065771	3.642075	0.9287479	0	11.74485
0.1081771	3.609895	0.9303609	0	11.74485
0.1113771	3.548379	0.9334017	0	11.74485
0.1129771	3.518952	0.9348365	0	11.74485
0.1145771	3.490354	0.9362186	0	11.74485
0.1161771	3.462546	0.9375507	0	11.74485
0.1193771	3.409159	0.9400759	0	11.74485
0.1209771	3.383513	0.9412737	0	11.74485
0.1225771	3.358524	0.9424313	0	11.74485
0.1241771	3.334165	0.9435506	0	11.74485
0.1273771	3.287228	0.9456818	0	11.74485
0.1289771	3.264601	0.9466972	0	11.74485
0.1305771	3.242506	0.9476811	0	11.74485
0.1321771	3.220922	0.948635	0	11.74485
0.1353771	3.179205	0.9504581	0	11.74485
0.1369771	3.159036	0.9513298	0	11.74485
0.1385771	3.139303	0.9521764	0	11.74485
0.1401771	3.119992	0.952999	0	11.74485

TABLE 5A.1   Tabulated Solutions to Eqs. (5.3) and (5.4) (*Continued*)

t	Y0	Y2	Y1	Y
0.1433771	3.08257	0.9545762	0	11.74485
0.1449771	3.064433	0.9553326	0	11.74485
0.1465771	3.046659	0.9560687	0	11.74485
0.1481771	3.029237	0.9567853	0	11.74485
0.1513771	2.995403	0.9581629	0	11.74485
0.1529771	2.978968	0.9588253	0	11.74485
0.1545771	2.96284	0.959471	0	11.74485
0.1561771	2.947011	0.9601006	0	11.74485
0.1593771	2.916208	0.9613139	0	11.74485
0.1609771	2.901218	0.9618986	0	11.74485
0.1625771	2.88649	0.9624695	0	11.74485
0.1641771	2.872017	0.9630268	0	11.74485
0.1673771	2.843806	0.9641031	0	11.74485
0.1689771	2.830054	0.9646229	0	11.74485
0.1705771	2.816528	0.9651309	0	11.74485
0.1721771	2.803223	0.9656276	0	11.74485
0.1753771	2.777249	0.9665884	0	11.74485
0.1769771	2.764568	0.9670532	0	11.74485
0.1785771	2.752085	0.967508	0	11.74485
0.1801771	2.739793	0.9679532	0	11.74485
0.1833771	2.715766	0.9688157	0	11.74485
0.1849771	2.70402	0.9692336	0	11.74485
0.1865771	2.692447	0.969643	0	11.74485
0.1881771	2.681043	0.9700441	0	11.74485
0.1913771	2.658723	0.9708223	0	11.74485
0.1929771	2.6478	0.9711999	0	11.74485
0.1945771	2.637029	0.9715701	0	11.74485
0.1961771	2.626406	0.9719331	0	11.74485
0.1993771	2.605595	0.9726383	0	11.74485
0.2	2.601609	0.9727725	0	11.74485

TABLE 5A.1 Tabulated Solutions to Eqs. (5.3) and (5.4) (*Continued*)

t	Y0	Y2	Y1	Y
0	1.	0	0	1.
0.004631	1.24728	0.0434794	0.0126345	1.25151
0.006231	1.34202	0.0601827	0.0176158	1.350416
0.009431	1.548129	0.096637	0.0287851	1.571504
0.011031	1.660356	0.1165682	0.0350907	1.695825
0.012631	1.779296	0.1377676	0.0419761	1.831102
0.014231	1.905399	0.1603427	0.0495364	1.978998
0.017431	2.180772	0.2100708	0.0671808	2.321126
0.019031	2.330714	0.2374498	0.0775963	2.520804
0.020631	2.489063	0.2666452	0.0893715	2.74424
0.022231	2.655737	0.29774	0.1028107	2.995974
0.025431	3.011979	0.3657786	0.1363904	3.607805
0.027031	3.199275	0.4026501	0.1577358	3.982587
0.028631	3.390097	0.4412313	0.1832409	4.414597
0.030231	3.581553	0.4812371	0.2140423	4.911748
0.0332707	3.933374	0.5595647	0.2919816	6.038297
0.0346195	4.078027	0.5944335	0.3363547	6.588248
0.0360362	4.218097	0.6304889	0.3894198	7.151461
0.0382526	4.406139	0.6845389	0.481232	7.862805
0.0401722	4.533055	0.7278618	0.5605716	8.183673
0.0423902	4.635244	0.7727565	0.6409317	8.201185
0.0445764	4.691047	0.8110117	0.7036966	7.976632
0.0467789	4.707775	0.8435287	0.7518	7.646972
0.049171	4.689984	0.8725292	0.7908948	7.266543
0.0504903	4.667778	0.8860246	0.808076	7.064167
0.0534692	4.594883	0.9109984	0.8386789	6.644725
0.0550692	4.546751	0.9217712	0.8516052	6.442804
0.0566692	4.494717	0.9310316	0.8627041	6.256583
0.0582692	4.440188	0.9390034	0.8723251	6.084911
0.0614692	4.327795	0.951827	0.8881532	5.780033
0.0630692	4.271415	0.9569849	0.894736	5.644332
0.0646692	4.215599	0.9614721	0.9006181	5.518285
0.0662692	4.16069	0.9653883	0.9059055	5.400909
0.0694692	4.054487	0.9718291	0.9150235	5.188774
0.0710692	4.003465	0.9744831	0.9189824	5.092562
0.0726692	3.953926	0.9768292	0.9226088	5.002094

TABLE 5A.2 Tabulated Solutions to Eqs. (5.3) and (5.4)

t	Y0	Y2	Y1	Y
0.0742692	3.905898	0.9789092	0.9259433	4.916841
0.0774692	3.814372	0.9824084	0.9318677	4.76016
0.0790692	3.770832	0.9838834	0.9345116	4.687951
0.0806692	3.728728	0.9852058	0.9369729	4.619378
0.0822692	3.688017	0.9863945	0.9392701	4.554152
0.0854692	3.610583	0.9884326	0.9434342	4.432716
0.0870692	3.573762	0.989308	0.9453274	4.376062
0.0886692	3.538137	0.9901019	0.9471096	4.321857
0.0902692	3.503658	0.9908236	0.9487905	4.269929
0.0934692	3.437948	0.9920802	0.9518809	4.172296
0.0950692	3.406624	0.9926281	0.9533048	4.12632
0.0966692	3.37626	0.9931298	0.9546561	4.082078
0.0982692	3.346815	0.9935898	0.9559403	4.039463
0.1014692	3.290522	0.994401	0.9583266	3.958725
0.1030692	3.263599	0.9947591	0.959437	3.920429
0.1046692	3.237444	0.9950894	0.9604974	3.883411
0.1062692	3.212024	0.9953945	0.9615109	3.847601
0.1094692	3.163267	0.995938	0.9634095	3.779346
0.1110692	3.139871	0.9961803	0.9642999	3.746784
0.1126692	3.117094	0.9964052	0.9651542	3.715196
0.1142692	3.09491	0.9966142	0.9659745	3.684532
0.1174692	3.052227	0.9969897	0.967521	3.625801
0.1190692	3.031682	0.9971585	0.9682507	3.597652
0.1206692	3.011642	0.9973159	0.9689535	3.570265
0.1222692	2.992085	0.997463	0.9696308	3.543604
0.1254692	2.954349	0.9977291	0.9709144	3.492335
0.1270692	2.936136	0.9978495	0.971523	3.467668
0.1286692	2.918338	0.9979623	0.9721111	3.44361
0.1302692	2.900939	0.9980682	0.9726796	3.420135
0.1334692	2.867282	0.9982608	0.9737615	3.374842
0.1350692	2.850997	0.9983485	0.9742766	3.35298
0.1366692	2.835057	0.998431	0.9747757	3.331614
0.1382692	2.819451	0.9985086	0.9752593	3.310725
0.1414692	2.789194	0.9986506	0.976183	3.270307
0.1430692	2.774522	0.9987156	0.9766243	3.250745
0.1446692	2.76014	0.998777	0.9770528	3.231593

TABLE 5A.2 Tabulated Solutions to Eqs. (5.3) and (5.4) (*Continued*)

t	Y0	Y2	Y1	Y
0.1462692	2.74604	0.9988349	0.977469	3.212837
0.1494692	2.718648	0.9989413	0.9782662	3.176458
0.1510692	2.705339	0.9989901	0.9786483	3.15881
0.1526692	2.692278	0.9990364	0.9790199	3.141505
0.1542692	2.679456	0.9990802	0.9793815	3.124534
0.1574692	2.654502	0.999161	0.9800761	3.091549
0.1590692	2.642357	0.9991983	0.9804098	3.075513
0.1606692	2.630423	0.9992337	0.9807349	3.059771
0.1622692	2.618696	0.9992672	0.9810518	3.044312
0.1654692	2.595836	0.9993293	0.9816618	3.014209
0.1670692	2.584693	0.9993581	0.9819555	2.99955
0.1686692	2.573732	0.9993854	0.9822421	2.985141
0.1702692	2.562951	0.9994114	0.9825217	2.970976
0.1734692	2.541904	0.9994597	0.9830613	2.94335
0.1750692	2.53163	0.9994821	0.9833216	2.929875
0.1766692	2.521515	0.9995034	0.9835759	2.916617
0.1782692	2.511557	0.9995238	0.9838243	2.903571
0.1814692	2.492092	0.9995617	0.9843045	2.87809
0.1830692	2.482578	0.9995793	0.9845366	2.865644
0.1846692	2.473204	0.9995961	0.9847635	2.853388
0.1862692	2.463967	0.9996121	0.9849855	2.841317
0.1894692	2.445892	0.9996421	0.9854152	2.81771
0.1910692	2.437046	0.9996561	0.9856232	2.806165
0.1926692	2.428325	0.9996695	0.9858269	2.794787
0.1942692	2.419725	0.9996822	0.9860262	2.783572
0.1974692	2.402878	0.9997061	0.9864127	2.761613
0.1990692	2.394625	0.9997173	0.9866	2.750863
0.2	2.389874	0.9997236	0.9867073	2.744677

TABLE 5A.2   Tabulated Solutions to Eqs. (5.3) and (5.4) (Continued)

t	Y0	Y2	Y1	Y
0	1.	0	0	1.
0.0051431	1.272855	0.075412	0.0487235	1.277042
0.0067431	1.365658	0.1011426	0.0657339	1.373468
0.0083431	1.462435	0.1280327	0.0837581	1.475428
0.0115431	1.668168	0.1854429	0.1232109	1.697676
0.0131431	1.777118	0.2160134	0.1448387	1.81886
0.0147431	1.889994	0.2478359	0.1678777	1.947349
0.0163431	2.006569	0.2808906	0.1924411	2.083562
0.0195431	2.249238	0.3504706	0.2465929	2.380484
0.0211431	2.374128	0.3867904	0.2763885	2.54152
0.0227431	2.50027	0.4239198	0.3081039	2.710791
0.0243431	2.626572	0.4616395	0.3417755	2.887758
0.0275431	2.874339	0.5377244	0.4148246	3.260094
0.0291431	2.99278	0.5754187	0.453899	3.451472
0.0307431	3.105459	0.6123849	0.4942792	3.642391
0.0323431	3.210821	0.6482409	0.5355044	3.828953
0.0355431	3.394237	0.7152063	0.6180384	4.170917
0.0371431	3.470316	0.7457276	0.6579309	4.317201
0.0387431	3.535246	0.7739955	0.6959615	4.441904
0.0403431	3.588947	0.7998981	0.7315287	4.54263
0.0435431	3.664035	0.8445401	0.7936853	4.669947
0.0451431	3.686763	0.8634069	0.8199441	4.698853
0.0467431	3.7008	0.8801363	0.8430464	4.70778
0.0483431	3.707141	0.8948909	0.8631862	4.699759
0.0515431	3.700745	0.9191895	0.8956637	4.645126
0.0531431	3.689897	0.9290929	0.9085955	4.60406
0.0547431	3.675084	0.9377268	0.9197065	4.556927
0.0563431	3.657043	0.9452471	0.929256	4.505563
0.0595431	3.613765	0.9574973	0.9445592	4.39573
0.0611431	3.589555	0.9624656	0.9506834	4.339313
0.0627431	3.564187	0.9667981	0.9559919	4.282871
0.0643431	3.537991	0.9705803	0.9606071	4.226908
0.0675431	3.484168	0.9767803	0.9681544	4.117783
0.0691431	3.45695	0.9793177	0.9712461	4.065061
0.0707431	3.429738	0.9815459	0.9739688	4.013743
0.0723431	3.402652	0.983506	0.976374	3.963898

TABLE 5A.3   Tabulated Solutions to Eqs. (5.3) and (5.4)

t	Y0	Y2	Y1	Y
0.0755431	3.34922	0.9867578	0.9803992	3.868741
0.0771431	3.323006	0.988106	0.9820872	3.823424
0.0787431	3.297193	0.9893002	0.9835959	3.779587
0.0803431	3.271812	0.9903598	0.9849478	3.737193
0.0835431	3.222437	0.9921403	0.987256	3.656566
0.0851431	3.198471	0.9928881	0.9882433	3.618237
0.0867431	3.174995	0.993556	0.9891365	3.581166
0.0883431	3.152009	0.9941536	0.9899463	3.545301
0.0915431	3.107503	0.9951694	0.9913517	3.476993
0.0931431	3.085973	0.9956011	0.9919624	3.444453
0.0947431	3.064913	0.9959896	0.9925204	3.412927
0.0963431	3.044317	0.9963397	0.9930311	3.382371
0.0995431	3.004474	0.996941	0.9939291	3.324
0.1011431	2.985206	0.9971992	0.9943244	3.296105
0.1027431	2.966359	0.997433	0.9946885	3.269021
0.1043431	2.947921	0.9976451	0.9950242	3.242713
0.1075431	2.91223	0.9980126	0.9956209	3.192289
0.1091431	2.894955	0.9981719	0.9958864	3.168113
0.1107431	2.878044	0.9983169	0.9961324	3.144588
0.1123431	2.861488	0.9984491	0.9963607	3.121686
0.1155431	2.829397	0.9986801	0.9967701	3.077655
0.1171431	2.813841	0.9987809	0.9969538	3.056477
0.1187431	2.798599	0.9988732	0.997125	3.035827
0.1203431	2.783662	0.9989578	0.9972847	3.015685
0.1235431	2.754662	0.9991065	0.9975732	2.976847
0.1251431	2.740584	0.9991718	0.9977035	2.958113
0.1267431	2.726774	0.9992319	0.9978256	2.939813
0.1283431	2.713225	0.9992872	0.9979399	2.921931
0.1315431	2.686881	0.9993849	0.9981477	2.887361
0.1331431	2.674071	0.9994281	0.9982422	2.870643
0.1347431	2.661492	0.999468	0.998331	2.854286
0.1363431	2.649138	0.9995049	0.9984146	2.838278
0.1395431	2.625081	0.9995703	0.9985672	2.807256
0.1411431	2.613364	0.9995994	0.9986369	2.79222
0.1427431	2.601848	0.9996264	0.9987027	2.777487
0.1443431	2.590527	0.9996513	0.9987647	2.763048

TABLE 5A.3 Tabulated Solutions to Eqs. (5.3) and (5.4) (*Continued*)

t	Y0	Y2	Y1	Y
0.1475431	2.568448	0.9996959	0.9988787	2.735009
0.1491431	2.55768	0.9997158	0.9989309	2.721391
0.1507431	2.547086	0.9997343	0.9989804	2.708031
0.1523431	2.536662	0.9997514	0.9990272	2.69492
0.1555431	2.516305	0.9997822	0.9991134	2.669413
0.1571431	2.506364	0.999796	0.9991532	2.657004
0.1587431	2.496575	0.9998089	0.9991909	2.644814
0.1603431	2.486935	0.9998209	0.9992266	2.632838
0.1635431	2.468086	0.9998424	0.9992927	2.609502
0.1651431	2.45887	0.9998521	0.9993233	2.59813
0.1667431	2.449788	0.9998612	0.9993524	2.586948
0.1683431	2.440837	0.9998696	0.99938	2.575951
0.1715431	2.423316	0.9998849	0.9994312	2.554491
0.1731431	2.414739	0.9998918	0.999455	2.544018
0.1747431	2.406281	0.9998982	0.9994776	2.53371
0.1763431	2.39794	0.9999042	0.9994992	2.523564
0.1795431	2.381594	0.9999152	0.9995393	2.503737
0.1811431	2.373585	0.9999201	0.999558	2.494048
0.1827431	2.365682	0.9999247	0.9995758	2.484505
0.1843431	2.357882	0.9999291	0.9995927	2.475103
0.1875431	2.342583	0.999937	0.9996244	2.456708
0.1891431	2.335079	0.9999406	0.9996392	2.44771
0.1907431	2.32767	0.9999439	0.9996533	2.438839
0.1923431	2.320354	0.9999471	0.9996668	2.430092
0.1955431	2.305991	0.9999528	0.999692	2.412963
0.1971431	2.298941	0.9999555	0.9997038	2.404573
0.1987431	2.291976	0.9999579	0.999715	2.396298
0.2	2.286563	0.9999598	0.9997236	2.389874

TABLE 5A.3   Tabulated Solutions to Eqs. (5.3) and (5.4) (Continued)

t	Y0	Y2	Y1	Y
0	1.	0	0	1.
0.0048899	1.233664	0.2348916	0.0714435	1.258526
0.0064899	1.30494	0.3068212	0.0969946	1.35071
0.0080899	1.372758	0.3754482	0.1236983	1.446852
0.0112899	1.496308	0.5011417	0.18072	1.651295
0.0128899	1.551411	0.5576024	0.211092	1.759611
0.0144899	1.601823	0.6095689	0.2427171	1.871878
0.0160899	1.647494	0.6569919	0.2755792	1.987888
0.0192899	1.7249	0.7384783	0.344809	2.22965
0.0208899	1.756984	0.7728726	0.3809839	2.354255
0.0224899	1.784992	0.8033533	0.4179985	2.480267
0.0240899	1.809225	0.830208	0.4556414	2.606624
0.0272899	1.847665	0.8742781	0.5317177	2.855174
0.0288899	1.862529	0.8921187	0.5694922	2.974379
0.0304899	1.874911	0.9075651	0.6065992	3.088076
0.0320899	1.885105	0.920898	0.6426559	3.194695
0.0352899	1.899991	0.9422374	0.7101813	3.3812
0.0368899	1.905154	0.9506924	0.7410437	3.459012
0.0384899	1.909069	0.9579312	0.7696777	3.525721
0.0400899	1.911912	0.9641208	0.7959588	3.58119
0.0432899	1.914973	0.9739202	0.8413491	3.659579
0.0448899	1.915441	0.9777691	0.8605674	3.683771
0.0464899	1.915341	0.9810502	0.8776248	3.699117
0.0480899	1.914759	0.9838465	0.8926804	3.706608
0.0512899	1.912432	0.9882584	0.917495	3.702044
0.0528899	1.910806	0.9899872	0.9276144	3.691898
0.0544899	1.908936	0.9914597	0.9364385	3.677661
0.0560899	1.906862	0.9927141	0.9441252	3.660085
0.0592899	1.902231	0.9946932	0.9566465	3.617464
0.0608899	1.899726	0.9954691	0.9617241	3.593473
0.0624899	1.897124	0.9961305	0.9661512	3.568264
0.0640899	1.894444	0.9966944	0.9700152	3.54218
0.0672899	1.888901	0.9975854	0.9763474	3.488466
0.0688899	1.886063	0.9979353	0.9789379	3.46126
0.0704899	1.883192	0.998234	0.9812121	3.434039
0.0720899	1.880298	0.998489	0.9832121	3.406926

TABLE 5A.4 Tabulated Solutions to Eqs. (5.3) and (5.4)

t	Y0	Y2	Y1	Y
0.0752899	1.87446	0.9988928	0.9865289	3.353402
0.0768899	1.871528	0.9990517	0.9879034	3.327129
0.0784899	1.868592	0.9991876	0.9891206	3.30125
0.0800899	1.865655	0.9993038	0.9902004	3.275798
0.0832899	1.859791	0.9994882	0.9920139	3.226275
0.0848899	1.856868	0.999561	0.9927753	3.202231
0.0864899	1.853953	0.9996233	0.9934552	3.178677
0.0880899	1.851048	0.9996766	0.9940634	3.155614
0.0912899	1.845272	0.9997615	0.9950968	3.110954
0.0928899	1.842402	0.9997951	0.9955358	3.089348
0.0944899	1.839545	0.9998239	0.9959308	3.068215
0.0960899	1.836702	0.9998486	0.9962867	3.047546
0.0992899	1.831059	0.999888	0.9968977	3.007562
0.1008899	1.82826	0.9999036	0.99716	2.988227
0.1024899	1.825475	0.9999171	0.9973975	2.969314
0.1040899	1.822706	0.9999286	0.9976129	2.950812
0.1072899	1.817212	0.999947	0.997986	2.914999
0.1088899	1.814488	0.9999543	0.9981476	2.897664
0.1104899	1.81178	0.9999606	0.9982948	2.880696
0.1120899	1.809087	0.9999661	0.998429	2.864085
0.1152899	1.803746	0.9999747	0.9986633	2.831888
0.1168899	1.801098	0.9999782	0.9987656	2.816282
0.1184899	1.798465	0.9999812	0.9988591	2.800991
0.1200899	1.795848	0.9999838	0.9989449	2.786006
0.1232899	1.790656	0.9999879	0.9990956	2.756916
0.1248899	1.788083	0.9999895	0.9991618	2.742793
0.1264899	1.785524	0.9999909	0.9992227	2.728942
0.1280899	1.782979	0.9999922	0.9992787	2.715352
0.1312899	1.777932	0.9999941	0.9993778	2.68893
0.1328899	1.77543	0.9999949	0.9994215	2.676082
0.1344899	1.772942	0.9999956	0.9994619	2.663467
0.1360899	1.770467	0.9999962	0.9994992	2.651079
0.1392899	1.765559	0.9999972	0.9995655	2.626953
0.1408899	1.763126	0.9999975	0.999595	2.615205
0.1424899	1.760705	0.9999979	0.9996222	2.603657
0.1440899	1.758298	0.9999981	0.9996475	2.592306

TABLE 5A.4 Tabulated Solutions to Eqs. (5.3) and (5.4) (*Continued*)

t	Y0	Y2	Y1	Y
0.1472899	1.753522	0.9999986	0.9996926	2.570168
0.1488899	1.751154	0.9999988	0.9997127	2.559372
0.1504899	1.748798	0.999999	0.9997314	2.548751
0.1520899	1.746455	0.9999991	0.9997488	2.5383
0.1552899	1.741806	0.9999993	0.99978	2.517892
0.1568899	1.7395	0.9999994	0.9997939	2.507927
0.1584899	1.737207	0.9999995	0.9998069	2.498114
0.1600899	1.734925	0.9999995	0.999819	2.488451
0.1632899	1.730397	0.9999997	0.9998408	2.469557
0.1648899	1.728151	0.9999997	0.9998506	2.460319
0.1664899	1.725916	0.9999997	0.9998598	2.451216
0.1680899	1.723693	0.9999998	0.9998683	2.442245
0.1712899	1.719281	0.9999998	0.9998837	2.424684
0.1728899	1.717091	0.9999999	0.9998907	2.416088
0.1744899	1.714913	0.9999999	0.9998972	2.407612
0.1760899	1.712746	0.9999999	0.9999033	2.399252
0.1792899	1.708445	0.9999999	0.9999143	2.382872
0.1808899	1.70631	0.9999999	0.9999193	2.374845
0.1824899	1.704186	0.9999999	0.999924	2.366925
0.1840899	1.702072	0.9999999	0.9999284	2.359109
0.1872899	1.697876	1.	0.9999364	2.343779
0.1888899	1.695794	1.	0.99994	2.33626
0.1904899	1.693722	1.	0.9999434	2.328837
0.1920899	1.69166	1.	0.9999466	2.321506
0.1952899	1.687565	1.	0.9999524	2.307115
0.1968899	1.685533	1.	0.9999551	2.300051
0.1984899	1.68351	1.	0.9999576	2.293073
0.2	1.68161	1.	0.9999598	2.286563

TABLE 5A.4   Tabulated Solutions to Eqs. (5.3) and (5.4) (*Continued*)

t	w	u	t	w	u
0	0	−10000.	0.0788636	0.0253718	−3.307154
0.0040345	0.0057164	−201.4752	0.0804636	0.025108	−3.152256
0.0062433	0.0091056	−127.3415	0.0820636	0.0248468	−3.005439
0.0083494	0.0121197	−92.8611	0.0852636	0.0243338	−2.734103
0.0103823	0.0147773	−72.77623	0.0868636	0.0240827	−2.608699
0.0120049	0.0167111	−61.61559	0.0884636	0.0238355	−2.489603
0.0149231	0.0197736	−47.65989	0.0900636	0.0235924	−2.376446
0.016039	0.0208085	−43.6719	0.0932636	0.0231195	−2.166615
0.0185069	0.0228451	−36.57616	0.0948636	0.02289	−2.069327
0.0213182	0.024773	−30.52067	0.0964636	0.0226651	−1.97675
0.0228636	0.0256702	−27.83829	0.0980636	0.0224451	−1.888626
0.0244636	0.0264876	−25.42684	0.1012636	0.0220198	−1.724787
0.0260636	0.0271997	−23.31998	0.1028636	0.0218146	−1.648635
0.0292636	0.0283385	−19.82053	0.1044636	0.0216143	−1.576059
0.0308636	0.0287797	−18.35401	0.1060636	0.021419	−1.506872
0.0324636	0.0291447	−17.0391	0.1092636	0.0210432	−1.377979
0.0340636	0.0294398	−15.85445	0.1108636	0.0208626	−1.317953
0.0372636	0.0298437	−13.80877	0.1124636	0.0206869	−1.260676
0.0388636	0.0299635	−12.92106	0.1140636	0.0205159	−1.206012
0.0404636	0.0300351	−12.10916	0.1172636	0.0201881	−1.104009
0.0420636	0.0300632	−11.36438	0.1188636	0.0200311	−1.056433
0.0452636	0.0300053	−10.04759	0.1204636	0.0198787	−1.010993
0.0468636	0.0299269	−9.463721	0.1220636	0.0197306	−0.9675861
0.0484636	0.0298201	−8.92292	0.1252636	0.0194475	−0.8864855
0.0500636	0.0296881	−8.421022	0.1268636	0.0193123	−0.8486125
0.0532636	0.0293595	−7.51976	0.1284636	0.0191812	−0.8124124
0.0548636	0.0291681	−7.114338	0.1300636	0.019054	−0.7778067
0.0564636	0.0289617	−6.735586	0.1332636	0.0188114	−0.7130848
0.0580636	0.0287422	−6.381262	0.1348636	0.0186957	−0.6828311
0.0612636	0.0282716	−5.738083	0.1364636	0.0185837	−0.6538962
0.0628636	0.0280238	−5.44582	0.1380636	0.0184752	−0.6262198
0.0644636	0.0277695	−5.171114	0.1412636	0.0182686	−0.5744154
0.0660636	0.0275102	−4.912653	0.1428636	0.0181702	−0.550181
0.0692636	0.0269807	−4.439813	0.1444636	0.018075	−0.5269919
0.0708636	0.0267127	−4.22337	0.1460636	0.0179829	−0.504801
0.0724636	0.0264436	−4.01902	0.1492636	0.0178077	−0.4632374
0.0740636	0.0261743	−3.825944	0.1508636	0.0177244	−0.4437816
0.0772636	0.0256378	−3.470671	0.1524636	0.0176439	−0.4251577

TABLE 5A.5   Tabulated Solutions to Eqs. (4.17) and (4.18)

t	w	u	t	w	u
0.1540636	0.017566	−0.407329	0.1780636	0.0166705	−0.214904
0.1572636	0.0174181	−0.3739182	0.1812636	0.0165829	−0.1974094
0.1588636	0.0173478	−0.3582709	0.1828636	0.0165414	−0.189208
0.1604636	0.0172799	−0.343288	0.1844636	0.0165014	−0.1813501
0.1620636	0.0172144	−0.3289404	0.1860636	0.0164628	−0.1738209
0.1652636	0.0170898	−0.3020421	0.1892636	0.0163896	−0.1596937
0.1668636	0.0170307	−0.2894397	0.1908636	0.0163549	−0.1530695
0.1684636	0.0169737	−0.2773694	0.1924636	0.0163215	−0.1467219
0.1700636	0.0169186	−0.2658082	0.1940636	0.0162892	−0.1406391
0.1732636	0.0168141	−0.2441262	0.1972636	0.0162281	−0.1292237
0.1748636	0.0167645	−0.2339645	0.1988636	0.0161992	−0.1238702
0.1764636	0.0167166	−0.2242298	0.2	0.0161793	−0.1202038

TABLE 5A.5   Tabulated Solutions to Eqs. (4.17) and (4.18) (*Continued*)

t	w	u	t	w	u
0	0	−10000.	0.0786141	254.1315	−3.332067
0.0042899	61.16479	−189.0702	0.0802141	251.4895	−3.175861
0.0061723	89.99676	−128.9107	0.0834141	246.2865	−2.887435
0.0082544	119.8931	−94.03992	0.0850141	243.7331	−2.754248
0.0102642	146.3009	−73.72568	0.0866141	241.2158	−2.627826
0.0127619	175.5633	−57.37999	0.0882141	238.7375	−2.507771
0.0147543	196.1071	−48.31584	0.0914141	233.9065	−2.285303
0.0170393	216.7481	−40.54521	0.0930141	231.5574	−2.182225
0.0183009	226.8786	−37.09442	0.0946141	229.2544	−2.084178
0.0210839	246.2723	−30.96251	0.0962141	226.9987	−1.990884
0.0226141	255.3268	−28.24587	0.0994141	222.6322	−1.817529
0.0242141	263.6732	−25.78124	0.1010141	220.5226	−1.736995
0.0274141	277.2449	−21.73937	0.1026141	218.4625	−1.660268
0.0290141	282.6237	−20.06446	0.1042141	216.452	−1.587146
0.0306141	287.1613	−18.57207	0.1074141	212.5797	−1.450981
0.0322141	290.9257	−17.23504	0.1090141	210.7175	−1.387595
0.0354141	296.3885	−14.94284	0.1106141	208.9042	−1.327127
0.0370141	298.2041	−13.95464	0.1122141	207.1394	−1.269431
0.0386141	299.4812	−13.05423	0.1154141	203.7528	−1.161808
0.0402141	300.2695	−12.23112	0.1170141	202.1299	−1.111626
0.0434141	300.5611	−10.78246	0.1186141	200.553	−1.063708
0.0450141	300.1471	−10.1428	0.1202141	199.0214	−1.017941
0.0466141	299.4103	−9.551804	0.1234141	196.0906	−0.9324565
0.0482141	298.3847	−9.004576	0.1250141	194.6899	−0.8925463
0.0514141	295.591	−8.024943	0.1266141	193.3311	−0.8544051
0.0530141	293.8789	−7.585531	0.1282141	192.0135	−0.8179494
0.0546141	291.9903	−7.175731	0.1314141	189.4977	−0.7497822
0.0562141	289.9477	−6.792976	0.1330141	188.2978	−0.7179246
0.0594141	285.4823	−6.099705	0.1346141	187.1354	−0.6874595
0.0610141	283.0959	−5.785324	0.1362141	186.0095	−0.6583231
0.0626141	280.6287	−5.490196	0.1394141	183.8637	−0.6037953
0.0642141	278.0953	−5.212843	0.1410141	182.842	−0.5782911
0.0674141	272.8812	−4.70625	0.1426141	181.8531	−0.5538893
0.0690141	270.2235	−4.474705	0.1442141	180.8964	−0.5305404
0.0706141	267.5454	−4.256298	0.1474141	179.0754	−0.4868137
0.0722141	264.8558	−4.050118	0.1490141	178.2095	−0.4663481
0.0754141	259.4736	−3.671188	0.1506141	177.3722	−0.4467592
0.0770141	256.7946	−3.496979	0.1522141	176.5626	−0.4280081

TABLE 5A.6   Tabulated Solutions to Eqs. (4.17) and (4.18)

t	w	u		t	w	u
0.1554141	175.0237	−0.3928728		0.1794141	166.3275	−0.2073373
0.1570141	174.2927	−0.3764195		0.1810141	165.8951	−0.1987199
0.1586141	173.5863	−0.360666		0.1826141	165.4779	−0.1904637
0.1602141	172.9038	−0.3455814		0.1842141	165.0753	−0.1825532
0.1634141	171.6075	−0.3173039		0.1874141	164.3122	−0.1677111
0.1650141	170.9924	−0.3040564		0.1890141	163.9508	−0.1607521
0.1666141	170.3983	−0.2913689		0.1906141	163.6021	−0.1540838
0.1682141	169.8245	−0.2792172		0.1922141	163.2659	−0.1476938
0.1714141	168.7358	−0.2564295		0.1954141	162.6287	−0.1357025
0.1730141	168.2194	−0.2457502		0.1970141	162.3271	−0.130079
0.1746141	167.721	−0.2355202		0.1986141	162.0362	−0.1246899
0.1762141	167.2399	−0.2257202		0.2	161.7926	−0.1202038

TABLE **5A.6**   Tabulated Solutions to Eqs. (4.17) and (4.18) (*Continued*)

t	w	u	t	w	u
0	0	−10000.	0.0786141	2541.315	−3.332067
0.0042899	611.6479	−189.0702	0.0802141	2514.895	−3.175861
0.0061723	899.9676	−128.9107	0.0834141	2462.865	−2.887435
0.0082544	1198.931	−94.03992	0.0850141	2437.331	−2.754248
0.0102642	1463.009	−73.72568	0.0866141	2412.158	−2.627826
0.0127619	1755.633	−57.37999	0.0882141	2387.375	−2.507771
0.0147543	1961.071	−48.31584	0.0914141	2339.065	−2.285303
0.0170393	2167.481	−40.54521	0.0930141	2315.574	−2.182225
0.0183009	2268.786	−37.09442	0.0946141	2292.544	−2.084178
0.0210839	2462.723	−30.96251	0.0962141	2269.987	−1.990884
0.0226141	2553.268	−28.24587	0.0994141	2226.322	−1.817529
0.0242141	2636.732	−25.78124	0.1010141	2205.226	−1.736995
0.0274141	2772.449	−21.73937	0.1026141	2184.625	−1.660268
0.0290141	2826.237	−20.06446	0.1042141	2164.52	−1.587146
0.0306141	2871.613	−18.57207	0.1074141	2125.797	−1.450981
0.0322141	2909.257	−17.23504	0.1090141	2107.175	−1.387595
0.0354141	2963.885	−14.94284	0.1106141	2089.042	−1.327127
0.0370141	2982.041	−13.95464	0.1122141	2071.394	−1.269431
0.0386141	2994.812	−13.05423	0.1154141	2037.528	−1.161808
0.0402141	3002.695	−12.23112	0.1170141	2021.299	−1.111626
0.0434141	3005.611	−10.78246	0.1186141	2005.53	−1.063708
0.0450141	3001.471	−10.1428	0.1202141	1990.214	−1.017941
0.0466141	2994.103	−9.551804	0.1234141	1960.906	−0.9324565
0.0482141	2983.847	−9.004576	0.1250141	1946.899	−0.8925463
0.0514141	2955.91	−8.024943	0.1266141	1933.311	−0.8544051
0.0530141	2938.789	−7.585531	0.1282141	1920.135	−0.8179494
0.0546141	2919.903	−7.175731	0.1314141	1894.977	−0.7497822
0.0562141	2899.477	−6.792976	0.1330141	1882.978	−0.7179246
0.0594141	2854.823	−6.099705	0.1346141	1871.354	−0.6874595
0.0610141	2830.959	−5.785324	0.1362141	1860.095	−0.6583231
0.0626141	2806.287	−5.490196	0.1394141	1838.637	−0.6037953
0.0642141	2780.953	−5.212843	0.1410141	1828.42	−0.5782911
0.0674141	2728.812	−4.70625	0.1426141	1818.531	−0.5538893
0.0690141	2702.235	−4.474705	0.1442141	1808.964	−0.5305404
0.0706141	2675.454	−4.256298	0.1474141	1790.754	−0.4868137
0.0722141	2648.558	−4.050118	0.1490141	1782.095	−0.4663481
0.0754141	2594.736	−3.671188	0.1506141	1773.722	−0.4467592
0.0770141	2567.946	−3.496979	0.1522141	1765.626	−0.4280081

TABLE 5A.7 Tabulated Solutions to Eqs. (4.17) and (4.18)

t	w	u	t	w	u
0.1554141	1750.237	−0.3928728	0.1794141	1663.275	−0.2073373
0.1570141	1742.927	−0.3764195	0.1810141	1658.951	−0.1987199
0.1586141	1735.863	−0.360666	0.1826141	1654.779	−0.1904637
0.1602141	1729.038	−0.3455814	0.1842141	1650.753	−0.1825532
0.1634141	1716.075	−0.3173039	0.1874141	1643.122	−0.1677111
0.1650141	1709.924	−0.3040564	0.1890141	1639.508	−0.1607521
0.1666141	1703.983	−0.2913689	0.1906141	1636.021	−0.1540838
0.1682141	1698.245	−0.2792172	0.1922141	1632.659	−0.1476938
0.1714141	1687.358	−0.2564295	0.1954141	1626.287	−0.1357025
0.1730141	1682.194	−0.2457502	0.1970141	1623.271	−0.130079
0.1746141	1677.21	−0.2355202	0.1986141	1620.362	−0.1246899
0.1762141	1672.399	−0.2257202	0.2	1617.926	−0.1202038

TABLE 5A.7 Tabulated Solutions to Eqs. (4.17) and (4.18) (*Continued*)

t	w	u	t	w	u
0	0	−10000.	0.0722141	2.649E+04	−4.050118
0.0042899	6116.479	−189.0702	0.0754141	2.595E+04	−3.671188
0.0061723	8999.676	−128.9107	0.0770141	2.568E+04	−3.496979
0.0082544	1.199E+04	−94.03992	0.0786141	2.541E+04	−3.332067
0.0102642	1.463E+04	−73.72568	0.0802141	2.515E+04	−3.175861
0.0127619	1.756E+04	−57.37999	0.0834141	2.463E+04	−2.887435
0.0147543	1.961E+04	−48.31584	0.0850141	2.437E+04	−2.754248
0.0170393	2.167E+04	−40.54521	0.0866141	2.412E+04	−2.627826
0.0183009	2.269E+04	−37.09442	0.0882141	2.387E+04	−2.507771
0.0210839	2.463E+04	−30.96251	0.0914141	2.339E+04	−2.285303
0.0226141	2.553E+04	−28.24587	0.0930141	2.316E+04	−2.182225
0.0242141	2.637E+04	−25.78124	0.0946141	2.293E+04	−2.084178
0.0274141	2.772E+04	−21.73937	0.0962141	2.27E+04	−1.990884
0.0290141	2.826E+04	−20.06446	0.0994141	2.226E+04	−1.817529
0.0306141	2.872E+04	−18.57207	0.1010141	2.205E+04	−1.736995
0.0322141	2.909E+04	−17.23504	0.1026141	2.185E+04	−1.660268
0.0354141	2.964E+04	−14.94284	0.1042141	2.165E+04	−1.587146
0.0370141	2.982E+04	−13.95464	0.1074141	2.126E+04	−1.450981
0.0386141	2.995E+04	−13.05423	0.1090141	2.107E+04	−1.387595
0.0402141	3.003E+04	−12.23112	0.1106141	2.089E+04	−1.327127
0.0434141	3.006E+04	−10.78246	0.1122141	2.071E+04	−1.269431
0.0450141	3.001E+04	−10.1428	0.1154141	2.038E+04	−1.161808
0.0466141	2.994E+04	−9.551804	0.1170141	2.021E+04	−1.111626
0.0482141	2.984E+04	−9.004576	0.1186141	2.006E+04	−1.063708
0.0514141	2.956E+04	−8.024943	0.1202141	1.99E+04	−1.017941
0.0530141	2.939E+04	−7.585531	0.1234141	1.961E+04	−0.9324565
0.0546141	2.92E+04	−7.175731	0.1250141	1.947E+04	−0.8925463
0.0562141	2.899E+04	−6.792976	0.1266141	1.933E+04	−0.8544051
0.0594141	2.855E+04	−6.099705	0.1282141	1.92E+04	−0.8179494
0.0610141	2.831E+04	−5.785324	0.1314141	1.895E+04	−0.7497822
0.0626141	2.806E+04	−5.490196	0.1330141	1.883E+04	−0.7179246
0.0642141	2.781E+04	−5.212843	0.1346141	1.871E+04	−0.6874595
0.0674141	2.729E+04	−4.70625	0.1362141	1.86E+04	−0.6583231
0.0690141	2.702E+04	−4.474705	0.1394141	1.839E+04	−0.6037953
0.0706141	2.675E+04	−4.256298	0.1410141	1.828E+04	−0.5782911

TABLE 5A.8 Tabulated Solutions to Eqs. (4.17) and (4.18)

t	w	u	t	w	u
0.1426141	1.819E+04	−0.5538893	0.1730141	1.682E+04	−0.2457502
0.1442141	1.809E+04	−0.5305404	0.1746141	1.677E+04	−0.2355202
0.1474141	1.791E+04	−0.4868137	0.1762141	1.672E+04	−0.2257202
0.1490141	1.782E+04	−0.4663481	0.1794141	1.663E+04	−0.2073373
0.1506141	1.774E+04	−0.4467592	0.1810141	1.659E+04	−0.1987199
0.1522141	1.766E+04	−0.4280081	0.1826141	1.655E+04	−0.1904637
0.1554141	1.75E+04	−0.3928728	0.1842141	1.651E+04	−0.1825532
0.1570141	1.743E+04	−0.3764195	0.1874141	1.643E+04	−0.1677111
0.1586141	1.736E+04	−0.360666	0.1890141	1.64E+04	−0.1607521
0.1602141	1.729E+04	−0.3455814	0.1906141	1.636E+04	−0.1540838
0.1634141	1.716E+04	−0.3173039	0.1922141	1.633E+04	−0.1476938
0.1650141	1.71E+04	−0.3040564	0.1954141	1.626E+04	−0.1357025
0.1666141	1.704E+04	−0.2913689	0.1970141	1.623E+04	−0.130079
0.1682141	1.698E+04	−0.2792172	0.1986141	1.62E+04	−0.1246899
0.1714141	1.687E+04	−0.2564295	0.2	1.618E+04	−0.1202038

TABLE 5A.8 Tabulated Solutions to Eqs. (4.17) and (4.18) (*Continued*)

t	w	u	t	w	u
0	0	−10000.	0.0720972	2.205E+04	−3.949E−05
0.0041332	4.035E+04	−20.09528	0.0741637	2.205E+04	−2.7E−05
0.0061122	4.296E+04	−11.02092	0.0762302	2.205E+04	−1.845E−05
0.0082397	4.115E+04	−6.448411	0.0782967	2.205E+04	−1.261E−05
0.0102382	3.798E+04	−4.109012	0.0803631	2.205E+04	−8.623E−06
0.0124595	3.433E+04	−2.575465	0.0824296	2.205E+04	−5.894E−06
0.0142424	3.175E+04	−1.79853	0.0844961	2.205E+04	−4.029E−06
0.0161023	2.952E+04	−1.248751	0.0865626	2.205E+04	−2.754E−06
0.0180209	2.768E+04	−0.8632175	0.088629	2.205E+04	−1.883E−06
0.0206443	2.582E+04	−0.5250177	0.0900067	2.205E+04	−1.461E−06
0.0226459	2.48E+04	−0.360657	0.0920732	2.205E+04	−9.986E−07
0.0246674	2.405E+04	−0.2473774	0.0941936	2.205E+04	−6.758E−07
0.0260231	2.366E+04	−0.1922821	0.0964853	2.205E+04	−4.432E−07
0.0280653	2.32E+04	−0.1316797	0.0981221	2.205E+04	−3.279E−07
0.0301151	2.288E+04	−0.0901257	0.1007671	2.205E+04	−2.015E−07
0.0321702	2.264E+04	−0.0616602	0.1026759	2.205E+04	−1.418E−07
0.0342288	2.247E+04	−0.0421738	0.1047197	2.205E+04	−9.733E−08
0.0362899	2.235E+04	−0.0288402	0.1069184	2.205E+04	−6.493E−08
0.0383527	2.226E+04	−0.0197196	0.1080833	2.205E+04	−5.24E−08
0.0404167	2.22E+04	−0.0134821	0.1105622	2.205E+04	−3.32E−08
0.0424815	2.216E+04	−0.0092171	0.1132694	2.205E+04	−2.017E−08
0.0445468	2.213E+04	−0.006301	0.1147221	2.205E+04	−1.544E−08
0.0466124	2.211E+04	−0.0043074	0.1162494	2.205E+04	−1.165E−08
0.0486784	2.209E+04	−0.0029445	0.1194494	2.205E+04	−6.465E−09
0.0500557	2.208E+04	−0.0022849	0.1210494	2.205E+04	−4.815E−09
0.0521219	2.207E+04	−0.0015619	0.1226494	2.205E+04	−3.587E−09
0.0541882	2.207E+04	−0.0010677	0.1242494	2.205E+04	−2.672E−09
0.0562545	2.206E+04	−0.0007298	0.1274494	2.205E+04	−1.482E−09
0.0583209	2.206E+04	−0.0004989	0.1290494	2.205E+04	−1.104E−09
0.0603873	2.206E+04	−0.000341	0.1306494	2.205E+04	−8.224E−10
0.0624538	2.206E+04	−0.0002331	0.1322494	2.205E+04	−6.126E−10
0.0645202	2.206E+04	−0.0001593	0.1354494	2.205E+04	−3.399E−10
0.0665867	2.205E+04	−0.0001089	0.1370494	2.205E+04	−2.531E−10
0.0686531	2.205E+04	−7.446E−05	0.1386494	2.205E+04	−1.886E−10
0.0700308	2.205E+04	−5.778E−05	0.1402494	2.205E+04	−1.404E−10

TABLE 5A.9 Tabulated Solutions to Eqs. (4.17) and (4.18)

t	w	u	t	w	u
0.1434494	2.205E+04	−7.792E−11	0.1722494	2.205E+04	−3.881E−13
0.1450494	2.205E+04	−5.804E−11	0.1754494	2.205E+04	−2.153E−13
0.1466494	2.205E+04	−4.323E−11	0.1770494	2.205E+04	−1.604E−13
0.1482494	2.205E+04	−3.22E−11	0.1786494	2.205E+04	−1.195E−13
0.1514494	2.205E+04	−1.787E−11	0.1802494	2.205E+04	−8.899E−14
0.1530494	2.205E+04	−1.331E−11	0.1834494	2.205E+04	−4.938E−14
0.1546494	2.205E+04	−9.912E−12	0.1850494	2.205E+04	−3.678E−14
0.1562494	2.205E+04	−7.383E−12	0.1866494	2.205E+04	−2.739E−14
0.1594494	2.205E+04	−4.096E−12	0.1882494	2.205E+04	−2.04E−14
0.1610494	2.205E+04	−3.051E−12	0.1914494	2.205E+04	−1.132E−14
0.1626494	2.205E+04	−2.273E−12	0.1930494	2.205E+04	−8.432E−15
0.1642494	2.205E+04	−1.693E−12	0.1946494	2.205E+04	−6.281E−15
0.1674494	2.205E+04	−9.392E−13	0.1962494	2.205E+04	−4.678E−15
0.1690494	2.205E+04	−6.996E−13	0.1994494	2.205E+04	−2.596E−15
0.1706494	2.205E+04	−5.211E−13	0.2	2.205E+04	−2.345E−15

TABLE 5A.9 Tabulated Solutions to Eqs. (4.17) and (4.18) (*Continued*)

t	w	u	t	w	u
0	0	−10000.	0.0723138	1.544E+05	−1.266E−07
0.0041817	3.006E+05	−11.47518	0.074549	1.544E+05	−7.032E−08
0.0063186	2.797E+05	−5.389015	0.0761882	1.544E+05	−4.57E−08
0.0081754	2.49E+05	−3.033202	0.0789222	1.544E+05	−2.226E−08
0.0102411	2.187E+05	−1.669818	0.0809688	1.544E+05	−1.3E−08
0.0124493	1.951E+05	−0.9053483	0.0820729	1.544E+05	−9.721E−09
0.0142816	1.817E+05	−0.5508832	0.0844722	1.544E+05	−5.172E−09
0.0161505	1.724E+05	−0.3338761	0.0871788	1.544E+05	−2.538E−09
0.0180426	1.661E+05	−0.201847	0.0886724	1.544E+05	−1.714E−09
0.0204271	1.611E+05	−0.1073742	0.0902724	1.544E+05	−1.125E−09
0.0223438	1.587E+05	−0.0647395	0.0934724	1.544E+05	−4.849E−10
0.0242653	1.571E+05	−0.0390135	0.0950724	1.544E+05	−3.184E−10
0.0261896	1.561E+05	−0.0235031	0.0966724	1.544E+05	−2.09E−10
0.0281157	1.555E+05	−0.0141564	0.0982724	1.544E+05	−1.372E−10
0.0300429	1.551E+05	−0.0085257	0.1014724	1.544E+05	−5.914E−11
0.0324526	1.548E+05	−0.0045229	0.1030724	1.544E+05	−3.883E−11
0.0343808	1.546E+05	−0.0027236	0.1046724	1.544E+05	−2.549E−11
0.0363092	1.545E+05	−0.0016401	0.1062724	1.544E+05	−1.674E−11
0.0382378	1.545E+05	−0.0009876	0.1094724	1.544E+05	−7.213E−12
0.0401664	1.544E+05	−0.0005947	0.1110724	1.544E+05	−4.736E−12
0.042095	1.544E+05	−0.0003581	0.1126724	1.544E+05	−3.109E−12
0.0440237	1.544E+05	−0.0002156	0.1142724	1.544E+05	−2.041E−12
0.0464345	1.544E+05	−0.0001144	0.1174724	1.544E+05	−8.797E−13
0.0483632	1.544E+05	−6.887E−05	0.1190724	1.544E+05	−5.776E−13
0.0502919	1.544E+05	−4.147E−05	0.1206724	1.544E+05	−3.792E−13
0.0522206	1.544E+05	−2.497E−05	0.1222724	1.544E+05	−2.489E−13
0.0541493	1.544E+05	−1.504E−05	0.1254724	1.544E+05	−1.073E−13
0.0560781	1.544E+05	−9.054E−06	0.1270724	1.544E+05	−7.044E−14
0.0580068	1.544E+05	−5.452E−06	0.1286724	1.544E+05	−4.625E−14
0.0604176	1.544E+05	−2.892E−06	0.1302724	1.544E+05	−3.036E−14
0.0623464	1.544E+05	−1.741E−06	0.1334724	1.544E+05	−1.309E−14
0.0642751	1.544E+05	−1.049E−06	0.1350724	1.544E+05	−8.591E−15
0.0662649	1.544E+05	−6.213E−07	0.1366724	1.544E+05	−5.64E−15
0.0684709	1.544E+05	−3.478E−07	0.1382724	1.544E+05	−3.703E−15
0.0703008	1.544E+05	−2.149E−07	0.1414724	1.544E+05	−1.596E−15

TABLE 5A.10   Tabulated Solutions to Eqs. (4.17) and (4.18)

t	w	u	t	w	u
0.1430724	1.544E+05	−1.048E−15	0.1734724	1.544E+05	−3.532E−19
0.1446724	1.544E+05	−6.879E−16	0.1750724	1.544E+05	−2.318E−19
0.1462724	1.544E+05	−4.516E−16	0.1766724	1.544E+05	−1.522E−19
0.1494724	1.544E+05	−1.947E−16	0.1782724	1.544E+05	−9.993E−20
0.1510724	1.544E+05	−1.278E−16	0.1814724	1.544E+05	−4.307E−20
0.1526724	1.544E+05	−8.39E−17	0.1830724	1.544E+05	−2.828E−20
0.1542724	1.544E+05	−5.508E−17	0.1846724	1.544E+05	−1.856E−20
0.1574724	1.544E+05	−2.374E−17	0.1862724	1.544E+05	−1.219E−20
0.1590724	1.544E+05	−1.559E−17	0.1894724	1.544E+05	−5.253E−21
0.1606724	1.544E+05	−1.023E−17	0.1910724	1.544E+05	−3.449E−21
0.1622724	1.544E+05	−6.718E−18	0.1926724	1.544E+05	−2.264E−21
0.1654724	1.544E+05	−2.896E−18	0.1942724	1.544E+05	−1.486E−21
0.1670724	1.544E+05	−1.901E−18	0.1974724	1.544E+05	−6.407E−22
0.1686724	1.544E+05	−1.248E−18	0.1990724	1.544E+05	−4.206E−22
0.1702724	1.544E+05	−8.193E−19	0.2	1.544E+05	−3.296E−22

TABLE 5A.10 Tabulated Solutions to Eqs. (4.17) and (4.18) (*Continued*)

t	w	u	t	w	u
0	0	−10000.	0.0732936	5.718E+05	−5.236E−22
0.0041575	7.672E+05	−1.268482	0.0748936	5.718E+05	−1.673E−22
0.0060272	6.367E+05	−0.3232498	0.0764936	5.718E+05	−5.347E−23
0.0081445	5.891E+05	−0.0711169	0.0780936	5.718E+05	−1.709E−23
0.0101037	5.767E+05	−0.0176551	0.0812936	5.718E+05	−1.745E−24
0.012067	5.732E+05	−0.0043776	0.0828936	5.718E+05	−5.577E−25
0.0140313	5.722E+05	−0.0010851	0.0844936	5.718E+05	−1.782E−25
0.0161746	5.719E+05	−0.0002369	0.0860936	5.718E+05	−5.695E−26
0.0181392	5.719E+05	−5.872E−05	0.0892936	5.718E+05	−5.817E−27
0.0201039	5.718E+05	−1.455E−05	0.0908936	5.718E+05	−1.859E−27
0.0220686	5.718E+05	−3.607E−06	0.0924936	5.718E+05	−5.94E−28
0.0240333	5.718E+05	−8.94E−07	0.0940936	5.718E+05	−1.898E−28
0.0261016	5.718E+05	−2.059E−07	0.0972936	5.718E+05	−1.939E−29
0.028303	5.718E+05	−4.313E−08	0.0988936	5.718E+05	−6.196E−30
0.0300951	5.718E+05	−1.208E−08	0.1004936	5.718E+05	−1.98E−30
0.0324308	5.718E+05	−2.301E−09	0.1020936	5.718E+05	−6.327E−31
0.0342551	5.718E+05	−6.301E−10	0.1052936	5.718E+05	−6.462E−32
0.0366075	5.718E+05	−1.186E−10	0.1068936	5.718E+05	−2.065E−32
0.0386426	5.718E+05	−2.794E−11	0.1084936	5.718E+05	−6.599E−33
0.0412936	5.718E+05	−4.242E−12	0.1100936	5.718E+05	−2.109E−33
0.0428936	5.718E+05	−1.356E−12	0.1132936	5.718E+05	−2.154E−34
0.0444936	5.718E+05	−4.332E−13	0.1148936	5.718E+05	−6.883E−35
0.0460936	5.718E+05	−1.384E−13	0.1164936	5.718E+05	−2.2E−35
0.0492936	5.718E+05	−1.414E−14	0.1180936	5.718E+05	−7.029E−36
0.0508936	5.718E+05	−4.519E−15	0.1212936	5.718E+05	−7.179E−37
0.0524936	5.718E+05	−1.444E−15	0.1228936	5.718E+05	−2.294E−37
0.0540936	5.718E+05	−4.615E−16	0.1244936	5.718E+05	−7.332E−38
0.0572936	5.718E+05	−4.713E−17	0.1260936	5.718E+05	−2.343E−38
0.0588936	5.718E+05	−1.506E−17	0.1292936	5.718E+05	−2.393E−39
0.0604936	5.718E+05	−4.813E−18	0.1308936	5.718E+05	−7.647E−40
0.0620936	5.718E+05	−1.538E−18	0.1324936	5.718E+05	−2.444E−40
0.0652936	5.718E+05	−1.571E−19	0.1340936	5.718E+05	−7.809E−41
0.0668936	5.718E+05	−5.02E−20	0.1372936	5.718E+05	−7.975E−42
0.0684936	5.718E+05	−1.604E−20	0.1388936	5.718E+05	−2.549E−42
0.0700936	5.718E+05	−5.127E−21	0.1404936	5.718E+05	−8.145E−43

TABLE **5A.11**   Tabulated Solutions to Eqs. (4.17) and (4.18)

t	w	u	t	w	u
0.1420936	5.718E+05	−2.603E–43	0.1724936	5.718E+05	−1.005E–52
0.1452936	5.718E+05	−2.658E–44	0.1740936	5.718E+05	−3.213E–53
0.1468936	5.718E+05	−8.495E–45	0.1772936	5.718E+05	−3.281E–54
0.1484936	5.718E+05	−2.715E–45	0.1788936	5.718E+05	−1.048E–54
0.1500936	5.718E+05	−8.676E–46	0.1804936	5.718E+05	−3.351E–55
0.1532936	5.718E+05	−8.86E–47	0.1820936	5.718E+05	−1.071E–55
0.1548936	5.718E+05	−2.831E–47	0.1852936	5.718E+05	−1.094E–56
0.1564936	5.718E+05	−9.049E–48	0.1868936	5.718E+05	−3.495E–57
0.1580936	5.718E+05	−2.892E–48	0.1884936	5.718E+05	−1.117E–57
0.1612936	5.718E+05	−2.953E–49	0.1900936	5.718E+05	−3.569E–58
0.1628936	5.718E+05	−9.438E–50	0.1932936	5.718E+05	−3.645E–59
0.1644936	5.718E+05	−3.016E–50	0.1948936	5.718E+05	−1.165E–59
0.1660936	5.718E+05	−9.638E–51	0.1964936	5.718E+05	−3.722E–60
0.1692936	5.718E+05	−9.843E–52	0.1980936	5.718E+05	−1.19E–60
0.1708936	5.718E+05	−3.146E–52	0.2	5.718E+05	−3.058E–61

TABLE 5A.11 Tabulated Solutions to Eqs. (4.17) and (4.18) (*Continued*)

t	w	u	t	w	u
0	0	−10000.	0.0724762	1.286E+06	−8.471E−14
0.0040288	1.287E+06	−6.919E−05	0.0743077	1.286E+06	−4.435E−15
0.006024	1.286E+06	−1.275E−07	0.0771965	1.286E+06	−3.83E−14
0.0081769	1.286E+06	−1.427E−10	0.0786122	1.286E+06	1.057E−13
0.0105186	1.286E+06	−7.901E−14	0.0805322	1.286E+06	9.377E−15
0.0121858	1.286E+06	1.296E−15	0.082983	1.286E+06	1.276E−14
0.0153858	1.286E+06	3.316E−14	0.084583	1.286E+06	−6.456E−14
0.0168428	1.286E+06	−1.058E−13	0.0868071	1.286E+06	−3.084E−14
0.0187868	1.286E+06	−1.06E−14	0.0888431	1.286E+06	−5.234E−15
0.0211779	1.286E+06	−1.082E−14	0.0904007	1.286E+06	2.323E−14
0.0227779	1.286E+06	5.472E−14	0.0930681	1.286E+06	8.278E−14
0.024096	1.286E+06	−1.049E−13	0.0948992	1.286E+06	4.289E−15
0.027061	1.286E+06	4.956E−15	0.0962054	1.286E+06	−7.849E−15
0.0285756	1.286E+06	−1.916E−14	0.0992109	1.286E+06	−1.057E−13
0.0301274	1.286E+06	8.35E−14	0.1002442	1.286E+06	5.354E−14
0.0322173	1.286E+06	1.959E−14	0.1021514	1.286E+06	4.424E−15
0.0343893	1.286E+06	6.922E−15	0.1051891	1.286E+06	6.681E−14
0.0374306	1.286E+06	1.058E−13	0.1064541	1.286E+06	−1.035E−13
0.0384868	1.286E+06	−6.099E−14	0.1082906	1.286E+06	−5.597E−15
0.0403657	1.286E+06	−4.213E−15	0.1110115	1.286E+06	−2.43E−14
0.0433787	1.286E+06	−5.826E−14	0.1125337	1.286E+06	9.631E−14
0.0446804	1.286E+06	1.047E−13	0.1145506	1.286E+06	1.472E−14
0.0465354	1.286E+06	6.39E−15	0.1168143	1.286E+06	8.132E−15
0.0491827	1.286E+06	2.06E−14	0.1184143	1.286E+06	−4.113E−14
0.0507259	1.286E+06	−8.737E−14	0.1208368	1.286E+06	−5.157E−14
0.052793	1.286E+06	−1.798E−14	0.1227524	1.286E+06	−4.483E−15
0.0549919	1.286E+06	−7.262E−15	0.1242	1.286E+06	1.384E−14
0.0565919	1.286E+06	3.673E−14	0.1270479	1.286E+06	1.019E−13
0.0580194	1.286E+06	−1.057E−13	0.128884	1.286E+06	5.517E−15
0.0609563	1.286E+06	4.294E−15	0.1300489	1.286E+06	−5.507E−15
0.0623794	1.286E+06	−1.217E−14	0.13313	1.286E+06	−9.714E−14
0.0652669	1.286E+06	−1.045E−13	0.1342505	1.286E+06	7.83E−14
0.0662237	1.286E+06	3.27E−14	0.1360886	1.286E+06	4.242E−15
0.0682454	1.286E+06	5.154E−15	0.139017	1.286E+06	4.282E−14
0.0713261	1.286E+06	9.096E−14	0.1404014	1.286E+06	−1.056E−13

TABLE 5A.12 Tabulated Solutions to Eqs. (4.17) and (4.18)

t	w	u	t	w	u
0.1423034	1.286E+06	−8.502E−15	0.1728135	1.286E+06	2.783E−14
0.1448029	1.286E+06	−1.449E−14	0.1743039	1.286E+06	−9.937E−14
0.1463894	1.286E+06	7.034E−14	0.1762905	1.286E+06	−1.288E−14
0.1485694	1.286E+06	2.682E−14	0.1786072	1.286E+06	−9.307E−15
0.1506455	1.286E+06	5.695E−15	0.1802072	1.286E+06	4.708E−14
0.1522177	1.286E+06	−2.645E−14	0.1825685	1.286E+06	4.456E−14
0.154831	1.286E+06	−7.561E−14	0.1845171	1.286E+06	4.707E−15
0.156675	1.286E+06	−4.256E−15	0.1875682	1.286E+06	7.509E−14
0.1580142	1.286E+06	8.856E−15	0.1887878	1.286E+06	−9.588E−14
0.1609862	1.286E+06	1.055E−13	0.1906221	1.286E+06	−5.187E−15
0.1628811	1.286E+06	8.167E−15	0.1934068	1.286E+06	−2.945E−14
0.1654008	1.286E+06	1.525E−14	0.1948847	1.286E+06	1.008E−13
0.1669808	1.286E+06	−7.256E−14	0.1968586	1.286E+06	1.216E−14
0.1682134	1.286E+06	9.804E−14	0.1991982	1.286E+06	9.821E−15
0.1700484	1.286E+06	5.313E−15	0.2	1.286E+06	−5.834E−16

TABLE 5A.12 Tabulated Solutions to Eqs. (4.17) and (4.18) (*Continued*)

t	Y0	Y2	Y1	Y
0	1.	0	0	1.
0.004359	1.47218	0.0129927	0	1.473998
0.007559	1.887874	0.0261945	0	1.895785
0.009159	2.125638	0.0344436	0	2.139863
0.010759	2.388529	0.0441308	0	2.41291
0.012359	2.681527	0.0556099	0	2.722127
0.015559	3.38291	0.0859702	0	3.49122
0.017159	3.806001	0.1062996	0	3.982331
0.0186034	4.237509	0.1287257	0	4.5123
0.0210299	5.073482	0.1779106	0	5.657177
0.022076	5.470579	0.2046193	0	6.276749
0.0249041	6.554646	0.2958113	0	8.359175
0.0269476	7.18246	0.3759752	0	9.950042
0.0285802	7.484542	0.4426141	0	10.89004
0.0300905	7.591702	0.5014594	0	11.39343
0.032047	7.537041	0.5691191	0	11.68333
0.0342838	7.320413	0.6325497	0	11.78975
0.0363293	7.066648	0.6787404	0	11.81749
0.0389026	6.741316	0.7242717	0	11.8267
0.0410777	6.48557	0.7545792	0	11.82855
0.042367	6.345555	0.7698417	0	11.82892
0.0454371	6.047731	0.8000437	0	11.82919
0.0470371	5.911127	0.8130893	0	11.82922
0.0486371	5.785962	0.8246892	0	11.82923
0.0502371	5.671172	0.8350736	0	11.82924
0.0534371	5.468716	0.8528996	0	11.82924
0.0550371	5.379262	0.8606128	0	11.82924
0.0566371	5.296616	0.8676674	0	11.82924
0.0582371	5.220104	0.8741467	0	11.82924
0.0614371	5.083157	0.8856474	0	11.82924
0.0630371	5.021725	0.8907773	0	11.82924
0.0646371	4.964417	0.8955528	0	11.82924
0.0662371	4.910864	0.9000103	0	11.82924
0.0694371	4.813746	0.9080933	0	11.82924
0.0710371	4.769628	0.91177	0	11.82924
0.0726371	4.72815	0.9152325	0	11.82924

TABLE **5A.13** Tabulated Solutions to Eqs. (5.33) and (5.34)

t	Y0	Y2	Y1	Y
0.0742371	4.689101	0.9184992	0	11.82924
0.0774371	4.617555	0.9245083	0	11.82924
0.0790371	4.584732	0.9272782	0	11.82924
0.0806371	4.553686	0.9299074	0	11.82924
0.0822371	4.524289	0.9324063	0	11.82924
0.0854371	4.469991	0.9370495	0	11.82924
0.0870371	4.444889	0.9392099	0	11.82924
0.0886371	4.421032	0.9412723	0	11.82924
0.0902371	4.398338	0.9432428	0	11.82924
0.0934371	4.356152	0.9469312	0	11.82924
0.0950371	4.336528	0.948659	0	11.82924
0.0966371	4.317806	0.9503154	0	11.82924
0.0982371	4.299931	0.9519043	0	11.82924
0.1014371	4.266529	0.9548944	0	11.82924
0.1030371	4.250914	0.9563024	0	11.82924
0.1046371	4.235969	0.9576564	0	11.82924
0.1062371	4.221656	0.9589591	0	11.82924
0.1094371	4.194796	0.961421	0	11.82924
0.1110371	4.182187	0.9625848	0	11.82924
0.1126371	4.170086	0.9637067	0	11.82924
0.1142371	4.15847	0.9647887	0	11.82924
0.1174371	4.136589	0.9668399	0	11.82924
0.1190371	4.126282	0.9678126	0	11.82924
0.1206371	4.116369	0.9687521	0	11.82924
0.1222371	4.106831	0.9696597	0	11.82924
0.1254371	4.088813	0.971385	0	11.82924
0.1270371	4.080299	0.9722051	0	11.82924
0.1286371	4.072097	0.9729984	0	11.82924
0.1302371	4.06419	0.973766	0	11.82924
0.1334371	4.049214	0.9752281	0	11.82924
0.1350371	4.042121	0.9759244	0	11.82924
0.1366371	4.035275	0.9765988	0	11.82924
0.1382371	4.028667	0.9772521	0	11.82924
0.1414371	4.016122	0.9784986	0	11.82924
0.1430371	4.010167	0.9790933	0	11.82924
0.1446371	4.004412	0.9796698	0	11.82924

TABLE 5A.13 Tabulated Solutions to Eqs. (5.33) and (5.34) (*Continued*)

t	Y0	Y2	Y1	Y
0.1462371	3.998849	0.9802288	0	11.82924
0.1494371	3.988269	0.9812968	0	11.82924
0.1510371	3.983237	0.981807	0	11.82924
0.1526371	3.978369	0.982302	0	11.82924
0.1542371	3.973658	0.9827824	0	11.82924
0.1574371	3.964683	0.9837013	0	11.82924
0.1590371	3.960408	0.9841407	0	11.82924
0.1606371	3.956268	0.9845673	0	11.82924
0.1622371	3.952257	0.9849817	0	11.82924
0.1654371	3.944606	0.9857749	0	11.82924
0.1670371	3.940956	0.9861546	0	11.82924
0.1686371	3.937418	0.9865235	0	11.82924
0.1702371	3.933989	0.9868819	0	11.82924
0.1734371	3.927437	0.9875686	0	11.82924
0.1750371	3.924309	0.9878976	0	11.82924
0.1766371	3.921274	0.9882174	0	11.82924
0.1782371	3.918329	0.9885282	0	11.82924
0.1814371	3.912699	0.9891242	0	11.82924
0.1830371	3.910007	0.9894099	0	11.82924
0.1846371	3.907395	0.9896877	0	11.82924
0.1862371	3.904858	0.9899579	0	11.82924
0.1894371	3.900004	0.9904762	0	11.82924
0.1910371	3.897681	0.9907247	0	11.82924
0.1926371	3.895425	0.9909666	0	11.82924
0.1942371	3.893234	0.9912018	0	11.82924
0.1974371	3.889036	0.9916533	0	11.82924
0.1990371	3.887027	0.99187	0	11.82924
0.2	3.885845	0.9919975	0	11.82924

TABLE **5A.13**   Tabulated Solutions to Eqs. (5.33) and (5.34) (*Continued*)

t	Y0	Y2	Y1	Y
0	1.	0	0	1.
0.0043706	1.469951	0.0448426	0.0130377	1.474452
0.0075706	1.87359	0.0885813	0.0262716	1.892724
0.0091706	2.099217	0.1149546	0.0345547	2.133097
0.0107706	2.343284	0.1449413	0.0443016	2.400235
0.0123706	2.607652	0.1790628	0.0558877	2.700158
0.0155706	3.202339	0.2618647	0.0868037	3.431502
0.0171706	3.532151	0.3114506	0.1078089	3.885396
0.0187706	3.879311	0.366721	0.1342	4.417361
0.0202451	4.207749	0.422344	0.1648947	4.990425
0.0228171	4.767525	0.5270979	0.2389056	6.198635
0.0250254	5.185178	0.6187002	0.3291907	7.36845
0.0261057	5.352493	0.6615904	0.3821291	7.901081
0.0285991	5.622704	0.7502051	0.5127836	8.709904
0.0303496	5.715824	0.8013117	0.5965414	8.825593
0.0320986	5.742239	0.8427961	0.665249	8.666575
0.0347417	5.69366	0.8895104	0.7412461	8.202449
0.036691	5.617331	0.9142106	0.7810006	7.829336
0.03893	5.510156	0.9351497	0.8151603	7.431823
0.0401949	5.446037	0.9443208	0.8305385	7.228602
0.0431046	5.299452	0.9601916	0.8584392	6.820126
0.0447046	5.222577	0.9666133	0.8705092	6.627292
0.0463046	5.149622	0.9718418	0.8808726	6.453951
0.0495046	5.016867	0.9796659	0.8977451	6.156647
0.0511046	4.957162	0.9826038	0.9047062	6.028732
0.0527046	4.901743	0.9850582	0.9109026	5.912426
0.0543046	4.850399	0.9871199	0.9164554	5.806349
0.0575046	4.758959	0.990337	0.9259983	5.62031
0.0591046	4.718358	0.9915948	0.9301305	5.53845
0.0607046	4.68084	0.9926707	0.9339104	5.462976
0.0623046	4.646172	0.9935943	0.9373818	5.393228
0.0655046	4.584517	0.9950774	0.9435401	5.268691
0.0671046	4.557137	0.995673	0.9462844	5.212958
0.0687046	4.531815	0.9961906	0.9488368	5.161048
0.0703046	4.508391	0.9966413	0.9512166	5.112618
0.0735046	4.46665	0.9973788	0.9555239	5.025018

TABLE 5A.14   Tabulated Solutions to Eqs. (5.33) and (5.34)

t	Y0	Y2	Y1	Y
0.0751046	4.448073	0.9976804	0.9574789	4.985335
0.0767046	4.430866	0.9979451	0.9593168	4.948101
0.0783046	4.414925	0.9981779	0.9610478	4.913122
0.0815046	4.38646	0.9985636	0.964223	4.849246
0.0831046	4.373765	0.9987232	0.965682	4.820049
0.0847046	4.361992	0.9988642	0.9670639	4.792502
0.0863046	4.351072	0.998989	0.9683744	4.766488
0.0895046	4.331542	0.9991977	0.9708009	4.718638
0.0911046	4.322819	0.9992847	0.9719257	4.696616
0.0927046	4.314723	0.9993619	0.9729966	4.675749
0.0943046	4.307207	0.9994306	0.9740173	4.655963
0.0975046	4.293752	0.999546	0.9759198	4.619361
0.0991046	4.287736	0.9995945	0.9768074	4.602423
0.1007046	4.282149	0.9996376	0.9776557	4.58632
0.1023046	4.276961	0.999676	0.9784671	4.571
0.1055046	4.267666	0.9997409	0.9799873	4.542531
0.1071046	4.263508	0.9997682	0.9806998	4.529297
0.1087046	4.259645	0.9997926	0.9813828	4.516681
0.1103046	4.256057	0.9998144	0.9820378	4.504647
0.1135046	4.249627	0.9998512	0.9832698	4.482197
0.1151046	4.24675	0.9998668	0.9838492	4.471723
0.1167046	4.244077	0.9998807	0.984406	4.461715
0.1183046	4.241593	0.9998931	0.984941	4.452146
0.1215046	4.237142	0.9999142	0.9859502	4.43424
0.1231046	4.235151	0.9999231	0.9864262	4.42586
0.1247046	4.2333	0.9999311	0.9868843	4.417836
0.1263046	4.231581	0.9999383	0.9873253	4.41015
0.1295046	4.228499	0.9999504	0.988159	4.395729
0.1311046	4.227121	0.9999555	0.9885531	4.388961
0.1327046	4.22584	0.9999601	0.9889328	4.382471
0.1343046	4.22465	0.9999643	0.9892989	4.376244
0.1375046	4.222517	0.9999713	0.9899922	4.364532
0.1391046	4.221563	0.9999742	0.9903204	4.359024
0.1407046	4.220677	0.9999769	0.9906371	4.353733
0.1423046	4.219853	0.9999793	0.9909427	4.34865
0.1455046	4.218378	0.9999833	0.9915222	4.33907

TABLE 5A.14   Tabulated Solutions to Eqs. (5.33) and (5.34) (*Continued*)

t	Y0	Y2	Y1	Y
0.1471046	4.217718	0.9999851	0.991797	4.334555
0.1487046	4.217105	0.9999866	0.9920623	4.330214
0.1503046	4.216536	0.999988	0.9923185	4.326038
0.1535046	4.215516	0.9999903	0.9928049	4.318153
0.1551046	4.21506	0.9999913	0.9930359	4.314431
0.1567046	4.214636	0.9999922	0.993259	4.310848
0.1583046	4.214242	0.999993	0.9934746	4.307397
0.1615046	4.213538	0.9999944	0.9938844	4.300873
0.1631046	4.213222	0.999995	0.994079	4.297789
0.1647046	4.21293	0.9999955	0.9942673	4.294816
0.1663046	4.212658	0.9999959	0.9944492	4.291951
0.1695046	4.212171	0.9999967	0.9947954	4.286527
0.1711046	4.211953	0.9999971	0.9949599	4.28396
0.1727046	4.211751	0.9999974	0.9951191	4.281483
0.1743046	4.211564	0.9999976	0.995273	4.279095
0.1775046	4.211228	0.9999981	0.9955661	4.274567
0.1791046	4.211078	0.9999983	0.9957055	4.272421
0.1807046	4.210938	0.9999985	0.9958403	4.27035
0.1823046	4.210809	0.9999986	0.9959709	4.268351
0.1855046	4.210577	0.9999989	0.9962194	4.264558
0.1871046	4.210473	0.999999	0.9963377	4.262759
0.1887046	4.210377	0.9999991	0.9964522	4.261021
0.1903046	4.210288	0.9999992	0.996563	4.259343
0.1935046	4.210128	0.9999994	0.9967742	4.256156
0.1951046	4.210057	0.9999994	0.9968747	4.254642
0.1967046	4.20999	0.9999995	0.996972	4.25318
0.1983046	4.209929	0.9999995	0.9970663	4.251767
0.2	4.209868	0.9999996	0.9971628	4.250322

TABLE 5A.14   Tabulated Solutions to Eqs. (5.33) and (5.34) (*Continued*)

t	Y0	Y2	Y1	Y
0	1.	0	0	1.
0.0046011	1.493151	0.0737689	0.047643	1.497137
0.0062011	1.685166	0.1053327	0.0685188	1.69364
0.0094011	2.106149	0.1795487	0.1190398	2.133194
0.0110011	2.336194	0.2228151	0.1495891	2.380075
0.0126011	2.579161	0.2704761	0.1843511	2.647504
0.0142011	2.833878	0.3225584	0.2238718	2.936875
0.0174011	3.36679	0.4386693	0.3190652	3.58123
0.0190011	3.634531	0.5010244	0.3751356	3.930339
0.0206011	3.893041	0.5643437	0.4363586	4.287181
0.0222011	4.133242	0.6267003	0.5014446	4.637911
0.0253122	4.516576	0.7376005	0.6302785	5.232324
0.0267269	4.645653	0.7808953	0.6851951	5.435377
0.0281504	4.745684	0.8190618	0.7355507	5.586383
0.0312154	4.869718	0.8829183	0.8230064	5.736197
0.0326475	4.892952	0.9050488	0.853967	5.739494
0.0352915	4.895717	0.9356768	0.8972014	5.674584
0.0366206	4.883343	0.9470767	0.9134358	5.620447
0.0394139	4.839795	0.9647015	0.9388535	5.485756
0.0409105	4.810801	0.9714826	0.9488306	5.409556
0.0424935	4.778157	0.977171	0.9573585	5.329684
0.0440935	4.744395	0.9817062	0.9643174	5.25151
0.0472935	4.677992	0.9881329	0.974589	5.106686
0.0488935	4.646396	0.9903956	0.9783929	5.040835
0.0504935	4.616259	0.9922033	0.9815445	4.979449
0.0520935	4.587721	0.9936528	0.9841717	4.922416
0.0552935	4.535629	0.9957604	0.9882297	4.820597
0.0568935	4.512052	0.9965228	0.9898011	4.77533
0.0584935	4.490058	0.997142	0.9911377	4.733488
0.0600935	4.469578	0.9976463	0.9922793	4.694821
0.0632935	4.432845	0.9983951	0.9940998	4.626072
0.0648935	4.416428	0.9986716	0.9948265	4.595554
0.0664935	4.401201	0.9988988	0.9954554	4.567342
0.0680935	4.387085	0.999086	0.9960014	4.541254
0.0712935	4.361883	0.9993681	0.9968907	4.494796
0.0728935	4.350656	0.9994737	0.9972528	4.474131
0.0744935	4.340258	0.9995613	0.9975699	4.455

TABLE 5A.15 Tabulated Solutions to Eqs. (5.33) and (5.34)

t	Y0	Y2	Y1	Y
0.0760935	4.330628	0.999634	0.997848	4.437283
0.0792935	4.313452	0.9997446	0.9983076	4.405663
0.0808935	4.305803	0.9997864	0.9984972	4.391567
0.0824935	4.298721	0.9998213	0.9986645	4.3785
0.0840935	4.292161	0.9998503	0.9988124	4.366384
0.0872935	4.280458	0.9998949	0.999059	4.344721
0.0888935	4.275246	0.9999119	0.9991616	4.335049
0.0904935	4.270418	0.9999261	0.9992527	4.326074
0.0920935	4.265945	0.999938	0.9993335	4.317744
0.0952935	4.257961	0.9999563	0.9994692	4.302834
0.0968935	4.254403	0.9999632	0.999526	4.296169
0.0984935	4.251107	0.9999691	0.9995766	4.289981
0.1000935	4.248051	0.999974	0.9996217	4.284234
0.1032935	4.242596	0.9999816	0.9996977	4.27394
0.1048935	4.240164	0.9999846	0.9997296	4.269336
0.1064935	4.237909	0.999987	0.9997581	4.26506
0.1080935	4.235819	0.9999891	0.9997836	4.261087
0.1112935	4.232085	0.9999923	0.9998267	4.253968
0.1128935	4.23042	0.9999935	0.9998448	4.250783
0.1144935	4.228876	0.9999945	0.9998611	4.247824
0.1160935	4.227445	0.9999954	0.9998756	4.245075
0.1192935	4.224887	0.9999967	0.9999002	4.240147
0.1208935	4.223745	0.9999972	0.9999106	4.237942
0.1224935	4.222687	0.9999977	0.9999199	4.235894
0.1240935	4.221705	0.999998	0.9999282	4.233991
0.1272935	4.219951	0.9999986	0.9999423	4.23058
0.1288935	4.219168	0.9999988	0.9999483	4.229053
0.1304935	4.218442	0.999999	0.9999537	4.227635
0.1320935	4.217768	0.9999992	0.9999584	4.226318
0.1352935	4.216564	0.9999994	0.9999666	4.223957
0.1368935	4.216027	0.9999995	0.9999701	4.2229
0.1384935	4.215529	0.9999996	0.9999731	4.221919
0.1400935	4.215066	0.9999996	0.9999759	4.221007
0.1432935	4.214239	0.9999997	0.9999806	4.219374
0.1448935	4.21387	0.9999998	0.9999826	4.218643
0.1464935	4.213528	0.9999998	0.9999844	4.217964
0.1480935	4.21321	0.9999998	0.999986	4.217334

TABLE 5A.15 Tabulated Solutions to Eqs. (5.33) and (5.34) (*Continued*)

t	Y0	Y2	Y1	Y
0.1512935	4.212643	0.9999999	0.9999888	4.216204
0.1528935	4.212389	0.9999999	0.9999899	4.215699
0.1544935	4.212154	0.9999999	0.999991	4.21523
0.1560935	4.211936	0.9999999	0.9999919	4.214794
0.1592935	4.211546	1.	0.9999935	4.214013
0.1608935	4.211371	1.	0.9999941	4.213664
0.1624935	4.21121	1.	0.9999947	4.21334
0.1640935	4.21106	1.	0.9999953	4.213039
0.1672935	4.210792	1.	0.9999962	4.2125
0.1688935	4.210672	1.	0.9999966	4.212259
0.1704935	4.210561	1.	0.9999969	4.212035
0.1720935	4.210458	1.	0.9999973	4.211827
0.1752935	4.210274	1.	0.9999978	4.211455
0.1768935	4.210191	1.	0.999998	4.211288
0.1784935	4.210115	1.	0.9999982	4.211134
0.1800935	4.210044	1.	0.9999984	4.21099
0.1832935	4.209917	1.	0.9999987	4.210733
0.1848935	4.209861	1.	0.9999988	4.210619
0.1864935	4.209808	1.	0.999999	4.210512
0.1880935	4.20976	1.	0.9999991	4.210413
0.1912935	4.209673	1.	0.9999993	4.210236
0.1928935	4.209634	1.	0.9999993	4.210157
0.1944935	4.209598	1.	0.9999994	4.210083
0.1960935	4.209564	1.	0.9999995	4.210015
0.1992935	4.209504	1.	0.9999996	4.209893
0.2	4.209492	1.	0.9999996	4.209868

TABLE 5A.15 Tabulated Solutions to Eqs. (5.33) and (5.34) (*Continued*)

t	Y0	Y2	Y1	Y
0	1.	0	0	1.
0.0047591	1.482992	0.2503616	0.0767365	1.511603
0.0063591	1.647703	0.3377406	0.1086396	1.704774
0.0095591	1.970022	0.5078432	0.1836312	2.128284
0.0111591	2.123508	0.5869069	0.2273251	2.359629
0.0127591	2.269303	0.659721	0.2754252	2.603827
0.0143199	2.402669	0.7234922	0.3265957	2.853193
0.0170167	2.610084	0.8148414	0.4240127	3.30201
0.0182506	2.694991	0.8485416	0.4715515	3.50957
0.0206095	2.840645	0.8997696	0.5646758	3.894359
0.0229029	2.963426	0.9351446	0.6532143	4.23052
0.0240455	3.018617	0.9483939	0.6947025	4.376061
0.0263532	3.119912	0.9681803	0.7699656	4.61446
0.028726	3.212367	0.9811944	0.8329254	4.777981
0.0312048	3.299	0.9894526	0.8827369	4.869473
0.0324973	3.340867	0.9922841	0.9029298	4.891353
0.0352194	3.422866	0.9960981	0.9349907	4.896173
0.0366636	3.46338	0.997316	0.9474088	4.88283
0.0381561	3.503299	0.9981925	0.9576824	4.861636
0.041306	3.581617	0.9992367	0.9730331	4.802778
0.042906	3.618538	0.9995141	0.9784453	4.769481
0.044506	3.653646	0.9996934	0.982712	4.735686
0.046106	3.687013	0.9998082	0.9860871	4.702241
0.049306	3.748766	0.9999267	0.9909009	4.638478
0.050906	3.777266	0.9999552	0.992608	4.608744
0.052506	3.804252	0.9999729	0.993978	4.58063
0.054106	3.82978	0.9999837	0.9950812	4.554182
0.057306	3.876675	0.9999942	0.9966949	4.506232
0.058906	3.898152	0.9999966	0.997282	4.484635
0.060506	3.918386	0.999998	0.9977605	4.464533
0.062106	3.937433	0.9999988	0.9981515	4.445846
0.065306	3.972181	0.9999996	0.9987345	4.412391
0.066906	3.987987	0.9999998	0.9989506	4.397458
0.068506	4.002818	0.9999999	0.9991287	4.383616
0.070106	4.016723	0.9999999	0.9992757	4.370789
0.073306	4.041951	1.	0.9994979	4.347898
0.074906	4.053367	1.	0.9995813	4.337704

TABLE 5A.16 Tabulated Solutions to Eqs. (5.33) and (5.34)

t	Y0	Y2	Y1	Y
0.076506	4.064044	1.	0.9996506	4.328263
0.078106	4.074025	1.	0.9997082	4.31952
0.081306	4.09206	1.	0.999796	4.303925
0.082906	4.100189	1.	0.9998293	4.296981
0.084506	4.107775	1.	0.999857	4.29055
0.086106	4.11485	1.	0.9998802	4.284593
0.089306	4.127596	1.	0.9999158	4.273966
0.090906	4.133325	1.	0.9999293	4.269232
0.092506	4.138662	1.	0.9999407	4.264846
0.094106	4.143632	1.	0.9999502	4.260782
0.097306	4.152566	1.	0.9999649	4.253529
0.098906	4.156574	1.	0.9999705	4.250296
0.100506	4.160303	1.	0.9999752	4.247301
0.102106	4.163772	1.	0.9999791	4.244524
0.105306	4.169997	1.	0.9999852	4.239566
0.106906	4.172786	1.	0.9999876	4.237355
0.108506	4.175379	1.	0.9999895	4.235305
0.110106	4.177789	1.	0.9999912	4.233405
0.113306	4.182109	1.	0.9999938	4.230011
0.114906	4.184043	1.	0.9999947	4.228497
0.116506	4.185839	1.	0.9999956	4.227093
0.118106	4.187508	1.	0.9999963	4.225791
0.121306	4.190498	1.	0.9999974	4.223465
0.122906	4.191835	1.	0.9999978	4.222427
0.124506	4.193077	1.	0.9999981	4.221464
0.126106	4.19423	1.	0.9999984	4.220571
0.129306	4.196295	1.	0.9999989	4.218976
0.130906	4.197218	1.	0.9999991	4.218263
0.132506	4.198074	1.	0.9999992	4.217603
0.134106	4.19887	1.	0.9999993	4.21699
0.137306	4.200294	1.	0.9999995	4.215895
0.138906	4.20093	1.	0.9999996	4.215406
0.140506	4.201521	1.	0.9999997	4.214952
0.142106	4.202069	1.	0.9999997	4.214532
0.145306	4.20305	1.	0.9999998	4.21378
0.146906	4.203489	1.	0.9999998	4.213444

TABLE 5A.16 Tabulated Solutions to Eqs. (5.33) and (5.34) (*Continued*)

t	Y0	Y2	Y1	Y
0.148506	4.203895	1.	0.9999999	4.213132
0.150106	4.204273	1.	0.9999999	4.212843
0.153306	4.204948	1.	0.9999999	4.212327
0.154906	4.20525	1.	0.9999999	4.212096
0.156506	4.20553	1.	0.9999999	4.211882
0.158106	4.20579	1.	0.9999999	4.211684
0.161306	4.206255	1.	1.	4.211328
0.162906	4.206463	1.	1.	4.21117
0.164506	4.206655	1.	1.	4.211023
0.166106	4.206834	1.	1.	4.210886
0.169306	4.207154	1.	1.	4.210642
0.170906	4.207297	1.	1.	4.210534
0.172506	4.207429	1.	1.	4.210432
0.174106	4.207552	1.	1.	4.210339
0.177306	4.207772	1.	1.	4.210171
0.178906	4.207871	1.	1.	4.210096
0.180506	4.207962	1.	1.	4.210027
0.182106	4.208046	1.	1.	4.209962
0.185306	4.208198	1.	1.	4.209847
0.186906	4.208265	1.	1.	4.209795
0.188506	4.208328	1.	1.	4.209748
0.190106	4.208386	1.	1.	4.209703
0.193306	4.20849	1.	1.	4.209624
0.194906	4.208537	1.	1.	4.209589
0.196506	4.20858	1.	1.	4.209556
0.198106	4.20862	1.	1.	4.209525
0.2	4.208664	1.	1.	4.209492

TABLE 5A.16 Tabulated Solutions to Eqs. (5.33) and (5.34) (*Continued*)

t	w	u	t	w	u
0	0	−10000.	0.0721463	2.474E+05	−0.0002072
0.00424	4.249E+05	−28.47461	0.0744646	2.474E+05	−0.0001416
0.0063122	4.681E+05	−15.77001	0.0760101	2.474E+05	−0.0001099
0.0080809	4.646E+05	−10.36682	0.0783284	2.474E+05	−7.511E−05
0.010158	4.394E+05	−6.677972	0.0806467	2.474E+05	−5.134E−05
0.0125124	4.025E+05	−4.223888	0.0821923	2.474E+05	−3.984E−05
0.0144305	3.735E+05	−2.965798	0.0845106	2.474E+05	−2.723E−05
0.0164527	3.466E+05	−2.068068	0.0860562	2.474E+05	−2.113E−05
0.0185564	3.234E+05	−1.434262	0.0883745	2.474E+05	−1.445E−05
0.0207215	3.046E+05	−0.9906103	0.0906928	2.474E+05	−9.874E−06
0.0221909	2.943E+05	−0.77267	0.0922384	2.474E+05	−7.662E−06
0.0244241	2.818E+05	−0.5312233	0.0945567	2.474E+05	−5.238E−06
0.0266832	2.725E+05	−0.3645854	0.0961023	2.474E+05	−4.064E−06
0.0281998	2.676E+05	−0.2834686	0.0984206	2.474E+05	−2.778E−06
0.0304862	2.62E+05	−0.1941896	0.1007389	2.474E+05	−1.899E−06
0.0320161	2.591E+05	−0.1508471	0.1022845	2.474E+05	−1.474E−06
0.0343173	2.558E+05	−0.1032335	0.1046028	2.474E+05	−1.007E−06
0.0366239	2.534E+05	−0.0706228	0.1061671	2.474E+05	−7.793E−07
0.0381638	2.522E+05	−0.0548233	0.1086665	2.474E+05	−5.171E−07
0.0404759	2.508E+05	−0.0374911	0.1104483	2.474E+05	−3.86E−07
0.0420184	2.501E+05	−0.0290984	0.1123348	2.474E+05	−2.832E−07
0.0443334	2.493E+05	−0.0198951	0.1143388	2.474E+05	−2.039E−07
0.0466495	2.488E+05	−0.0136016	0.1164754	2.474E+05	−1.436E−07
0.0481939	2.485E+05	−0.0105554	0.1187628	2.474E+05	−9.864E−08
0.050511	2.482E+05	−0.0072159	0.1212231	2.474E+05	−6.587E−08
0.052056	2.48E+05	−0.0055996	0.1225264	2.474E+05	−5.319E−08
0.0543737	2.478E+05	−0.0038278	0.1252994	2.474E+05	−3.374E−08
0.0566916	2.477E+05	−0.0026166	0.1267786	2.474E+05	−2.647E−08
0.0582369	2.476E+05	−0.0020305	0.1283271	2.474E+05	−2.053E−08
0.060555	2.476E+05	−0.001388	0.1315271	2.474E+05	−1.214E−08
0.0621005	2.475E+05	−0.0010771	0.1331271	2.474E+05	−9.339E−09
0.0644187	2.475E+05	−0.0007362	0.1347271	2.474E+05	−7.183E−09
0.0667369	2.475E+05	−0.0005033	0.1363271	2.474E+05	−5.524E−09
0.0682824	2.475E+05	−0.0003905	0.1395271	2.474E+05	−3.267E−09
0.0706007	2.474E+05	−0.000267	0.1411271	2.474E+05	−2.513E−09

TABLE 5A.17   Tabulated Solutions to Eqs. (5.35) and (5.36)

t	w	u	t	w	u
0.1427271	2.474E+05	−1.933E−09	0.1731271	2.474E+05	−1.317E−11
0.1443271	2.474E+05	−1.486E−09	0.1747271	2.474E+05	−1.013E−11
0.1475271	2.474E+05	−8.792E−10	0.1763271	2.474E+05	−7.79E−12
0.1491271	2.474E+05	−6.761E−10	0.1795271	2.474E+05	−4.608E−12
0.1507271	2.474E+05	−5.2E−10	0.1811271	2.474E+05	−3.544E−12
0.1523271	2.474E+05	−3.999E−10	0.1827271	2.474E+05	−2.725E−12
0.1555271	2.474E+05	−2.365E−10	0.1843271	2.474E+05	−2.096E−12
0.1571271	2.474E+05	−1.819E−10	0.1875271	2.474E+05	−1.24E−12
0.1587271	2.474E+05	−1.399E−10	0.1891271	2.474E+05	−9.535E−13
0.1603271	2.474E+05	−1.076E−10	0.1907271	2.474E+05	−7.333E−13
0.1635271	2.474E+05	−6.365E−11	0.1923271	2.474E+05	−5.64E−13
0.1651271	2.474E+05	−4.895E−11	0.1955271	2.474E+05	−3.336E−13
0.1667271	2.474E+05	−3.765E−11	0.1971271	2.474E+05	−2.566E−13
0.1683271	2.474E+05	−2.895E−11	0.1987271	2.474E+05	−1.973E−13
0.1715271	2.474E+05	−1.713E−11	0.2	2.474E+05	−1.601E−13

TABLE 5A.17   Tabulated Solutions to Eqs. (5.35) and (5.36) (*Continued*)

t	w	u	t	w	u
0	0	−10000.	0.0721971	9.163E+05	−3.664E−13
0.0042143	1.557E+06	−5.24456	0.0753971	9.163E+05	−8.871E−14
0.0061261	1.279E+06	−2.036174	0.0769971	9.163E+05	−4.365E−14
0.0082535	1.088E+06	−0.7604266	0.0785971	9.163E+05	−2.148E−14
0.0101987	9.985E+05	−0.3163509	0.0801971	9.163E+05	−1.057E−14
0.0121774	9.544E+05	−0.1308013	0.0833971	9.163E+05	−2.559E−15
0.0141705	9.336E+05	−0.05394	0.0849971	9.163E+05	−1.259E−15
0.0161696	9.24E+05	−0.0222193	0.0865971	9.163E+05	−6.195E−16
0.0181713	9.197E+05	−0.0091485	0.0881971	9.163E+05	−3.048E−16
0.0201739	9.178E+05	−0.0037661	0.0913971	9.163E+05	−7.381E−17
0.022177	9.169E+05	−0.0015502	0.0929971	9.163E+05	−3.632E−17
0.0241802	9.166E+05	−0.0006381	0.0945971	9.163E+05	−1.787E−17
0.0261835	9.164E+05	−0.0002626	0.0961971	9.163E+05	−8.793E−18
0.0281868	9.163E+05	−0.0001081	0.0993971	9.163E+05	−2.129E−18
0.0301902	9.163E+05	−4.45E−05	0.1009971	9.163E+05	−1.048E−18
0.0321936	9.163E+05	−1.832E−05	0.1025971	9.163E+05	−5.155E−19
0.0341969	9.163E+05	−7.539E−06	0.1041971	9.163E+05	−2.536E−19
0.0362003	9.163E+05	−3.103E−06	0.1073971	9.163E+05	−6.141E−20
0.0382036	9.163E+05	−1.277E−06	0.1089971	9.163E+05	−3.022E−20
0.0402856	9.163E+05	−5.077E−07	0.1105971	9.163E+05	−1.487E−20
0.0420159	9.163E+05	−2.359E−07	0.1121971	9.163E+05	−7.317E−21
0.0440399	9.163E+05	−9.62E−08	0.1153971	9.163E+05	−1.771E−21
0.0464732	9.163E+05	−3.273E−08	0.1169971	9.163E+05	−8.717E−22
0.0482064	9.163E+05	−1.518E−08	0.1185971	9.163E+05	−4.289E−22
0.0502231	9.163E+05	−6.213E−09	0.1201971	9.163E+05	−2.11E−22
0.0526278	9.163E+05	−2.141E−09	0.1233971	9.163E+05	−5.11E−23
0.0545271	9.163E+05	−9.226E−10	0.1249971	9.163E+05	−2.514E−23
0.0567531	9.163E+05	−3.441E−10	0.1265971	9.163E+05	−1.237E−23
0.0580248	9.163E+05	−1.958E−10	0.1281971	9.163E+05	−6.088E−24
0.0609971	9.163E+05	−5.246E−11	0.1313971	9.163E+05	−1.474E−24
0.0625971	9.163E+05	−2.581E−11	0.1329971	9.163E+05	−7.253E−25
0.0641971	9.163E+05	−1.27E−11	0.1345971	9.163E+05	−3.569E−25
0.0673971	9.163E+05	−3.075E−12	0.1361971	9.163E+05	−1.756E−25
0.0689971	9.163E+05	−1.513E−12	0.1393971	9.163E+05	−4.252E−26
0.0705971	9.163E+05	−7.446E−13	0.1409971	9.163E+05	−2.092E−26

TABLE 5A.18   Tabulated Solutions to Eqs. (5.35) and (5.36)

t	w	u	t	w	u
0.1425971	9.163E+05	–1.029E–26	0.1729971	9.163E+05	–1.448E–32
0.1441971	9.163E+05	–5.065E–27	0.1745971	9.163E+05	–7.127E–33
0.1473971	9.163E+05	–1.226E–27	0.1761971	9.163E+05	–3.507E–33
0.1489971	9.163E+05	–6.035E–28	0.1793971	9.163E+05	–8.491E–34
0.1505971	9.163E+05	–2.969E–28	0.1809971	9.163E+05	–4.178E–34
0.1521971	9.163E+05	–1.461E–28	0.1825971	9.163E+05	–2.056E–34
0.1553971	9.163E+05	–3.538E–29	0.1841971	9.163E+05	–1.012E–34
0.1569971	9.163E+05	–1.741E–29	0.1873971	9.163E+05	–2.449E–35
0.1585971	9.163E+05	–8.565E–30	0.1889971	9.163E+05	–1.205E–35
0.1601971	9.163E+05	–4.215E–30	0.1905971	9.163E+05	–5.93E–36
0.1633971	9.163E+05	–1.02E–30	0.1921971	9.163E+05	–2.918E–36
0.1649971	9.163E+05	–5.021E–31	0.1953971	9.163E+05	–7.065E–37
0.1665971	9.163E+05	–2.471E–31	0.1969971	9.163E+05	–3.476E–37
0.1681971	9.163E+05	–1.216E–31	0.1985971	9.163E+05	–1.711E–37
0.1713971	9.163E+05	–2.944E–32	0.2	9.163E+05	–9.186E–38

TABLE **5A.18**   Tabulated Solutions to Eqs. (5.35) and (5.36) (*Continued*)

t	w	u	t	w	u
0	0	−10000.	0.0720528	2.061E+06	−3.789E−28
0.0040428	2.069E+06	−0.0099871	0.0752528	2.061E+06	−6.436E−29
0.0060381	2.061E+06	−0.000196	0.0768528	2.061E+06	2.652E−29
0.0080336	2.061E+06	−3.846E−06	0.0784528	2.061E+06	−1.093E−29
0.0100389	2.061E+06	−7.4E−08	0.0800528	2.061E+06	4.505E−30
0.0121101	2.061E+06	−1.251E−09	0.0832528	2.061E+06	7.653E−31
0.0142516	2.061E+06	−1.839E−11	0.0848528	2.061E+06	−3.154E−31
0.016765	2.061E+06	−1.206E−13	0.0864528	2.061E+06	1.3E−31
0.0192528	2.061E+06	1.914E−15	0.0880528	2.061E+06	−5.358E−32
0.0208528	2.061E+06	−7.888E−16	0.0912528	2.061E+06	−9.101E−33
0.0224528	2.061E+06	3.251E−16	0.0928528	2.061E+06	3.751E−33
0.0240528	2.061E+06	−1.34E−16	0.0944528	2.061E+06	−1.546E−33
0.0272528	2.061E+06	−2.276E−17	0.0960528	2.061E+06	6.371E−34
0.0288528	2.061E+06	9.38E−18	0.0992528	2.061E+06	1.082E−34
0.0304528	2.061E+06	−3.866E−18	0.1008528	2.061E+06	−4.46E−35
0.0320528	2.061E+06	1.593E−18	0.1024528	2.061E+06	1.838E−35
0.0352528	2.061E+06	2.706E−19	0.1040528	2.061E+06	−7.576E−36
0.0368528	2.061E+06	−1.115E−19	0.1072528	2.061E+06	−1.287E−36
0.0384528	2.061E+06	4.597E−20	0.1088528	2.061E+06	5.304E−37
0.0400528	2.061E+06	−1.895E−20	0.1104528	2.061E+06	−2.186E−37
0.0432528	2.061E+06	−3.218E−21	0.1120528	2.061E+06	9.009E−38
0.0448528	2.061E+06	1.326E−21	0.1152528	2.061E+06	1.53E−38
0.0464528	2.061E+06	−5.467E−22	0.1168528	2.061E+06	−6.307E−39
0.0480528	2.061E+06	2.253E−22	0.1184528	2.061E+06	2.599E−39
0.0512528	2.061E+06	3.827E−23	0.1200528	2.061E+06	−1.071E−39
0.0528528	2.061E+06	−1.577E−23	0.1232528	2.061E+06	−1.82E−40
0.0544528	2.061E+06	6.501E−24	0.1248528	2.061E+06	7.5E−41
0.0560528	2.061E+06	−2.679E−24	0.1264528	2.061E+06	−3.091E−41
0.0592528	2.061E+06	−4.551E−25	0.1280528	2.061E+06	1.274E−41
0.0608528	2.061E+06	1.876E−25	0.1312528	2.061E+06	2.164E−42
0.0624528	2.061E+06	−7.731E−26	0.1328528	2.061E+06	−8.919E−43
0.0640528	2.061E+06	3.186E−26	0.1344528	2.061E+06	3.676E−43
0.0672528	2.061E+06	5.412E−27	0.1360528	2.061E+06	−1.515E−43
0.0688528	2.061E+06	−2.231E−27	0.1392528	2.061E+06	−2.573E−44
0.0704528	2.061E+06	9.193E−28	0.1408528	2.061E+06	1.061E−44

TABLE **5A.19** Tabulated Solutions to Eqs. (5.35) and (5.36)

t	w	u	t	w	u
0.1424528	2.061E+06	−4.371E−45	0.1728528	2.061E+06	2.121E−52
0.1440528	2.061E+06	1.801E−45	0.1744528	2.061E+06	−8.74E−53
0.1472528	2.061E+06	3.06E−46	0.1760528	2.061E+06	3.602E−53
0.1488528	2.061E+06	−1.261E−46	0.1792528	2.061E+06	6.119E−54
0.1504528	2.061E+06	5.198E−47	0.1808528	2.061E+06	−2.522E−54
0.1520528	2.061E+06	−2.142E−47	0.1824528	2.061E+06	1.039E−54
0.1552528	2.061E+06	−3.639E−48	0.1840528	2.061E+06	−4.284E−55
0.1568528	2.061E+06	1.5E−48	0.1872528	2.061E+06	−7.276E−56
0.1584528	2.061E+06	−6.181E−49	0.1888528	2.061E+06	2.999E−56
0.1600528	2.061E+06	2.547E−49	0.1904528	2.061E+06	−1.236E−56
0.1632528	2.061E+06	4.327E−50	0.1920528	2.061E+06	5.094E−57
0.1648528	2.061E+06	−1.783E−50	0.1952528	2.061E+06	8.653E−58
0.1664528	2.061E+06	7.35E−51	0.1968528	2.061E+06	−3.566E−58
0.1680528	2.061E+06	−3.029E−51	0.1984528	2.061E+06	1.47E−58
0.1712528	2.061E+06	−5.146E−52	0.2	2.061E+06	−4.893E−59

TABLE 5A.19 Tabulated Solutions to Eqs. (5.35) and (5.36) (*Continued*)

t	w	u	t	w	u
0	0	–10000.	0.0728718	2.912E+05	–0.0018739
0.0042744	4.286E+05	–59.38565	0.0746897	2.912E+05	–0.0014542
0.0060196	4.939E+05	–36.88367	0.0765078	2.911E+05	–0.0011284
0.0083106	5.192E+05	–22.21471	0.0783258	2.911E+05	–0.0008756
0.0105586	5.082E+05	–14.46715	0.0801439	2.911E+05	–0.0006795
0.0124765	4.863E+05	–10.35285	0.082871	2.911E+05	–0.0004645
0.0145722	4.58E+05	–7.335154	0.084689	2.911E+05	–0.0003604
0.0160557	4.38E+05	–5.801449	0.0865071	2.911E+05	–0.0002797
0.0183896	4.087E+05	–4.055683	0.0883252	2.911E+05	–0.000217
0.020006	3.909E+05	–3.183671	0.0901433	2.911E+05	–0.0001684
0.0225025	3.675E+05	–2.205218	0.0928704	2.911E+05	–0.0001151
0.0242051	3.544E+05	–1.72261	0.0946885	2.91E+05	–8.933E–05
0.0268024	3.382E+05	–1.186341	0.0965066	2.91E+05	–6.932E–05
0.0285564	3.295E+05	–0.9239841	0.0983247	2.91E+05	–5.379E–05
0.0303242	3.223E+05	–0.7190781	0.1001428	2.91E+05	–4.174E–05
0.032103	3.163E+05	–0.5592616	0.10287	2.91E+05	–2.853E–05
0.0347866	3.093E+05	–0.3832522	0.1046881	2.91E+05	–2.214E–05
0.0365834	3.058E+05	–0.2977616	0.1065062	2.91E+05	–1.718E–05
0.038385	3.028E+05	–0.2312785	0.1083243	2.91E+05	–1.333E–05
0.0401903	3.005E+05	–0.1796015	0.1101424	2.91E+05	–1.035E–05
0.0429033	2.978E+05	–0.1228691	0.1128696	2.91E+05	–7.071E–06
0.0447145	2.964E+05	–0.095382	0.1146877	2.91E+05	–5.487E–06
0.0465273	2.953E+05	–0.0740376	0.1165058	2.91E+05	–4.258E–06
0.0483413	2.945E+05	–0.0574656	0.1183239	2.91E+05	–3.304E–06
0.0501561	2.938E+05	–0.0446006	0.120142	2.91E+05	–2.564E–06
0.0528798	2.93E+05	–0.0304935	0.1228691	2.91E+05	–1.753E–06
0.0546962	2.926E+05	–0.0236647	0.1246872	2.91E+05	–1.36E–06
0.0565129	2.923E+05	–0.0183648	0.1265053	2.91E+05	–1.055E–06
0.05833	2.92E+05	–0.0142516	0.1283368	2.91E+05	–8.174E–07
0.0601473	2.918E+05	–0.0110595	0.1302532	2.91E+05	–6.257E–07
0.0628736	2.916E+05	–10.0075603	0.1322725	2.91E+05	–4.721E–07
0.0646913	2.915E+05	–0.0058668	0.1344061	2.91E+05	–3.505E–07
0.0665091	2.914E+05	–0.0045526	0.1366675	2.91E+05	–2.557E–07
0.0683269	2.913E+05	–0.0035328	0.1390723	2.91E+05	–1.828E–07
0.0701448	2.913E+05	–0.0027414	0.1403343	2.91E+05	–1.533E–07

TABLE 5A.20 Tabulated Solutions to Eqs. (5.35) and (5.36)

t	w	u	t	w	u
0.1429909	2.91E+05	−1.058E−07	0.1729298	2.91E+05	−1.625E−09
0.1443918	2.91E+05	−8.705E−08	0.1745298	2.91E+05	−1.3E−09
0.147357	2.91E+05	−5.756E−08	0.1761298	2.91E+05	−1.04E−09
0.1489298	2.91E+05	−4.622E−08	0.1793298	2.91E+05	−6.654E−10
0.1505298	2.91E+05	−3.697E−08	0.1809298	2.91E+05	−5.323E−10
0.1521298	2.91E+05	−2.958E−08	0.1825298	2.91E+05	−4.258E−10
0.1553298	2.91E+05	−1.893E−08	0.1841298	2.91E+05	−3.406E−10
0.1569298	2.91E+05	−1.514E−08	0.1873298	2.91E+05	−2.18E−10
0.1585298	2.91E+05	−1.211E−08	0.1889298	2.91E+05	−1.744E−10
0.1601298	2.91E+05	−9.689E−09	0.1905298	2.91E+05	−1.395E−10
0.1633298	2.91E+05	−6.2E−09	0.1921298	2.91E+05	−1.116E−10
0.1649298	2.91E+05	−4.96E−09	0.1953298	2.91E+05	−7.14E−11
0.1665298	2.91E+05	−3.968E−09	0.1969298	2.91E+05	−5.712E−11
0.1681298	2.91E+05	−3.174E−09	0.1985298	2.91E+05	−4.569E−11
0.1713298	2.91E+05	−2.031E−09	0.2	2.91E+05	−3.722E−11

TABLE 5A.20 Tabulated Solutions to Eqs. (5.35) and (5.36) (*Continued*)

t	w	u	t	w	u
0	0	–10000.	0.0721564	1.078E+06	–7.636E–11
0.0041511	1.858E+06	–12.88304	0.0753564	1.078E+06	–2.287E–11
0.0062437	1.583E+06	–5.121836	0.0769564	1.078E+06	–1.252E–11
0.0080379	1.387E+06	–2.478102	0.0785564	1.078E+06	–6.85E–12
0.0102664	1.235E+06	–1.04048	0.0801564	1.078E+06	–3.749E–12
0.0122361	1.161E+06	–0.4899796	0.0833564	1.078E+06	–1.123E–12
0.0142321	1.121E+06	–0.2298011	0.0849564	1.078E+06	–6.146E–13
0.0162406	1.1E+06	–0.1075652	0.0865564	1.078E+06	–3.363E–13
0.0182552	1.089E+06	–0.0503021	0.0881564	1.078E+06	–1.841E–13
0.0202726	1.083E+06	–0.0235131	0.0913564	1.078E+06	–5.513E–14
0.0222912	1.081E+06	–0.0109887	0.0929564	1.078E+06	–3.017E–14
0.0243105	1.079E+06	–0.005135	0.0945564	1.078E+06	–1.651E–14
0.0263301	1.078E+06	–0.0023994	0.0961564	1.078E+06	–9.037E–15
0.0280132	1.078E+06	–0.0012728	0.0993564	1.078E+06	–2.707E–15
0.030033	1.078E+06	–0.0005947	0.1009564	1.078E+06	–1.481E–15
0.0320528	1.078E+06	–0.0002779	0.1025564	1.078E+06	–8.108E–16
0.0340727	1.078E+06	–0.0001298	0.1041564	1.078E+06	–4.437E–16
0.0360925	1.078E+06	–6.067E–05	0.1073564	1.078E+06	–1.329E–16
0.0381124	1.078E+06	–2.835E–05	0.1089564	1.078E+06	–7.273E–17
0.0401322	1.078E+06	–1.325E–05	0.1105564	1.078E+06	–3.981E–17
0.0421521	1.078E+06	–6.189E–06	0.1121564	1.078E+06	–2.179E–17
0.0441719	1.078E+06	–2.892E–06	0.1153564	1.078E+06	–6.525E–18
0.0461917	1.078E+06	–1.351E–06	0.1169564	1.078E+06	–3.571E–18
0.0482543	1.078E+06	–6.213E–07	0.1185564	1.078E+06	–1.954E–18
0.0502074	1.078E+06	–2.977E–07	0.1201564	1.078E+06	–1.07E–18
0.0524773	1.078E+06	–1.266E–07	0.1233564	1.078E+06	–3.204E–19
0.0540379	1.078E+06	–7.033E–08	0.1249564	1.078E+06	–1.753E–19
0.0564263	1.078E+06	–2.86E–08	0.1265564	1.078E+06	–9.596E–20
0.0585199	1.078E+06	–1.3E–08	0.1281564	1.078E+06	–5.251E–20
0.0601044	1.078E+06	–7.156E–09	0.1313564	1.078E+06	–1.573E–20
0.0628554	1.078E+06	–2.539E–09	0.1329564	1.078E+06	–8.608E–21
0.0650186	1.078E+06	–1.124E–09	0.1345564	1.078E+06	–4.711E–21
0.0662282	1.078E+06	–7.125E–10	0.1361564	1.078E+06	–2.578E–21
0.0689766	1.078E+06	–2.53E–10	0.1393564	1.078E+06	–7.723E–22
0.0705564	1.078E+06	–1.395E–10	0.1409564	1.078E+06	–4.226E–22

TABLE 5A.21   Tabulated Solutions to Eqs. (5.35) and (5.36)

t	w	u	t	w	u
0.1425564	1.078E+06	−2.313E−22	0.1729564	1.078E+06	−2.456E−27
0.1441564	1.078E+06	−1.266E−22	0.1745564	1.078E+06	−1.344E−27
0.1473564	1.078E+06	−3.792E−23	0.1761564	1.078E+06	−7.356E−28
0.1489564	1.078E+06	−2.075E−23	0.1793564	1.078E+06	−2.203E−28
0.1505564	1.078E+06	−1.136E−23	0.1809564	1.078E+06	−1.206E−28
0.1521564	1.078E+06	−6.215E−24	0.1825564	1.078E+06	−6.599E−29
0.1553564	1.078E+06	−1.862E−24	0.1841564	1.078E+06	−3.612E−29
0.1569564	1.078E+06	−1.019E−24	0.1873564	1.078E+06	−1.082E−29
0.1585564	1.078E+06	−5.576E−25	0.1889564	1.078E+06	−5.92E−30
0.1601564	1.078E+06	−3.052E−25	0.1905564	1.078E+06	−3.24E−30
0.1633564	1.078E+06	−9.14E−26	0.1921564	1.078E+06	−1.773E−30
0.1649564	1.078E+06	−5.002E−26	0.1953564	1.078E+06	−5.311E−31
0.1665564	1.078E+06	−2.738E−26	0.1969564	1.078E+06	−2.907E−31
0.1681564	1.078E+06	−1.498E−26	0.1985564	1.078E+06	−1.591E−31
0.1713564	1.078E+06	−4.487E−27	0.2	1.078E+06	−9.234E−32

TABLE 5A.21 Tabulated Solutions to Eqs. (5.35) and (5.36) (*Continued*)

t	w	u	t	w	u
0	0	−10000.	0.0727606	2.425E+06	−1.74E−44
0.004028	2.454E+06	−0.0575058	0.0743606	2.425E+06	2.101E−45
0.0060722	2.427E+06	−0.0018753	0.0775606	2.425E+06	3.062E−47
0.0080418	2.425E+06	−6.937E−05	0.0791606	2.425E+06	−3.697E−48
0.0100114	2.425E+06	−2.566E−06	0.0807606	2.425E+06	4.463E−49
0.012032	2.425E+06	−8.715E−08	0.0823606	2.425E+06	−5.389E−50
0.0141316	2.425E+06	−2.593E−09	0.0855606	2.425E+06	−7.855E−52
0.0162487	2.425E+06	−7.492E−11	0.0871606	2.425E+06	9.484E−53
0.0184383	2.425E+06	−1.904E−12	0.0887606	2.425E+06	−1.145E−53
0.0202561	2.425E+06	−8.233E−14	0.0903606	2.425E+06	1.382E−54
0.0231606	2.425E+06	5.056E−16	0.0935606	2.425E+06	2.015E−56
0.0247606	2.425E+06	−6.104E−17	0.0951606	2.425E+06	−2.433E−57
0.0263606	2.425E+06	7.37E−18	0.0967606	2.425E+06	2.937E−58
0.0295606	2.425E+06	1.074E−19	0.0983606	2.425E+06	−3.546E−59
0.0311606	2.425E+06	−1.297E−20	0.1015606	2.425E+06	−5.169E−61
0.0327606	2.425E+06	1.566E−21	0.1031606	2.425E+06	6.241E−62
0.0343606	2.425E+06	−1.891E−22	0.1047606	2.425E+06	−7.535E−63
0.0375606	2.425E+06	−2.756E−24	0.1063606	2.425E+06	9.098E−64
0.0391606	2.425E+06	3.327E−25	0.1095606	2.425E+06	1.326E−65
0.0407606	2.425E+06	−4.017E−26	0.1111606	2.425E+06	−1.601E−66
0.0423606	2.425E+06	4.85E−27	0.1127606	2.425E+06	1.933E−67
0.0455606	2.425E+06	7.07E−29	0.1143606	2.425E+06	−2.334E−68
0.0471606	2.425E+06	−8.536E−30	0.1175606	2.425E+06	−3.402E−70
0.0487606	2.425E+06	1.031E−30	0.1191606	2.425E+06	4.107E−71
0.0503606	2.425E+06	−1.244E−31	0.1207606	2.425E+06	−4.959E−72
0.0535606	2.425E+06	−1.814E−33	0.1223606	2.425E+06	5.987E−73
0.0551606	2.425E+06	2.19E−34	0.1255606	2.425E+06	8.727E−75
0.0567606	2.425E+06	−2.644E−35	0.1271606	2.425E+06	−1.054E−75
0.0583606	2.425E+06	3.192E−36	0.1287606	2.425E+06	1.272E−76
0.0615606	2.425E+06	4.653E−38	0.1303606	2.425E+06	−1.536E−77
0.0631606	2.425E+06	−5.617E−39	0.1335606	2.425E+06	−2.239E−79
0.0647606	2.425E+06	6.782E−40	0.1351606	2.425E+06	2.703E−80
0.0663606	2.425E+06	−8.188E−41	0.1367606	2.425E+06	−3.263E−81
0.0695606	2.425E+06	−1.194E−42	0.1383606	2.425E+06	3.94E−82
0.0711606	2.425E+06	1.441E−43	0.1415606	2.425E+06	5.743E−84

TABLE 5A.22   Tabulated Solutions to Eqs. (5.35) and (5.36)

t	w	u	t	w	u
0.1431606	2.425E+06	−6.934E–85	0.1735606	2.425E+06	2.487E–102
0.1447606	2.425E+06	8.372E–86	0.1751606	2.425E+06	−3.003E–103
0.1463606	2.425E+06	−1.011E–86	0.1767606	2.425E+06	3.626E–104
0.1495606	2.425E+06	−1.473E–88	0.1783606	2.425E+06	−4.378E–105
0.1511606	2.425E+06	1.779E–89	0.1815606	2.425E+06	−6.381E–107
0.1527606	2.425E+06	−2.148E–90	0.1831606	2.425E+06	7.704E–108
0.1543606	2.425E+06	2.593E–91	0.1847606	2.425E+06	−9.301E–109
0.1575606	2.425E+06	3.78E–93	0.1863606	2.425E+06	1.123E–109
0.1591606	2.425E+06	−4.563E–94	0.1895606	2.425E+06	1.637E–111
0.1607606	2.425E+06	5.51E–95	0.1911606	2.425E+06	−1.976E–112
0.1623606	2.425E+06	−6.652E–96	0.1927606	2.425E+06	2.386E–113
0.1655606	2.425E+06	−9.696E–98	0.1943606	2.425E+06	−2.881E–114
0.1671606	2.425E+06	1.171E–98	0.1975606	2.425E+06	−4.199E–116
0.1687606	2.425E+06	−1.413E–99	0.1991606	2.425E+06	5.07E–117
0.1703606	2.425E+06	1.706E–100	0.2	2.425E+06	1.218E–117

TABLE 5A.22   Tabulated Solutions to Eqs. (5.35) and (5.36) (*Continued*)

CHAPTER 6

Fire Load and Severity of Fires

Notation

$q = \dfrac{\sum M_i \Delta H_{ci}}{A_t}$	Total fire load per unit area
M_i	Total weight of each single combustible item in the fire compartment (kg)
ΔH_{ci}	Effective calorific value of each combustible item (MJ/kg)
A_t	Total internal surface area of the fire compartment (m^2)
α	Design rate of fire growth for t^2 fires
k	Thermal conductivity, which must have the dimensions W/m K or J/m s K
T	Temperature
d	Thickness in the direction if heat flow
ρ	Air density
c	Specific heat capacity
K	Number of collisions that result in a reaction per second
A	Total number of collisions
E	Activation energy
R	Ideal gas constant
P	Losses of heat owing to thermal radiation
e	Emissivity factor

σ	Boltzmann constant ($\sigma = 5.6703 \times 10^{-8}$ W/ m^2 K^4)
T_o	Ambient temperature
A_v	Area of openings
c_p	Average specific heat at constant pressure
t	Time
$\vec{v}(u;v;w)$	Velocity vector
M	Molecular weight
i and k	Gas component numbers
C_{mi}	Concentrations of mass fractions
D	Diffusion coefficient (m^2/s)
k_1	Portion of a chemical reaction velocity that is a function of temperature only
$m = m_A + m_B + \cdots$	Order of a chemical reaction
p	Pressure
ν	Kinematic viscosity; $\nu = \mu/\rho$
θ	Dimensionless temperature
τ	Dimensionless time
h	Height of the compartment (m)
a	Thermal diffusivity (m^2/s)
τ_e	Effective dimensionless time
AR	Area under the temperature-time curve (above a baseline of 300°C)
Time	$t = \dfrac{h^2}{a}\tau \text{(s)}$
Temperature	$T = \dfrac{RT_*^2}{E}\theta + T_*\text{(K)}$, where $T_* = 600$ K is the baseline temperature
θ_{st}	Dimensionless temperature (E119 standard fire exposure)
Coordinates	$\bar{x} = x/h$, and $\bar{z} = z/h$, where x and z are dimensionless coordinates

Velocities	$\bar{u} = \dfrac{v}{h}u$ (m/s) and $\bar{w} = \dfrac{v}{h}w$ (m/s), horizontal and vertical components of velocity, accordingly; v is kinematic viscosity (m^2/s) and u and w are dimensionless velocities
$Pr = v/a$	Prandtl number
$Fr = \dfrac{gh^3}{va}$	Froude number
g	Gravitational acceleration
$Le = a/d = Sc/Pr$	Lewis number
$Sc = v/D$	Schmidt number
$\beta = \dfrac{RT_*}{E}$	Dimensionless parameter
$\gamma = \dfrac{c_p RT_*^2}{QE}$	Dimensionless parameter
$P = \dfrac{e\sigma K_v(\beta T_*)^3 h}{\lambda}$	Thermal radiation dimensionless coefficient
$K_v = A_o h/V$	Dimensionless opening factor
A_o	Total area of vertical and horizontal openings
$\delta = \left(\dfrac{E}{RT_*^2}\right)Qz\left[\exp\left(-\dfrac{E}{RT_*}\right)\right]$	Frank-Kamenetskii's parameter
$C = [1 - P(t)/P_o]$	Concentration of the burned fuel product in the fire compartment
$\bar{W} = \dfrac{v}{h}W$	Vertical component of gas velocity
$\bar{U} = \dfrac{v}{h}U$	Horizontal component of gas velocity
$b = L/h$	Length L (width) and height h of fire compartment, accordingly
W, U	Dimensionless velocities

6.1   Fire-Load Survey Data

Fire loads historically have been established by surveys of typical buildings in various use categories. Live loads used in structural engineering design were established in the identical manner. The relationship between fuel load and fire severity was established by

Fuel load (lb/ft²)	5	10	15	20	30	40	50	60	70
Fire severity (hours)	½	1	1½	2	3	4½	7	8	9

TABLE 6.1   Fire Severity for Various Fuel Loads

Ingberg in 1928 (see Chap. 2). To achieve this goal, several survey programs were conducted to establish data for representative use categories of buildings. The National Bureau of Standards (NBS) published fuel-load surveys of residential buildings [1] and office buildings [2]. A very detailed study was carried out in Switzerland in 1967–1969 for the Swiss Fire Prevention Association for Industry and Trade [3].

The first very important finding by Ingberg [4] was the correlation between fuel load and fire duration shown in Table 6.1. The duration of the fire was referred to as the *fire severity* (in hours) and represented the time needed to consume most of the fuel in the compartment. The effects of ventilation, the form of the fuel, and heat losses to boundaries were not taken into account. Ingberg had considered the worst possible condition of a fire scenario: no suppression, meaning that all the combustibles in the compartment should be consumed without causing failure of any structural element. The second very important finding by Ingberg was the ability to compare and equate the severity of two fires (the so-called *t*-equivalence method). As described by Law [5], "The term *t-equivalent* is usually taken to be the exposure time in the standard fire resistance test which gives the same heating effect on a structure as a given compartment fire." The duration and severity of a real fire are not defined well as the standard fire test curve. The time-equivalence concept makes use of fire-load and ventilation data in a real compartment fire to produce a value that would be "equivalent" to the exposure time in the standard test (for a detailed explanation and application in a structural fire load design case, see below). Formulating equivalent fire exposures traditionally has been achieved by gathering data from room-burn experiments where protected steel temperatures were recorded and variables relating to the fire severity were changed systematically (e.g., ventilation, fire load, and compartment shape).

The fuel load data were provided by NBS (now the National Institute of Standards and Technology [NIST]) fuel-load surveys conducted in residences, offices, schools, hospitals, and warehouses.

Because the expected fire severity is one of the bases for the fire-resistance requirements of structures, knowledge of it is essential. At present, predictions of fire severities can be made as a function of the amount of combustible material present in the room, the size and geometry of the room, the dimensions of the ventilation available, the heat losses through the openings, the emissivity of the flames in the room, and the thermal properties of the room surfaces. However, in predicting fire performance, two critical parameters—the fire load and the opening factor of the room—must be known. Besides the ventilation characteristics, fire load is the starting point for estimating the potential size and severity of a fire. Previous surveys have measured only mass and calorific value of the fuel load. However, at present, a need has been identified for fuel-load data that also include the exposed surface area of the fuel items so that the rate and duration of fuel-controlled burning can be better assessed. The intensity and duration of fire in different building occupancies vary greatly depending on the amount and surface area of the combustible material present, as well as the characteristics of the available ventilation. Therefore, an accurate prediction of the possible fire load in a certain building occupancy will assist the engineer in better estimating the likely fire severity and thus help to provide adequate and cost-effective fire protection.

Fire load is the starting point for estimating the potential size and severity of a fire and thus the endurance required of walls, columns, doors, floor-ceiling assemblies, and other parts of the enclosing compartment. The term *fire load* is defined as the total heat content on complete combustion of all the combustible materials contained inside a building or the fire compartment. In this case, the heat content per unit area is called the *fire-load density*. The higher its value, the greater is the potential fire severity and damage because the duration of the burning period of the fire is considered proportional to fire load. Based on Pettersson [6], fire load per unit area is given by the total internal surface area of the fire compartment and is calculated from the relationship

$$q = \frac{\sum M_i \Delta H_{ci}}{A_t} \quad (MJ/m^2) \qquad \text{Total surface area} \qquad (6.1)$$

where M_i = total weight of each single combustible item in the fire
        compartment (kg)
     ΔH_{ci} = effective calorific value of each combustible item (MJ/kg)
     A_t = total internal surface area of the fire compartment (m²)

Fire load density in buildings is most often expressed in units of mega joules per square meter (floor area). It must be noted that many European references, such as Pettersson [6], use mega joules per square meter of total bounding internal surface of the room, which can cause major errors if the distinction is not clear.

However, the total internal surface area and the floor area can be converted from one to the other by using the following formula:

$$A_t = 2[A_f + h(L + B)] \quad (\text{m}^2) \tag{6.2}$$

where A_t = total internal surface area (m²)
   A_f = floor area (m²)
   h = height of the fire compartment (m)
   L = length of the fire compartment (m)
   B = width of the fire compartment (m)

In estimating fire loads in fire compartments, the combustion properties of the fire load, such as its nature, weight, thickness, surface area, and location, play a critical role. Therefore, fire loads are commonly divided into two categories:

1. *Fixed fire load* that consists of exposed combustible materials permanently affixed to walls, ceilings, or floors plus any other built-in fixtures.

2. *Movable fire* load that consists of combustible furniture and other contents that are brought into the building for use of the occupant.

In practice, the fire load will vary with the occupancy, with the location in the building, and with time. However, it is possible to determine by means of statistical surveys the probability of the presence of a certain fire-load density in various occupancies such as motels, hotels, offices, schools, and hospitals. Although many surveys have been conducted in the past, most of them have made similar assumptions in order to simplify the fire-load estimation. These assumptions include the following:

1. Combustible materials are uniformly distributed throughout the building.

2. All combustible materials would be involved in the fire.

3. All combustible materials in the fire compartment would undergo total combustion during the fire.

4. Fire load can be measured as the sum of the heat contents of the different materials (i.e., cellulosic and noncellulosic) but will be given in a more conventional way by being evaluated directly on a wood-equivalent basis.

Thus, the total fire load will be expressed in mass of wood equivalent. For combustible items, especially fixed-in-place combustibles that could not be weighed easily, their dimensions were measured, and volumes calculated and multiplied by specific density to estimate the equivalent weight of the combustible materials. These valves then were used to derive the total calorific value in mega joules of the combustibles within the building. Finally, the fire loads were divided by the floor area to obtain the fire-load density (MJ/m²).

To determine the magnitude of the fire load, many countries have conducted statistical investigations on residential buildings, office buildings, schools, hospitals, hotels, and so on. To ease comparisons between the results from each survey, all the results are then multiplied by a net calorific value of 16.7 MJ/kg (for wood). These fire-load surveys have been conducted since 1891. For detailed information regarding the statistical approaches of these surveys, see Heaney [7]. As an example here, Table 6.2 presents data from the *International Fire Engineering Guidelines* [8] in both mass and calorific values by different occupancies for comparison. One can see that the data collected from different countries over nearly half a century are reasonably consistent, therefore, for the purpose of defining the structural fire load (SFL), these data are assumed to be provided.

However, the fuel load by itself does not fully describe the severity of fire. The rate at which energy is released can vary greatly for the same fuel load depending on the fuel. It was discovered by Heskestad [9] in 1978 that the growth phase of flaming fires generally followed a polynomial curve, with most fuels reasonably described by the so-called t-squared form (αt^2). It became possible to select design fires based on the potential energy release of the fuel rather than on the fuel types of a "typical" group. An understanding of preflashover fires is very important for life safety. The prime objective of a fire safety engineer is to prevent or delay flashover, thus providing adequate time for the occupants of the building to escape. In multistory structures, this is achieved by designing early detection and sprinkler systems. Activation times of sprinklers and detectors need to be calculated. Thus a design rate of fire growth α has to be obtained based on the type, and amount of fuel and available ventilation. These calculations often are performed by two-zone computer models. Several empirical and theoretical correlations exist [10] for defining parameter α (see Table 6.3).

Below are the step-by-step procedures [based on tabular/numerical solutions of dimensionless differential equations (5.3) and (5.4) for ultrafast, fast, medium, and slow fire scenarios] for the following main objectives:

- Defining parameter α for t-square fire growth period
- Functional relationship: maximum temperature versus opening factor of fire compartment
- Use of time-equivalence method to finalize and verify the approximate formulas of SFL (for postflashover fires)
- Comparison of parametric fire curves and SFL curves
- Thermal analyses
- Examples

Occupancy	Mean MJ/m², IFEG (1983)	95% Fract. MJ/m², IFEG (1983)	95% Fract. kg/m², IFEG (1983)	95% Fract. lb/ft², IFEG (1983)	90% Fract. MJ/m², Swiss (1969)	90% Fract. lb/sq², Swiss (1969)
Dwelling	780	970	54	12	495	6
Hospital	230	520	29	6	495	6
Hospital storage	2,000	4,400	244	54	1,320	16
Hotel room	310	510	28	6	495	6
Office	420	760	42	9	1,320	16
Shop	600	1300	72	16	660	8
Manufacturing	300	720	40	9	660	8
Storage	1,180	2,690	149	33	825	10
Library	1,500	2,750	153	34	3,300	40
School	285	450	25	6	495	6

TABLE 6.2 Comparison of Fuel (Fire) Load Data

Growth Rate	Typical Scenario	α (kW/s)
Slow	Densely packed paper products	0.00293
Medium	Traditional mattress or armchair	0.01172
Fast	PU mattress (horizontal) or PE pallets stacked 1 m high	0.0469
Ultrafast	High rack storage, polyethylene foam, known as PE rigid foam stacked 5 m high	0.1876

TABLE 6.3   Parameters Used for t^2 Fires

6.2   Defining Parameter α for *t*-Square Fire Growth Period

We now can compare the fire growth period parameter α with the results from Cases 1 through 4 (see Chap. 5). The end of the fire growth period has been estimated at the temperature $T = 400°C$ (dimensionless time $\tau = 0.03$).

Case 1: Ultrafast Fire

1. General data:

$$h = 3.0 \text{ m} \quad \text{Height of fire compartment}$$
$$B = 4.0 \text{ m} \quad \text{Width of fire compartment}$$
$$L = 5.0 \text{ m} \quad \text{Length of fire compartment}$$

Prandtl number $\text{Pr} = 0.68$; $\rho = 0.524$ (kg/m³), air density; $\nu = 62.53 \times 10^{-6}$ (m²/s), kinematic viscosity; $a = 92 \times 10^{-6}$ (m²/s), thermal diffusivity; $c_p = 1.068$, specific heat capacity (kJ/kg K); and $T_* = 600$ K (or approximately 300°C); $\beta = RT_*/E = 0.1$.

2. From Table 5.3 [or Eq. (5.5)], calculate $d\theta/d\tau$ at $\tau = 0.03$:

$$\frac{d\theta}{d\tau} = 489 \tag{6.3}$$

3. Calculate the first derivative of the real temperature-time function using scale parameters from Chap. 5:

$$\frac{dT}{dt} = \beta T_* \frac{a}{h^2} \frac{d\theta}{d\tau} = 0.3 \tag{6.4}$$

4. Calculate parameter α:

$$\alpha = c_p \rho \frac{dT}{dt} = 0.524(1.068)0.3 = 0.1678 \tag{6.5}$$

Case 2: Fast Fire

1. General data: Same as Case 1.
2. From Table 5.4 [or Eq. (5.8)], calculate $d\theta/d\tau$ at $\tau = 0.03$:

$$\frac{d\theta}{d\tau} = 120 \tag{6.6}$$

3. Calculate the first derivative of the real temperature-time function using scale parameters from Chap. 5:

$$\frac{dT}{dt} = \beta T_* \frac{a}{h^2} \frac{d\theta}{d\tau} = 0.0736 \tag{6.7}$$

4. Calculate parameter α:

$$\alpha = c_p \rho \frac{dT}{dt} = 0.524(1.068)0.0736 = 0.0412 \tag{6.8}$$

Case 3: Medium Fire

1. General data: Same as Case 1.
2. From Table 5.5 [or Eq. (5.11)], calculate $d\theta/d\tau$ at $\tau = 0.03$:

$$\frac{d\theta}{d\tau} = 70 \tag{6.9}$$

3. Calculate the first derivative of the real temperature-time function using scale parameters from Chap. 5:

$$\frac{dT}{dt} = \beta T_* \frac{a}{h^2} \frac{d\theta}{d\tau} = 0.0429 \tag{6.10}$$

4. Calculate parameter α:

$$\alpha = c_p \rho \frac{dT}{dt} = 0.524(1.068)0.0429 = 0.024 \tag{6.11}$$

Case 4: Slow Fire

1. General data: Same as Case 1.
2. From Table 5.6 [or Eq. (5.14)], calculate $d\theta/d\tau$ at $\tau = 0.03$:

$$\frac{d\theta}{d\tau} = 7.37 \tag{6.12}$$

3. Calculate the first derivative of the real temperature-time function using scale parameters from Chap. 5:

$$\frac{dT}{dt} = \beta T_* \frac{a}{h^2} \frac{d\theta}{d\tau} = 0.00452 \tag{6.13}$$

4. Calculate parameter α:

$$\alpha = c_p \rho \frac{dT}{dt} = 0.524(1.068)0.00452 = 0.00253 \qquad (6.14)$$

Let's summarize all these results (see Table 6.4) and compare with data from Table 6.3.

Category	Fire Growth Constant [1] (kJ/s²)	$\dot{\theta}$ at $T = 400°C$	Fire Growth Constant (Calculated) (kJ/s²)	%
Ultrafast	$\alpha = 0.1876$	489	$\alpha = 0.1678$	11.8
Fast	$\alpha = 0.0469$	120	$\alpha = 0.0412$	13.8
Medium	$\alpha = 0.01172$	70	$\alpha = 0.02402$	−48.8
Slow	$\alpha = 0.00293$	7.37	$\alpha = 0.002529$	15.8

TABLE 6.4 Parameter α for t–Square Fire Development

The most important cases from a structural design point of view (Cases 1 and 2) have shown very good agreement.

6.3 Maximum Temperature versus Opening Factor

An analysis of the solutions to differential equations (5.1) through (5.4) (as a function of dimensionless parameter P and consequently the opening factor K_v; see "Notation") can provide very valuable information for the preliminary and approximate structural design process. For example, the temperature-time functions provided earlier are given for opening factor $K_v = 0.05$. The relationships between maximum fire temperature in the compartment and different opening factor values are very important information that a structural engineer would require for any preliminary design. This information is provided below. In order to build this relationship (maximum temperature in a compartment T_{max} versus opening factor K_v), let's go back to differential equations (5.1) through (5.4) and reanalyze them with the new set of parameters P that corresponds to new opening factors K_v from the interval $0.02 < K_v < 0.2$. This time around, only maximum dimensionless temperature will be computed (not the full temperature-time relationship) for each given value of opening factor K_v. The procedure itself is very similar to one used in Chap. 5; therefore, for example, notation of dimensionless temperatures is the same:

y is the dimensionless temperature θ with the corresponding parameter γ.

y0 is the dimensionless temperature θ with the corresponding parameter γ.

y1 is the concentration of the product of the first-order chemical reaction with corresponding parameter γ.

y2 is the concentration of the product of the first-order chemical reaction with corresponding parameter γ, and so on.

Case 1: Ultrafast Fire
$K_v = 0.02$ $P = 0.0932$

	Variable	Initial Value	Minimal Value	Maximal Value	Final Value
1	t	0	0	0.2	0.2
2	y	1	1	20.59544	20.59544
3	y0	1	1	9.926769	3.023292
4	y1	0	0	0	0
5	y2	0	0	0.9961146	0.9961146

TABLE 6.5 Calculated Values of DEQ Variables

Differential Equations

1. $d(y0)/d(t) = 20*(1–y2)*exp(y0/(1+.1*y0)) –.0932*y0^4$
2. $d(y2)/d(t) = 1.0*(1–y2)*exp(y0/(1+.1*y0))$
3. $d(y1)/d(t) = 0*(1–y1)^1.0*exp(y/(1+.1*y))$
4. $d(y)/d(t) = (1)* 20*(1–y1)^1.0*exp(y/(1+.1*y)) –.0932*y^4$

$$(6.15)$$

$K_v = 0.03$ $P = 0.134$

	Variable	Initial Value	Minimal Value	Maximal Value	Final Value
1	t	0	0	0.2	0.2
2	y	1	1	16.68685	16.68685
3	y0	1	1	8.690773	2.816653
4	y1	0	0	0	0
5	y2	0	0	0.9904837	0.9904837

TABLE 6.6 Calculated Values of DEQ Variables

Differential Equations

1. $d(y0)/d(t) = 20*(1–y2)*exp(y0/(1+.1*y0)) –.134*y0^4$
2. $d(y2)/d(t) = 1.0*(1–y2)*exp(y0/(1+.1*y0))$

3. $d(y1)/d(t) = 0*(1-y1)^{1.0}*exp(y/(1+.1*y))$

4. $d(y)/d(t) = (1)* 20*(1-y1)^{1.0}*exp(y/(1+.1*y)) -.134*y^4$

(6.16)

$K_v = 0.04$ $P = 0.186$

	Variable	Initial Value	Minimal Value	Maximal Value	Final Value
1	t	0	0	0.2	0.2
2	y	1	1	13.60152	13.60152
3	y0	1	1	7.698541	2.677698
4	y1	0	0	0	0
5	y2	0	0	0.9815785	0.9815785

TABLE **6.7**   Calculated Values of DEQ Variables

Differential Equations

1. $d(y0)/d(t) = 20*(1-y2)*exp(y0/(1+.1*y0)) -.186*y0^4$

2. $d(y2)/d(t) = 1.0*(1-y2)*exp(y0/(1+.1*y0))$

3. $d(y1)/d(t) = 0*(1-y1)^{1.0}*exp(y/(1+.1*y))$

4. $d(y)/d(t) = (1)* 20*(1-y1)^{1.0}*exp(y/(1+.1*y)) -.186*y^4$

(6.17)

$K_v = 0.05$ $P = 0.233$

	Variable	Initial Value	Minimal Value	Maximal Value	Final Value
1	t	0	0	0.2	0.2
2	y	1	1	11.74485	11.74485
3	y0	1	1	7.049816	2.601609
4	y1	0	0	0	0
5	y2	0	0	0.9727725	0.9727725

TABLE **6.8**   Calculated Values of DEQ Variables

Differential Equations

1. $d(y0)/d(t) = 20*(1-y2)*exp(y0/(1+.1*y0)) -.233*y0^4$

2. $d(y2)/d(t) = 1.0*(1-y2)*exp(y0/(1+.1*y0))$

3. $d(y1)/d(t) = 0*(1-y1)^{1.0}*exp(y/(1+.1*y))$

4. $d(y)/d(t) = (1)* 20*(1-y1)^{1.0}*exp(y/(1+.1*y)) -.233*y^4$

(6.18)

$K_v = 0.06$ $P = 0.278$

	Variable	Initial Value	Minimal Value	Maximal Value	Final Value
1	t	0	0	0.2	0.2
2	y	1	1	10.4464	10.4464
3	y0	1	1	6.561706	2.549676
4	y1	0	0	0	0
5	y2	0	0	0.9640877	0.9640877

TABLE 6.9   Calculated Values of DEQ Variables

Differential Equations

1. d(y0)/d(t) = 20*(1–y2)*exp(y0/(1+.1*y0)) –.278*y0^4
2. d(y2)/d(t) = 1.0*(1–y2)*exp(y0/(1+.1*y0))
3. d(y1)/d(t) = 0*(1–y1)^1.0*exp(y/(1+.1*y))
4. d(y)/d(t) = (1)* 20*(1–y1)^1.0*exp(y/(1+.1*y)) –.278*y^4

$$(6.19)$$

$K_v = 0.08$ $P = 0.373$

	Variable	Initial Value	Minimal Value	Maximal Value	Final Value
1	t	0	0	0.2	0.2
2	y	1	1	8.591577	8.591577
3	y0	1	1	5.808178	2.472316
4	y1	0	0	0	0
5	y2	0	0	0.9457345	0.9457345

TABLE 6.10   Calculated Values of DEQ Variables

Differential Equations

1. d(y0)/d(t) = 20*(1–y2)*exp(y0/(1+.1*y0)) –.373*y0^4
2. d(y2)/d(t) = 1.0*(1–y2)*exp(y0/(1+.1*y0))
3. d(y1)/d(t) = 0*(1–y1)^1.0*exp(y/(1+.1*y))
4. d(y)/d(t) = (1)* 20*(1–y1)^1.0*exp(y/(1+.1*y)) –.373*y^4

$$(6.20)$$

$K_v = 0.10$ $P = 0.466$

	Variable	Initial Value	Minimal Value	Maximal Value	Final Value
1	t	0	0	0.2	0.2
2	y	1	1	7.429827	7.429827
3	y0	1	1	5.290364	2.416911
4	y1	0	0	0	0
5	y2	0	0	0.928318	0.928318

TABLE **6.11**   Calculated Values of DEQ Variables

Differential Equations

1. $d(y0)/d(t) = 20*(1-y2)*exp(y0/(1+.1*y0)) -.466*y0^4$
2. $d(y2)/d(t) = 1.0*(1-y2)*exp(y0/(1+.1*y0))$
3. $d(y1)/d(t) = 0*(1-y1)^1.0*exp(y/(1+.1*y))$
4. $d(y)/d(t) = (1)* 20*(1-y1)^1.0*exp(y/(1+.1*y)) -.466*y^4$

$$(6.21)$$

$K_v = 0.12$ $P = 0.559$

	Variable	Initial Value	Minimal Value	Maximal Value	Final Value
1	t	0	0	0.2	0.2
2	y	1	1	6.620452	6.620452
3	y0	1	1	4.902166	2.371268
4	y1	0	0	0	0
5	y2	0	0	0.9117181	0.9117181

TABLE **6.12**   Calculated Values of DEQ Variables

Differential Equations

1. $d(y0)/d(t) = 20*(1-y2)*exp(y0/(1+.1*y0)) -.559*y0^4$
2. $d(y2)/d(t) = 1.0*(1-y2)*exp(y0/(1+.1*y0))$
3. $d(y1)/d(t) = 0*(1-y1)^1.0*exp(y/(1+.1*y))$
4. $d(y)/d(t) = (1)* 20*(1-y1)^1.0*exp(y/(1+.1*y)) -.559*y^4$

$$(6.22)$$

$K_v = 0.15$ $P = 0.699$

	Variable	Initial Value	Minimal Value	Maximal Value	Final Value
1	t	0	0	0.2	0.2
2	y	1	1	5.776856	5.776856
3	y0	1	1	4.462002	2.312724
4	y1	0	0	0	0
5	y2	0	0	0.8884133	0.8884133

TABLE 6.13   Calculated Values of DEQ Variables

Differential Equations

1. $d(y0)/d(t) = 20*(1-y2)*exp(y0/(1+.1*y0)) -.699*y0^4$
2. $d(y2)/d(t) = 1.0*(1-y2)*exp(y0/(1+.1*y0))$
3. $d(y1)/d(t) = 0*(1-y1)^1.0*exp(y/(1+.1*y))$
4. $d(y)/d(t) = (1)* 20*(1-y1)^1.0*exp(y/(1+.1*y)) -.699*y^4$

$$(6.23)$$

$K_v = 0.20$ $P = 0.932$

	Variable	Initial Value	Minimal Value	Maximal Value	Final Value
1	t	0	0	0.2	0.2
2	y	1	1	4.894005	4.894005
3	y0	1	1	3.96438	2.230826
4	y1	0	0	0	0
5	y2	0	0	0.853936	0.853936

TABLE 6.14   Calculated Values of DEQ Variables

Differential Equations

1. $d(y0)/d(t) = 20*(1-y2)*exp(y0/(1+.1*y0)) -.932*y0^4$
2. $d(y2)/d(t) = 1.0*(1-y2)*exp(y0/(1+.1*y0))$
3. $d(y1)/d(t) = 0*(1-y1)^1.0*exp(y/(1+.1*y))$
4. $d(y)/d(t) = (1)* 20*(1-y1)^1.0*exp(y/(1+.1*y)) -.932*y^4$

$$(6.24)$$

Case 2: Fast Fire
$K_v = 0.02$ $P = 0.0628$

	Variable	Initial Value	Minimal Value	Maximal Value	Final Value
1	t	0	0	0.2	0.2
2	y	1	1	11.27461	3.325467
3	y0	1	1	5.445123	3.131239
4	y1	0	0	0.9988736	0.9988736
5	y2	0	0	0.9999933	0.9999933

TABLE 6.15   Calculated Values of DEQ Variables

Differential Equations

1. $d(y0)/d(t) = 20*(1–y2)*exp(y0/(1+.1*y0)) –.0628*y0^4$
2. $d(y2)/d(t) = 3.5*(1–y2)*exp(y0/(1+.1*y0))$
3. $d(y1)/d(t) = 1.0*(1–y1)^1.0*exp(y/(1+.1*y))$
4. $d(y)/d(t) = (1)* 20*(1–y1)^1.0*exp(y/(1+.1*y)) –.0628*y^4$

$$(6.25)$$

$K_v = 0.03$ $P = 0.0942$

	Variable	Initial Value	Minimal Value	Maximal Value	Final Value
1	t	0	0	0.2	0.2
2	y	1	1	9.886612	3.016296
3	y0	1	1	5.140411	2.778713
4	y1	0	0	0.9959991	0.9959991
5	y2	0	0	0.999959	0.999959

TABLE 6.16   Calculated Values of DEQ Variables

Differential Equations

1. $d(y0)/d(t) = 20*(1–y2)*exp(y0/(1+.1*y0)) –.0942*y0^4$
2. $d(y2)/d(t) = 3.5*(1–y2)*exp(y0/(1+.1*y0))$
3. $d(y1)/d(t) = 1.0*(1–y1)^1.0*exp(y/(1+.1*y))$
4. $d(y)/d(t) = (1)* 20*(1–y1)^1.0*exp(y/(1+.1*y)) –.0942*y^4$

$$(6.26)$$

$K_v = 0.04$ $P = 0.1256$

	Variable	Initial Value	Minimal Value	Maximal Value	Final Value
1	t	0	0	0.2	0.2
2	y	1	1	8.926764	2.849018
3	y0	1	1	4.898414	2.551832
4	y1	0	0	0.9917736	0.9917736
5	y2	0	0	0.9998736	0.9998736

TABLE 6.17   Calculated Values of DEQ Variables

Differential Equations

1. d(y0)/d(t) = 20*(1–y2)*exp(y0/(1+.1*y0)) –.1256*y0^4
2. d(y2)/d(t) = 3.5*(1–y2)*exp(y0/(1+.1*y0))
3. d(y1)/d(t) = 1.0*(1–y1)^1.0*exp(y/(1+.1*y))
4. d(y)/d(t) = (1)* 20*(1–y1)^1.0*exp(y/(1+.1*y)) –.1256*y^4

(6.27)

$K_v = 0.05$ $P = 0.157$

	Variable	Initial Value	Minimal Value	Maximal Value	Final Value
1	t	0	0	0.2	0.2
2	y	1	1	8.201185	2.744677
3	y0	1	1	4.707775	2.389874
4	y1	0	0	0.9867073	0.9867073
5	y2	0	0	0.9997236	0.9997236

TABLE 6.18   Calculated Values of DEQ Variables

Differential Equations

1. d(y0)/d(t) = 20*(1–y2)*exp(y0/(1+.1*y0)) –.157*y0^4
2. d(y2)/d(t) = 3.5*(1–y2)*exp(y0/(1+.1*y0))
3. d(y1)/d(t) = 1.0*(1–y1)^1.0*exp(y/(1+.1*y))
4. d(y)/d(t) = (1)* 20*(1–y1)^1.0*exp(y/(1+.1*y)) –.157*y^4

(6.28)

$K_v = 0.06$ $P = 0.1884$

	Variable	Initial Value	Minimal Value	Maximal Value	Final Value
1	t	0	0	0.2	0.2
2	y	1	1	7.650644	2.673004
3	y0	1	1	4.54247	2.266796
4	y1	0	0	0.9811408	0.9811408
5	y2	0	0	0.9995043	0.9995043

TABLE 6.19   Calculated Values of DEQ Variables

Differential Equations

1. d(y0)/d(t) = 20*(1–y2)*exp(y0/(1+.1*y0)) –.1884*y0^4
2. d(y2)/d(t) = 3.5*(1–y2)*exp(y0/(1+.1*y0))
3. d(y1)/d(t) = 1.0*(1–y1)^1.0*exp(y/(1+.1*y))
4. d(y)/d(t) = (1)* 20*(1–y1)^1.0*exp(y/(1+.1*y)) –.1884*y^4

$$(6.29)$$

$K_v = 0.08$ $P = 0.2512$

	Variable	Initial Value	Minimal Value	Maximal Value	Final Value
1	t	0	0	0.2	0.2
2	y	1	1	6.820811	2.578807
3	y0	1	1	4.276035	2.089453
4	y1	0	0	0.9692756	0.9692756
5	y2	0	0	0.9988674	0.9988674

TABLE 6.20   Calculated Values of DEQ Variables

Differential Equations

1. d(y0)/d(t) = 20*(1–y2)*exp(y0/(1+.1*y0)) –.2512*y0^4
2. d(y2)/d(t) = 3.5*(1–y2)*exp(y0/(1+.1*y0))
3. d(y1)/d(t) = 1.0*(1–y1)^1.0*exp(y/(1+.1*y))
4. d(y)/d(t) = (1)* 20*(1–y1)^1.0*exp(y/(1+.1*y)) –.2512*y^4

$$(6.30)$$

$K_v = 0.10$ $P = 0.314$

	Variable	Initial Value	Minimal Value	Maximal Value	Final Value
1	t	0	0	0.2	0.2
2	y	1	1	6.240278	2.516616
3	y0	1	1	4.067243	1.96577
4	y1	0	0	0.9571003	0.9571003
5	y2	0	0	0.9980031	0.9980031

TABLE 6.21   Calculated Values of DEQ Variables

Differential Equations

1. $d(y0)/d(t) = 20*(1–y2)*exp(y0/(1+.1*y0)) –.314*y0^4$
2. $d(y2)/d(t) = 3.5*(1–y2)*exp(y0/(1+.1*y0))$
3. $d(y1)/d(t) = 1.0*(1–y1)^1.0*exp(y/(1+.1*y))$
4. $d(y)/d(t) = (1)* 20*(1–y1)^1.0*exp(y/(1+.1*y)) –.314*y^4$

$$(6.31)$$

$K_v = 0.12$ $P = 0.3768$

	Variable	Initial Value	Minimal Value	Maximal Value	Final Value
1	t	0	0	0.2	0.2
2	y	1	1	5.783721	2.469767
3	y0	1	1	3.894373	1.873313
4	y1	0	0	0.9450093	0.9450093
5	y2	0	0	0.9969614	0.9969614

TABLE 6.22   Calculated Values of DEQ Variables

Differential Equations

1. $d(y0)/d(t) = 20*(1+.1*y2)*exp(y0/(1+.1*y0)) –.3768*y0^4$
2. $d(y2)/d(t) = 3.5*(1–y2)*exp(y0/(1+.1*y0))$
3. $d(y1)/d(t) = 1.0*(1–y1)^1.0*exp(y/(1+.1*y))$
4. $d(y)/d(t) = (1)* 20*(1–y1)^1.0*exp(y/(1+.1*y)) –.3768*y^4$

$$(6.32)$$

$K_v = 0.15$ $P = 0.471$

	Variable	Initial Value	Minimal Value	Maximal Value	Final Value
1	t	0	0	0.2	0.2
2	y	1	1	5.268394	2.41426
3	y0	1	1	3.686598	1.770034
4	y1	0	0	0.9274035	0.9274035
5	y2	0	0	0.9951605	0.9951605

TABLE 6.23   Calculated Values of DEQ Variables

Differential Equations

1. $d(y0)/d(t) = 20*(1-y2)*exp(y0/(1+.1*y0)) -.471*y0^4$
2. $d(y2)/d(t) = 3.5*(1-y2)*exp(y0/(1+.1*y0))$
3. $d(y1)/d(t) = 1.0*(1-y1)^{1.0}*exp(y/(1+.1*y))$
4. $d(y)/d(t) = (1)* 20*(1-y1)^{1.0}*exp(y/(1+.1*y)) -.471*y^4$

$$(6.33)$$

$K_v = 0.20$ $P = 0.628$

	Variable	Initial Value	Minimal Value	Maximal Value	Final Value
1	t	0	0	0.2	0.2
2	y	1	1	4.669768	2.341232
3	y0	1	1	3.421334	1.651479
4	y1	0	0	0.8999782	0.8999782
5	y2	0	0	0.9917781	0.9917781

TABLE 6.24   Calculated Values of DEQ Variables

Differential Equations

1. $d(y0)/d(t) = 20*(1-y2)*exp(y0/(1+.1*y0)) -.628*y0^4$
2. $d(y2)/d(t) = 3.5*(1-y2)*exp(y0/(1+.1*y0))$
3. $d(y1)/d(t) = 1.0*(1-y1)^{1.0}*exp(y/(1+.1*y))$
4. $d(y)/d(t) = (1)* 20*(1-y1)^{1.0}*exp(y/(1+.1*y)) -.628*y^4$

$$(6.34)$$

Case 3: Medium Fire.

$K_v = 0.02$ $P = 0.0628$

	Variable	Initial Value	Minimal Value	Maximal Value	Final Value
1	t	0	0	0.2	0.2
2	y	1	1	5.445447	3.131239
3	y0	1	1	4.086852	2.923196
4	y1	0	0	0.9999933	0.9999933
5	y2	0	0	0.9999985	0.9999985

TABLE 6.25   Calculated Values of DEQ Variables

Differential Equations

1. $d(y0)/d(t) = 20*(1–y2)*\exp(y0/(1+.1*y0)) –.0628*y0^4$
2. $d(y2)/d(t) = 5.5*(1–y2)*\exp(y0/(1+.1*y0))$
3. $d(y1)/d(t) = 3.5*(1–y1)^{1.0}*\exp(y/(1+.1*y))$
4. $d(y)/d(t) = (1)* 20*(1–y1)^{1.0}*\exp(y/(1+.1*y)) –.0628*y^4$

$$(6.35)$$

$K_v = 0.03$ $P = 0.0942$

	Variable	Initial Value	Minimal Value	Maximal Value	Final Value
1	t	0	0	0.2	0.2
2	y	1	1	5.136693	2.778713
3	y0	1	1	3.934803	2.633277
4	y1	0	0	0.999959	0.999959
5	y2	0	0	0.9999933	0.9999933

TABLE 6.26   Calculated Values of DEQ Variables

Differential Equations

1. $d(y0)/d(t) = 20*(1–y2)*\exp(y0/(1+.1*y0)) –.0942*y0^4$
2. $d(y2)/d(t) = 5.5*(1–y2)*\exp(y0/(1+.1*y0))$
3. $d(y1)/d(t) = 3.5*(1–y1)^{1.0}*\exp(y/(1+.1*y))$
4. $d(y)/d(t) = (1)* 20*(1–y1)^{1.0}*\exp(y/(1+.1*y)) –.0942*y^4$

$$(6.36)$$

$K_v = 0.04$ $P = 0.1256$

	Variable	Initial Value	Minimal Value	Maximal Value	Final Value
1	t	0	0	0.2	0.2
2	y	1	1	4.901465	2.551832
3	y0	1	1	3.811134	2.434501
4	y1	0	0	0.9998736	0.9998736
5	y2	0	0	0.9999813	0.9999813

TABLE 6.27   Calculated Values of DEQ Variables

Differential Equations

1. $d(y0)/d(t) = 20*(1-y2)*exp(y0/(1+.1*y0)) -.1256*y0^4$
2. $d(y2)/d(t) = 5.5*(1-y2)*exp(y0/(1+.1*y0))$
3. $d(y1)/d(t) = 3.5*(1-y1)^{1.0}*exp(y/(1+.1*y))$
4. $d(y)/d(t) = (1)* 20*(1-y1)^{1.0}*exp(y/(1+.1*y)) -.1256*y^4$

(6.37)

$K_v = 0.05$ $P = 0.157$

	Variable	Initial Value	Minimal Value	Maximal Value	Final Value
1	t	0	0	0.2	0.2
2	y	1	1	4.70778	2.389874
3	y0	1	1	3.707141	2.286563
4	y1	0	0	0.9997236	0.9997236
5	y2	0	0	0.9999598	0.9999598

TABLE 6.28   Calculated Values of DEQ Variables

Differential Equations

1. $d(y0)/d(t) = 20*(1-y2)*exp(y0/(1+.1*y0)) -.157*y0^4$
2. $d(y2)/d(t) = 5.5*(1-y2)*exp(y0/(1+.1*y0))$
3. $d(y1)/d(t) = 3.5*(1-y1)^{1.0}*exp(y/(1+.1*y))$
4. $d(y)/d(t) = (1)* 20*(1-y1)^{1.0}*exp(y/(1+.1*y)) -.157*y^4$

(6.38)

$K_v = 0.06$ $P = 0.1884$

	Variable	Initial Value	Minimal Value	Maximal Value	Final Value
1	t	0	0	0.2	0.2
2	y	1	1	4.543338	2.266796
3	y0	1	1	3.614527	2.170579
4	y1	0	0	0.9995043	0.9995043
5	y2	0	0	0.9999271	0.9999271

TABLE **6.29**   Calculated Values of DEQ Variables

Differential Equations

1. d(y0)/d(t) = 20*(1–y2)*exp(y0/(1+.1*y0)) –.1884*y0^4
2. d(y2)/d(t) = 5.5*(1–y2)*exp(y0/(1+.1*y0))
3. d(y1)/d(t) = 3.5*(1–y1)^1.0*exp(y/(1+.1*y))
4. d(y)/d(t) = (1)* 20*(1–y1)^1.0*exp(y/(1+.1*y)) –.1884*y^4

(6.39)

$K_v = 0.08$ $P = 0.2512$

	Variable	Initial Value	Minimal Value	Maximal Value	Final Value
1	t	0	0	0.2	0.2
2	y	1	1	4.276746	2.089453
3	y0	1	1	3.465593	1.997618
4	y1	0	0	0.9988674	0.9988674
5	y2	0	0	0.9998249	0.9998249

TABLE **6.30**   Calculated Values of DEQ Variables

Differential Equations

1. d(y0)/d(t) = 20*(1–y2)*exp(y0/(1+.1*y0)) –.2512*y0^4
2. d(y2)/d(t) = 5.5*(1–y2)*exp(y0/(1+.1*y0))
3. d(y1)/d(t) = 3.5*(1–y1)^1.0*exp(y/(1+.1*y))
4. d(y)/d(t) = (1)* 20*(1–y1)^1.0*exp(y/(1+.1*y)) –.2512*y^4

(6.40)

$K_v = 0.10$ $P = 0.314$

	Variable	Initial Value	Minimal Value	Maximal Value	Final Value
1	t	0	0	0.2	0.2
2	y	1	1	4.067125	1.96577
3	y0	1	1	3.341476	1.872557
4	y1	0	0	0.9980031	0.9980031
5	y2	0	0	0.9996726	0.9996726

TABLE 6.31   Calculated Values of DEQ Variables

Differential Equations

1. $d(y0)/d(t) = 20*(1-y2)*\exp(y0/(1+.1*y0)) -.314*y0^4$
2. $d(y2)/d(t) = 5.5*(1-y2)*\exp(y0/(1+.1*y0))$
3. $d(y1)/d(t) = 3.5*(1-y1)^{1.0}*\exp(y/(1+.1*y))$
4. $d(y)/d(t) = (1)* 20*(1-y1)^{1.0}*\exp(y/(1+.1*y)) -.314*y^4$

$$(6.41)$$

$K_v = 0.12$ $P = 0.3768$

	Variable	Initial Value	Minimal Value	Maximal Value	Final Value
1	t	0	0	0.2	0.2
2	y	1	1	3.894847	1.873313
3	y0	1	1	3.236463	1.77651
4	y1	0	0	0.9969614	0.9969614
5	y2	0	0	0.999473	0.999473

TABLE 6.32   Calculated Values of DEQ Variables

Differential Equations

1. $d(y0)/d(t) = 20*(1-y2)*\exp(y0/(1+.1*y0)) -.3768*y0^4$
2. $d(y2)/d(t) = 5.5*(1-y2)*\exp(y0/(1+.1*y0))$
3. $d(y1)/d(t) = 3.5*(1-y1)^{1.0}*\exp(y/(1+.1*y))$
4. $d(y)/d(t) = (1)* 20*(1-y1)^{1.0}*\exp(y/(1+.1*y)) -.3768*y^4$

$$(6.42)$$

$K_v = 0.15$ $P = 0.471$

	Variable	Initial Value	Minimal Value	Maximal Value	Final Value
1	t	0	0	0.2	0.2
2	y	1	1	3.686517	1.770034
3	y0	1	1	3.105627	1.66655
4	y1	0	0	0.9951605	0.9951605
5	y2	0	0	0.9990955	0.9990955

TABLE 6.33   Calculated Values of DEQ Variables

Differential Equations

1. d(y0)/d(t) = 20*(1–y2)*exp(y0/(1+.1*y0)) –.471*y0^4
2. d(y2)/d(t) = 5.5*(1–y2)*exp(y0/(1+.1*y0))
3. d(y1)/d(t) = 3.5*(1–y1)^1.0*exp(y/(1+.1*y))
4. d(y)/d(t) = (1)* 20*(1–y1)^1.0*exp(y/(1+.1*y)) –.471*y^4

$$(6.43)$$

$K_v = 0.20$ $P = 0.628$

	Variable	Initial Value	Minimal Value	Maximal Value	Final Value
1	t	0	0	0.2	0.2
2	y	1	1	3.418942	1.651479
3	y0	1	1	2.930774	1.536984
4	y1	0	0	0.9917781	0.9917781
5	y2	0	0	0.9983021	0.9983021

TABLE 6.34   Calculated Values of DEQ Variables

Differential Equations

1. d(y0)/d(t) = 20*(1–y2)*exp(y0/(1+.1*y0)) –.628*y0^4
2. d(y2)/d(t) = 5.5*(1–y2)*exp(y0/(1+.1*y0))
3. d(y1)/d(t) = 3.5*(1–y1)^1.0*exp(y/(1+.1*y))
4. d(y)/d(t) = (1)* 20*(1–y1)^1.0*exp(y/(1+.1*y)) –.628*y^4

$$(6.44)$$

Case 4: Slow Fire

$K_v = 0.02$ $P = 0.0628$

	Variable	Initial Value	Minimal Value	Maximal Value	Final Value
1	t	0	0	0.2	0.2
2	y	1	1	4.086847	2.923196
3	y0	1	1	1.957302	1.845307
4	y1	0	0	0.9999985	0.9999985
5	y2	0	0	1	1

TABLE 6.35   Calculated Values of DEQ Variables

Differential Equations

1. $d(y0)/d(t) = 20*(1-y2)*exp(y0/(1+.1*y0)) -2.53*0-.0628*y0^4$
2. $d(y2)/d(t) = 20*(1-y2)*exp(y0/(1+.1*y0))$
3. $d(y1)/d(t) = 5.5*(1-y1)^{1.0}*exp(y/(1+.1*y))$
4. $d(y)/d(t) = (1)* 20*(1-y1)^{1.0}*exp(y/(1+.1*y)) -2.53*0-.0628*y^4$

$$(6.45)$$

$K_v = 0.03$ $P = 0.0942$

	Variable	Initial Value	Minimal Value	Maximal Value	Final Value
1	t	0	0	0.2	0.2
2	y	1	1	3.934713	2.633277
3	y0	1	1	1.941882	1.783726
4	y1	0	0	0.9999933	0.9999933
5	y2	0	0	1	1

TABLE 6.36   Calculated Values of DEQ Variables

Differential Equations

1. $d(y0)/d(t) = 20*(1-y2)*exp(y0/(1+.1*y0)) -2.53*0-.0942*y0^4$
2. $d(y2)/d(t) = 20*(1-y2)*exp(y0/(1+.1*y0))$
3. $d(y1)/d(t) = 5.5*(1-y1)^{1.0}*exp(y/(1+.1*y))$
4. $d(y)/d(t) = (1)* 20*(1-y1)^{1.0}*exp(y/(1+.1*y)) -2.53*0-.0942*y^4$

$$(6.46)$$

$K_v = 0.04$ $P = 0.1256$

	Variable	Initial Value	Minimal Value	Maximal Value	Final Value
1	t	0	0	0.2	0.2
2	y	1	1	3.811001	2.434501
3	y0	1	1	1.928026	1.729648
4	y1	0	0	0.9999813	0.9999813
5	y2	0	0	1	1

TABLE 6.37   Calculated Values of DEQ Variables

Differential Equations

1. d(y0)/d(t) = 20*(1–y2)*exp(y0/(1+.1*y0)) –2.53*0–.1256*y0^4
2. d(y2)/d(t) = 20*(1–y2)*exp(y0/(1+.1*y0))
3. d(y1)/d(t) = 5.5*(1–y1)^1.0*exp(y/(1+.1*y))
4. d(y)/d(t) = (1)* 20*(1–y1)^1.0*exp(y/(1+.1*y)) –2.53*0–.1256*y^4

$$(6.47)$$

$K_v = 0.05$ $P = 0.157$

	Variable	Initial Value	Minimal Value	Maximal Value	Final Value
1	t	0	0	0.2	0.2
2	y	1	1	3.706608	2.286563
3	y0	1	1	1.915441	1.68161
4	y1	0	0	0.9999598	0.9999598
5	y2	0	0	1	1

TABLE 6.38   Calculated Values of DEQ Variables

Differential Equations

1. d(y0)/d(t) = 20*(1–y2)*exp(y0/(1+.1*y0)) –2.53*0 –.157*y0^4
2. d(y2)/d(t) = 20*(1–y2)*exp(y0/(1+.1*y0))
3. d(y1)/d(t) = 5.5*(1–y1)^1.0*exp(y/(1+.1*y))
4. d(y)/d(t) = (1)* 20*(1–y1)^1.0*exp(y/(1+.1*y)) –2.53*0 –.157*y^4

$$(6.48)$$

$K_v = 0.06$ $P = 0.1884$

	Variable	Initial Value	Minimal Value	Maximal Value	Final Value
1	t	0	0	0.2	0.2
2	y	1	1	3.617213	2.170579
3	y0	1	1	1.903739	1.638522
4	y1	0	0	0.9999271	0.9999271
5	y2	0	0	1	1

TABLE 6.39   Calculated Values of DEQ Variables

Differential Equations

1. $d(y0)/d(t) = 20*(1-y2)*exp(y0/(1+.1*y0))-2.53*0-.1884*y0^4$
2. $d(y2)/d(t) = 20*(1-y2)*exp(y0/(1+.1*y0))$
3. $d(y1)/d(t) = 5.5*(1-y1)^1.0*exp(y/(1+.1*y))$
4. $d(y)/d(t) = (1)* 20*(1-y1)^1.0*exp(y/(1+.1*y))-2.53*0-.1884*y^4$

$$(6.49)$$

$K_v = 0.08$ $P = 0.2512$

	Variable	Initial Value	Minimal Value	Maximal Value	Final Value
1	t	0	0	0.2	0.2
2	y	1	1	3.465595	1.997618
3	y0	1	1	1.882226	1.564071
4	y1	0	0	0.9998249	0.9998249
5	y2	0	0	1	1

TABLE 6.40   Calculated Values of DEQ Variables

Differential Equations

1. $d(y0)/d(t) = 20*(1-y2)*exp(y0/(1+.1*y0))-2.53*0-.2512*y0^4$
2. $d(y2)/d(t) = 20*(1-y2)*exp(y0/(1+.1*y0))$
3. $d(y1)/d(t) = 5.5*(1-y1)^1.0*exp(y/(1+.1*y))$
4. $d(y)/d(t) = (1)* 20*(1-y1)^1.0*exp(y/(1+.1*y))-2.53*0-.2512*y^4$

$$(6.50)$$

$K_v = 0.10$ $P = 0.314$

	Variable	Initial Value	Minimal Value	Maximal Value	Final Value
1	t	0	0	0.2	0.2
2	y	1	1	3.341377	1.872557
3	y0	1	1	1.863281	1.501603
4	y1	0	0	0.9996726	0.9996726
5	y2	0	0	0.9999999	0.9999999

TABLE 6.41   Calculated Values of DEQ Variables

Differential Equations

1. $d(y0)/d(t) = 20*(1–y2)*exp(y0/(1+.1*y0))–2.53*0–.314*y0^4$
2. $d(y2)/d(t) = 20*(1–y2)*exp(y0/(1+.1*y0))$
3. $d(y1)/d(t) = 5.5*(1–y1)^{\wedge}1.0*exp(y/(1+.1*y))$
4. $d(y)/d(t) = (1)* 20*(1–y1)^{\wedge}1.0*exp(y/(1+.1*y))–2.53*0–.314*y^4$

(6.51)

$K_v = 0.12$ $P = 0.3768$

	Variable	Initial Value	Minimal Value	Maximal Value	Final Value
1	t	0	0	0.2	0.2
2	y	1	1	3.236587	1.77651
3	y0	1	1	1.846273	1.448107
4	y1	0	0	0.999473	0.999473
5	y2	0	0	0.9999999	0.9999999

TABLE 6.42   Calculated Values of DEQ Variables

Differential Equations

1. $d(y0)/d(t) = 20*(1–y2)*exp(y0/(1+.1*y0))–2.53*0–.3768*y0^4$
2. $d(y2)/d(t) = 20*(1–y2)*exp(y0/(1+.1*y0))$
3. $d(y1)/d(t) = 5.5*(1–y1)^{\wedge}1.0*exp(y/(1+.1*y))$
4. $d(y)/d(t) = (1)* 20*(1–y1)^{\wedge}1.0*exp(y/(1+.1*y))–2.53*0–.3768*y^4$

(6.52)

$K_v = 0.15$ $P = 0.471$

	Variable	Initial Value	Minimal Value	Maximal Value	Final Value
1	t	0	0	0.2	0.2
2	y	1	1	3.105377	1.66655
3	y0	1	1	1.822705	1.38041
4	y1	0	0	0.9990955	0.9990955
5	y2	0	0	0.9999999	0.9999999

TABLE **6.43** Calculated Values of DEQ Variables

Differential Equations

1. $d(y0)/d(t) = 20*(1-y2)*exp(y0/(1+.1*y0)) -2.53*0-.471*y0^4$
2. $d(y2)/d(t) = 20*(1-y2)*exp(y0/(1+.1*y0))$
3. $d(y1)/d(t) = 5.5*(1-y1)^{1.0}*exp(y/(1+.1*y))$
4. $d(y)/d(t) = (1)* 20*(1-y1)^{1.0}*exp(y/(1+.1*y)) -2.53*0-.471*y^4$

$$(6.53)$$

$K_v = 0.20$ $P = 0.628$

	Variable	Initial Value	Minimal Value	Maximal Value	Final Value
1	t	0	0	0.2	0.2
2	y	1	1	2.933169	1.536984
3	y0	1	1	1.788613	1.290999
4	y1	0	0	0.9983021	0.9983021
5	y2	0	0	0.9999997	0.9999997

TABLE **6.44** Calculated Values of DEQ Variables

Differential Equations

1. $d(y0)/d(t) = 20*(1-y2)*exp(y0/(1+.1*y0)) -2.53*0-.628*y0^4$
2. $d(y2)/d(t) = 20*(1-y2)*exp(y0/(1+.1*y0))$
3. $d(y1)/d(t) = 5.5*(1-y1)^{1.0}*exp(y/(1+.1*y))$
4. $d(y)/d(t) = (1)* 20*(1-y1)^{1.0}*exp(y/(1+.1*y)) -2.53*0-.628*y^4$

$$(6.54)$$

The summary of all the preceding information is provided in Table 6.45 or Fig. 6.1 and analytical formulas (6.55) through (6.58).

Opening Factor	Case 1	Case 2	Case 3	Case 4
K_v	$\theta_{1\,max}$	$\theta_{2\,max}$	$\theta_{3\,max}$	$\theta_{4\,max}$
0.02	20.6	11.27	5.44	4.09
0.03	16.7	9.89	5.14	3.93
0.04	13.6	8.93	4.9	3.81
0.05	11.74	8.2	4.71	3.71
0.06	10.45	7.65	4.54	3.62
0.08	8.59	6.84	4.28	3.46
0.1	7.43	6.24	4.07	3.34
0.12	6.62	5.78	3.89	3.24
0.15	5.78	5.27	3.69	3.1
0.2	4.89	4.67	3.42	2.93

TABLE 6.45   Dimensionless Temperature-Opening Factor K_v Data

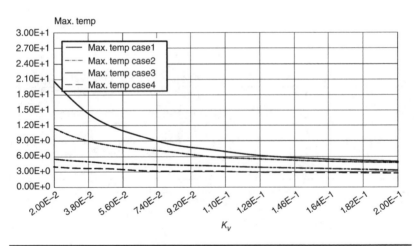

FIGURE 6.1   Dimensionless temperature-opening factor K_v curves.

The correspondent analytical formulas are as follows.

Case 1: Ultrafast Fire

Model $T1m = \theta_{max1} = a0 + a1*Kv + a2*Kv^2 + a3*Kv^3 + a4*Kv^4$

Variable	Value
a0	31.66894
a1	−701.2023
a2	7924.765
a3	−4.117E+04
a4	7.866E+04

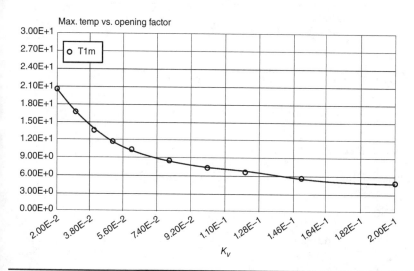

FIGURE 6.2   Maximum temperatures versus opening factor.

$$\theta_{max1} = 31.67 - 701.2K_v + 7{,}924.8K_v^2 - 41{,}170K_v^3 + 78{,}660K_v^4 \quad (6.55)$$

Case 2: Fast Fire

Model $T2m = \theta_{max2} = a0 + a1*Kv + a2*Kv^2 + a3*Kv^3 + a4*Kv^4$

Variable	Value
a0	14.78789
a1	−219.6985
a2	2281.948
a3	−1.157E+04
a4	2.196E+04

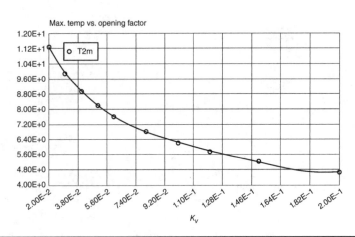

Figure 6.3 Maximum temperatures versus opening factor.

$$\theta_{\max 2} = 14.79 - 219.7K_v + 2,282K_v^2 - 11,570K_v^3 + 21,960K_v^4 \quad (6.56)$$

Case 3: Medium Fire

Model $T3m = \theta_{\max 3} = a0 + a1*Kv + a2*Kv^2 + a3*Kv^3 + a4*Kv^4$

Variable	Value
a0	6.177982
a1	−43.9206
a2	366.5795
a3	−1685.105
a4	3027.828

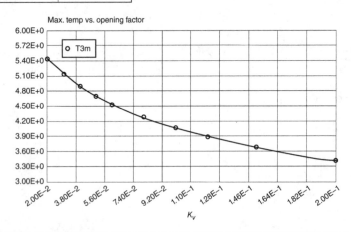

Figure 6.4   Maximum temperatures versus Opening factor.

$$\theta_{\max 3} = 6.178 - 43.92K_v + 366.6K_v^2 - 1685.1K_v^3 + 3027.8K_v^4 \quad (6.57)$$

Case 4: Slow Fire

Model   T4m= θ_{max4} = a0 + a1*Kv + a2*Kv^2 + a3*Kv^3 + a4*Kv^4

Variable	Value
a0	4.460537
a1	−21.92555
a2	173.2939
a3	−807.203
a4	1487.832

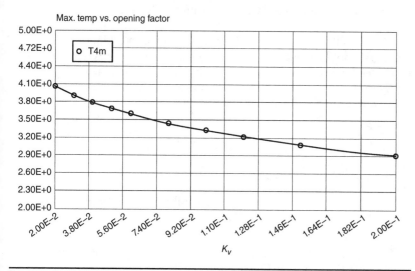

FIGURE 6.5   Maximum temperatures versus opening factor.

$$\theta_{max4} = 4.461 - 21.92K_v + 173.3K_v^2 - 807.2K_v^3 + 1487.8K_v^4 \quad (6.58)$$

6.4   Time-Equivalence Method

The fire severity could be related to the fire load of a room and expressed as an area under the temperature-time curve. The time-equivalence concept makes use of the fire load (see Table 6.1) to produce a value that would be "equivalent" to the exposure time in the standard test [4]. The *t-equivalence* has been defined as the exposure time in the standard fire-resistance test that gives the same heating effect on a structure as a given real-life compartment fire. Thus any fire temperature-time history could be compared with the standard curve. The severities of two fires are equal if the areas under the temperature-time curves are equal (above a baseline of 300°C). In order to achieve this goal, let's once again use the optimal control method: obtain the solutions to Eqs. (5.3) and (5.4) and the time τ_e (effective dimensionless time) if the payoff functional (the dimensionless area under the dimensionless temperature-time curve) is

$$\int_0^{\tau_e} \theta(\tau)d\tau = AR \tag{6.59}$$

where AR is the area under the temperature-time curve (above a baseline of 300°C) from the standard test for a given fire rating exposure time.

This is called a *fixed-endpoint, free-time problem* in optimal control method theory. First, let's rewrite the standard fire exposure curve data [11] in terms of dimensionless temperature and time (see Table 6.46).

The corresponding best-to-fit curve for the dimensionless temperature-time standard fire exposure is

$$\theta_{st} = 4.12 + 7.5\log(102\tau + 1) \tag{6.60}$$

Second, let's now calculate the effective time τ_e:

1. For each fire severity case (see Table 5.2) and a given fire rating (see Table 6.1), read the number AR (Table 6.46, last column).

2. Substitute AR into Eq. (6.59), and find the solution of Eqs. (5.3) and (5.4) with the functional Eq. (6.59).

3. The corresponding dimensionless time (upper limit of the integral) of the real-life compartment fire (see Table 6.47) transfers into the real duration time (see Table 6.47).

Now the time interval $(0 < \tau < \tau_e)$ of the SFL application is defined completely. Obviously, the time-equivalence method works both ways: If for some reason the structural engineer or the owner has defined the time interval of the SFL application, then the structural elements should have a corresponding fire rating. In some cases, the abnormal fire might spread from one floor to the next, and the compromised building system will be subjected to the SFL for much longer period of time; therefore, for all practical purposes, the time in Eqs. (5.3) and (5.4) should be assumed to be $\tau \to \infty$. The same assumption must be made if the whole compromised building structure is checked for global stability or excessive deformations owing to a decrease in stiffness with the temperature rise (or cool-off).

Temp., °F	Temp., °C	Temp., K	Time t, hours	Dimensionless time τ	Dimensionless temp. θ_{st}	Dimensionless area AR
1,000	538	811	0.083	0.00305	3.5	—
1,300	704	977	0.167	0.00615	6.28	0.01534
1,550	843	1,116	.5	0.0184	8.6	0.11616
1,700	927	1,200	1.0	0.0368	10.0	0.28359
1,850	1,010	1,283	2.0	0.0736	11.38	0.68780
2,000	1,093	1,366	4.0	0.1472	12.77	1.56709
2,300	1,260	1,533	8.0	0.2944	15.55	3.68035

Table 6.46 Dimensionless Standard Fire Exposure Curve Data

233

Category	Fire Severity, ½ hour	Fire Severity, 1 hour	Fire Severity, 2 hours	Fire Severity, 3 hours
Very fast	1.04	1.77	4.89	8.15
Fast	1.17	2.23	6.20	10.87
Medium	1.27	2.6	7.09	12.23
Slow	1.82	4.35	11.68	14.95

TABLE 6.47   Effective Time t_e

6.5   Comparison of Parametric and Structural Fire-Load Curves

The natural fire models take into account the main parameters that influence the growth and development of fires. Natural fires depend substantially on fire loads, openings, and thermal properties of the surrounding structure. The gas temperature in the compartment can be determined with parametric temperature-time curves. These curves consider the ventilation by an opening factor and the design value of the fire-load density [7] (see Chap. 2). The comparison will be limited here to the Eurocode model only.

Let's rewrite the Eq. (2.32) in a different form for computer-aided calculations:

$$T_t = 20 + 1325C$$

$$C = (1 - 0.324e^{-0.2t^*} - 0.204e^{-1.7t^*} - 0.472e^{-19t^*}) \qquad (6.61)$$

where

$$t^* = t(\Gamma)$$

$$\Gamma = \frac{(F_v/0.04)^2}{(b/1,160)^2} \qquad (6.62)$$

The duration of the fire is determined by the fire load:

$$t^* = 0.20(10^{-3})q_{t,d}\Gamma/F_v \qquad (6.63)$$

where

$$q_{t,d} = q_{f,d}A_f/A_t \qquad (6.64)$$

and $50 \leq q_{t,d} \leq 1,000$, $q_{t,d}$ is the design value of the fuel-load density (MJ/m²) related to the surface area A_t (m²) of the enclosure, and $q_{f,d}$ is the design value of the fuel-load density related to the surface area of the floor (MJ/m²) (see Table 5.2).

The parametric curve in this case is valid for compartments up to a floor area of 100 m² and a height of 4.5 m; therefore, the compartment size is as follows: $h = 3$ m, $B = 6$ m, and $L = 12$ m.

Data

Floor area: $A_f = 6(12) = 72$ m²
Total area: $A_t = 2(72) + 2(18) + 2(36) = 252$ m²
Ratio: $A_f/A_t = 0.286$
Boundary thermal conductivity: $b = 1,160$
Opening factor: $0.02 < F_v < 0.2$

Essential parameters in advanced SFL models are the maximum gas temperature in the compartment, the maximum duration of the fire, and the maximum temperature versus opening factor relationship. The main goal, consequently, is to compare the results in these areas. All computations are self-explanatory and based on the preceding formulas. They are presented below in tabulated form.

The following formula will be used to compare these maximum temperature data with the theoretical data (see Sec. 6.2):

$$T_{max} = \frac{RT_*^2}{E}\theta_{max} + T_* = 600\theta_{max} + 600 \qquad (6.65)$$

where dimension of T is in kelvins. Temperatures T_{max} from Table 6.50 also will be transferred into kelvins by adding 273°.

The dimensionless opening factor K_v should not be confused with the opening factor F_v, which has dimensions m$^{-1/2}$. In fact, the relationship between two of them is as follows:

$$K_v = \frac{F_v}{\sqrt{h}} \cdot \frac{A_t}{A_f} \qquad (6.66)$$

Assuming, for example, that $h = 1$ m and $A_f/A_t = 0.286$, we now have: $K_v = F_v/0.286$. Obviously, this will change if one (or both) of these parameters changes. In this respect, one might call a comparison of results "conditional." It provides the general information (maximum temperatures versus opening factors) but not an exact comparison. The range of parameters K_v was limited from 0.02 to 0.2, which corresponds [based on Eq. (6.66)] to the range of parameters F_v: $0.02 < F_v < 0.06$ (see Table 6.51). The values of parameters K_v are provided in parentheses in the second column of Table 6.51.

It can be seen from Table 6.51 that the maximum temperatures from parametric Eurocode curves are higher (between 3 and 26 percent) than the values from dimensionless analysis. The parametric curves also were compared directly with measured compartment fire-test temperatures. The data were gathered from experiments carried

F_v	0.02	0.03	0.04	0.05	0.06	0.08	0.10	0.12	0.15	0.20
Γ	0.25	0.562	1.0	1.562	2.25	4.0	6.25	9.0	14.06	25.0
Γ/F_v	12.5	18.73	25	31.24	37.5	50	62.5	75	93.73	125
C, Case 1	0.62	0.664	0.698	0.723	0.744	0.776	0.801	0.821	0.847	0.881
C, Case 2	0.587	0.627	0.659	0.684	0.7056	0.7387	0.7636	0.7842	0.8083	0.841
C, Case 3	0.5401	0.577	0.604	0.627	0.647	0.680	0.7056	0.7266	0.7518	0.7842
C, Case 4	0.4752	0.5265	0.554	0.5743	0.5912	0.6197	0.6436	0.6643	0.6899	0.7233

TABLE 6.48 Parameters F_v, Γ, Γ/F_v, and C

Category	Fuel Load, MJ/m²	$q_{t,d}$
Ultrafast	700	200
Fast	500	143
Medium	300	85.8
Slow	100	50 (minimum)

TABLE **6.49**   Fuel Load

out in the United States, France, and the United Kingdom. The conclusions were similar to the results in Table 6.51: Parametric temperature-time curves overestimate the temperatures achieved in real fire compartments.

6.6   Passive Fire Protection Design

A typical fire-resistance calculation involves estimating an expected natural-fire exposure, conducting a heat-transfer analysis to calculate the temperature-time relationship of the structural elements, and then calculating the ultimate load capacity, taking into account material degradation at high temperatures. The main objective of this section is the dimensionless analysis of a heat-transfer problem.

Traditional fire protection materials have included concrete, clay or concrete block work, and plasterboard. Until the late 1970, concrete was the most common form of fire protection for steelwork. The major disadvantages are cost, the increase in weight to the structure, and the time it takes to apply on site. Nowadays, modern lightweight sprays and boards have replaced these heavy concrete applications.

Fire protection of steel can be achieved by three methods: (1) insulating the element with spray material or board-type protection, (2) shielding, or (3) hollow sections can be filled with concrete or liquid to form a heat sink. This section deals primarily with the first method of fire protection. Passive fire protection materials insulate the structure from high temperatures, and they can be classified as *nonreactive* (e.g., boards and sprays) or *reactive* (e.g., intumescing coatings). Boards are fixed dry usually to columns. Beams are more commonly protected with spray materials. The main advantages of spray coverings are that they are cheap and they easily cover complex details. However, application is wet and may delay other work on a construction site. Fire ratings of up to 4 hours can be achieved through use of these methods. Thickness for an I-section with Hp/A (perimeter over section area) = 150 m^{-1} is on the order of 20 to 25 mm (1 inch) for 1 hour and 30 to 50 mm for 2 hours of fire resistance. Thermal conductivities range from approximately 0.03 to 0.05 W/m K. Since there is a large

F_v	0.02	0.03	0.04	0.05	0.06	0.08	0.10	0.12	0.15	0.20
Ultrafast t^*	0.500	0.749	1.0	1.249	1.5	2.0	2.5	3.0	3.75	5.0
Ultrafast T_{max}	841	900	945	978	1,006	1,048	1,081	1,108	1,142	1,187
Fast t^*	0.357	0.535	0.715	0.894	1.072	1.431	1.788	2.154	2.68	3.575
Fast T_{max}	798	851	893	926	955	999	1,032	1,059	1,091	1,134
Medium t^*	0.215	0.322	0.429	0.535	0.643	0.858	1.072	1.286	1.607	2.154
Medium T_{max}	736	784	820	851	877	921	955	983	1,016	1,059
Slow t^*	0.125	0.188	0.249	0.312	0.375	0.500	0.625	0.751	0.037	1.25
Slow T_{max}	650	718	754	781	803	841	873	900	934	978

TABLE 6.50　Maximum Temperatures Results

238

F_v	Case 1, T_{max} (K_l) (theoretical)	Case 1, T_{max} (parametric)	Case 2, T_{max} (theoretical)	Case 2, T_{max} (parametric)	Case 3, T_{max} (theoretical)	Case 3, T_{max} (parametric)	Case 4, T_{max} (theoretical)	Case 4, T_{max} (parametric)
0.02	1,151 (0.07)	1,114	1034	1,071	865	1,009	810	923
0.03	1,051 (0.10)	1,173	974	1,124	844	1,057	800	931
0.04	966 (0.14)	1,218	930	1,166	830	1,093	790	1,027
0.05	888 (0.175)	1,251	898	1,199	810	1,124	780	1,054
0.06	893 (0.20)	1,279	880	1,228	805	1,150	776	1,076

Table 6.51 Maximum Temperature Comparison

239

variety of spray materials and passive fire protection solutions, these numbers are used in this chapter as examples only. Shown below are the dimensionless parameters that must be changed if the thickness of insulation is larger (or smaller) than 50 mm or the thermal diffusivity parameter is different from that assumed here. Spray coatings provide protection against hydrocarbon-based fires, and some of these systems can provide up to a 4-hour fire rating. Spray-applied fireproofing materials typically are cement-based products or gypsum with a lightweight aggregate (e.g., vermiculite, perlite, or expanded polystyrene beads) that has some type of cellulosic or glass-fiber reinforcement. Spray-applied fireproofing typically is one of the more inexpensive means to protect structural elements. Thicknesses required to achieve various ratings may be found on a generic basis in some publications [5], but typically, they are provided by the manufacturer. Test methods exist to assess the adhesion and cohesion characteristics of the material. However, additional research is needed to better understand their performance, including deformation, brittleness, adhesion, and so on at higher temperatures. Let's note here again that the physical parameters, such as, for example, thermal diffusivity of insulating material and the air-gas mixture, should be taken at the maximum temperature for a given fire-severity case. Spray-applied products typically are used more to protect beams than columns.

Thermal diffusivity indicates how rapidly the material changes temperature. Lightweight materials tend to have high diffusivities because they change temperature quickly and their temperature responds rapidly to that of their surroundings. Conversely, heavyweight materials have low diffusivities, which means that they have a slow response to the surrounding temperatures. Since all the analyses here are presented in a dimensionless form, only the ratio between the thermal diffusivity of the insulating material and the air-gas mixture is involved (see below).

Let's now consider the one-dimensional heat flow, which is governed by the following partial differential equation:

$$\frac{\partial T}{\partial t} = a_1 \frac{\partial^2 T}{\partial x^2} \tag{6.67}$$

where T is the temperature, t is the time in seconds, and a_1 is the thermal diffusivity in square meters per second, given by $\lambda/\rho c_p$. In this treatment, the thermal conductivity λ (in W/m K), the density ρ (in kg/m³), and the heat capacity c_p (in kJ/kg K) are all considered to be constant. Boundary conditions are defined by specifying a flow rate or temperature at each end of the one-dimensional domain. The solution to the heat-flow problem is the internal temperature history that satisfies the field equation (6.67), the initial conditions, and the prescribed boundary conditions.

Most numerical methods convert the continuous partial differential or integral equations of heat transfer to a set of linear differential equations. The complex partial differential equation is transformed into a system of simultaneous first-order differential equations, one at each node. The system of nodal equations then is solved by step-by-step integration over the time domain. The *method of lines* is a general technique for solving partial differential equations (PDEs) typically by using finite-difference relationships for the spatial derivatives and ordinary differential equations for the time derivative. William E. Schiesser [12] at Lehigh University has been a major proponent of the numerical method of lines (NMOL).

Consider that a slab of insulating material with a thickness t_{ins} is subjected to the air-gas compartment temperature on the exposed side. This slab is shown in Fig. 6.6. For a numerical problem solution, the slab is divided into N sections with $N + 1$ node points.

The Polymath software does not accept any Greek letters as variables; therefore, as done previously, the dimensionless temperature and time are renamed as follows: T1 = θ_1, T2 = θ_2, T11 = θ_{11} are the dimensionless temperature functions at each slab section, and $t = \tau$ is the independent dimensionless time variable.

If the exposed surface is held at a temperature of $\theta(\tau)$, the fire compartment's temperature from Chap. 5 (see Cases 1 through 4), then the boundary condition at node 1 is as follows:

$$T1 = \theta_1 = A \, \exp[-(\tau - a)^2 / 2\sigma^2] \qquad (6.68)$$

where parameters A, a, and σ are defined by corresponding formulas from Chap. 5. The other boundary condition is that the boundary at node $N + 1$ has minimum temperature value. Thus,

$$\frac{\partial T_{N+1}}{\partial x} = 0 \qquad \text{for } t > 0 \qquad (6.69)$$

The boundary condition (6.68) (not to be confused with the term *thermal boundary condition*) is presented by the dimensionless function

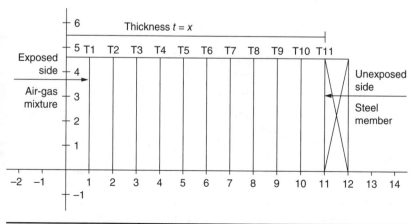

FIGURE 6.6 Non-steady-state heat conduction in a one-dimensional slab.

$\theta(\tau)$ (from Chap. 5); therefore, the original Eq. (6.67) has to be presented by the same set of dimensionless scale factors for temperature and time, that is,

$$T = \beta T_* \theta + T_* \qquad (6.70)$$

and

$$t = \frac{h^2}{a_2} \tau \qquad (6.71)$$

$$x = h\zeta$$

where a_2 is the thermal diffusivity of the air-gas mixture, h is the height of the compartment, $\beta = 0.1$, and $T_* = 600$ K is the base temperature. This, substituted into Eq. (6.67), yields

$$\frac{\partial \theta}{\partial \tau} = \alpha_o \frac{\partial^2 \theta}{\partial \zeta^2} \qquad (6.72)$$

where $\alpha_o = a_1 / a_2$ is the dimensionless thermal diffusivity parameter.

Boundary condition (6.68) is different (parameters A, a, and σ are different) for each fire-severity case: very fast, fast, medium, and slow. It is assumed here that the thermal diffusivity parameter a_1 is constant (e.g., for vermiculite insulation material, $a_1 = 6.77 \ 10^{-8}$ m^2/s, the thermal conductivity $\lambda = 0.13$ W/m K at elevated temperature, and $c_p = 525$ J/kg K per data from ISO 22007-2 [13]), and the air thermal diffusivity parameter a_2 changes with the maximum temperature [14], which is defined as follows:

$$a_2 = 9.1018 \times 10^{-11} \ T^2 + 8.8197 \times 10^{-8} \ T - 1.0654 \times 10^{-5} \qquad (6.73)$$

The advantages of vermiculite material are as follows:

- Vermiculite has reduced thermal conductivity.
- It is light in weight.
- It possesses improved workability.
- It is an excellent fire-resistance material.
- It has improved adhesion properties.
- It has increased resistance to cracking and shrinkage.
- It is easy to install or apply.

The thermal conductivity of vermiculite increases with temperature, but after reaching a temperature in the range of about 1,050 to 1,200°C, the value decreases again. The solutions to Eq. (6.67) are very sensitive to the thermal diffusivity values of the insulation material; therefore, the finite-difference equations below should be rerun if the input data differ substantially from the given data. Table 6.52 lists the data for some insulating materials that are used in our analyses.

Material ID	Property Name	Unit	Value	Multiplier	Unit	Value	Remarks
Steel	Density	lb/ft^3	480	16.02	kg/m^3	7690	
Steel	Specific heat	Btu/lb-F	0.172	4.19	kJ/kg K	0.721	
Steel	Thermal conductivity	Btu/ft h°F	20.0	1.73	W/m K	34.6	
Steel	Thermal diffusivity (10^{-6}) a_1	ft^2/s	242	0.02577	m^2/s	6.244	
Steel	Dimensionless thermal diffusivity α_o	None	0.00963		None	0.00963	
SFRM1	Density	lb/ft^3	19.6	16.02	kg/m^3	314.0	
SFRM1	Specific heat	Btu/lb°F	0.304	4.19	kJ/kg K	1.274	
SFRM1	Thermal conductivity	BTU/ft h°F	0.058	1.73	W/m K	0.10	
SFRM1	Thermal diffusivity (10^{-6}) a_1	ft^2/s	9.734	0.02577	m^2/s	0.2508	
SFRM1	Dimensionless thermal diffusivity α_o	None	0.0003867		None	0.0003867	
SFRM2	Density	lb/ft^3	16.9	16.02	kg/m^3	270.74	
SFRM2	Specific heat	Btu/lb°F	0.29	4.19	kJ/kg K	1.2151	
SFRM2	Thermal conductivity	Btu/ft h°F	0.069	1.73	W/m K	0.1194	
SFRM2	Thermal diffusivity (10^{-6}) a_1	ft^2/s	14.078	0.02577	m^2/s	0.3628	
SFRM2	Dimensionless thermal diffusivity α_o	None	0.000560		None	0.000560	
Ceramic blanket	Density	lb/ft^3	10	16.02	kg/m^3	160.2	

TABLE 6.52 Thermal Diffusivity Parameters

Material ID	Property Name	Unit	Value	Multiplier	Unit	Value	Remarks
Ceramic blanket	Specific heat	BTU/lb°F	0.234	4.19	kJ/kg K	0.98046	
Ceramic blanket	Thermal conductivity	Btu/ft h°F	0.0745	1.73	W/m K	0.1289	
Ceramic blanket	Thermal diffusivity (10^{-6}) a_1	ft²/s	31.84	0.02577	m²/s	0.82056	
Ceramic blanket	Dimensionless thermal diffusivity α_o	None	0.001265		None	0.001265	
Lightweight concrete	Density	lb/ft³	110	16.02	kg/m³	1762.2	
Lightweight concrete	Specific heat	Btu/lb°F	0.3	4.19	kJ/kg K	1.257	
Lightweight concrete	Thermal conductivity	Btu/ft h°F	0.3	1.73	W/m K	0.519	
Lightweight concrete	Thermal diffusivity (10^{-6}) a_1	ft²/s	9.0921	0.02577	m²/s	0.2343	
Lightweight concrete	Dimensionless thermal diffusivity α_o	None	0.000361		None	0.000361	
Air-gas mixture	Thermal diffusivity (10^{-4}) a_2	ft²/s	251.65	0.02577	m²/s	6.485	

TABLE 6.52 Thermal Diffusivity Parameters (Continued)

With $N = 10$ sections of length $\Delta\zeta = t_{ins}/10$ h = 50 mm/10(1,000) × 3 = 0.00167 and dimensionless thermal diffusivity $\alpha_o = 0.000262$ (vermiculite material), Eq. (6.66) can be rewritten using a central difference formula for the second derivative as

$$\frac{\partial\theta}{\partial\tau} = \frac{\alpha}{(\Delta\zeta)^2}(\theta_{n+1} - 2\theta_n + \theta_{n-1}) \qquad (6.74)$$

The boundary condition represented by Eq. (6.69) can be written using a second-order backward finite difference as

$$\frac{\partial\theta_{11}}{\partial\tau} = \frac{3\theta_{11} - 4\theta_{10} + \theta_9}{2(\Delta\zeta)} = 0 \qquad (6.75)$$

which can be solved for θ_{11} to yield

$$T11 = \theta_{11} = \frac{4\theta_{10} - \theta_9}{3} \qquad (6.76)$$

The problem then requires the solution of Eqs. (6.74), (6.75), and (6.76), which results in 9 simultaneous ordinary differential equations and 2 explicit algebraic equations for the 11 temperatures at the various nodes. The initial condition for each of the θ's is assumed to be equal to 1, and the independent dimensionless variable time τ varies from 0 to 0.2 [the same as it does in energy conservation Eqs. (5.3) and (5.4)]. The outputs of the dimensionless temperature-time functions for each severity case are presented below. The transients in temperature show an approach to non-steady state.

Case 1: 1,022 K < T_{max} < 1,305 K, Ultrafast Fire

Data $T_* = 600$ K; $\delta = 20$; $K_v = 0.05$; $\beta = 0.1$; $P = 0.157$; $a_2 = 2.58 \times 10^{-4}$; $0 < \tau < 0.2$; $\alpha = 0.000262$

	Variable	Initial Value	Minimal Value	Maximal Value	Final Value
1	deltax	0.00167	0.00167	0.00167	0.00167
2	t	0	0	0.2	0.2
3	T1	5.895865	5.386139	11.97944	5.386139
4	T10	1	1	8.576736	8.576736
5	T11	1	0.9978919	8.458245	8.458245
6	T2	1	1	10.75095	7.268025
7	T3	1	1	10.3175	8.406891
8	T4	1	1	10.24391	8.937297
9	T5	1	1	10.24391	8.937297
10	T6	1	1	9.333907	9.091417
11	T7	1	1	9.021585	9.020262
12	T8	1	1	8.932206	8.932206
13	T9	1	1	8.932206	8.932206

TABLE 6.53 Calculated Values of DEQ Variables

Differential Equations

1. $d(T2)/d(t) = 0.000262/deltax^2*(T3-2*T2+T1)$
2. $d(T3)/d(t) = 0.000262/deltax^2*(T4-2*T3+T2)$
3. $d(T4)/d(t) = 0.000262/deltax^2*(T5-2*T4+T3)$
4. $d(T5)/d(t) = 0.000262/deltax^2*(T5-2*T4+T3)$
5. $d(T6)/d(t) = 0.000262/deltax^2*(T7-2*T6+T5)$
6. $d(T7)/d(t) = 0.000262/deltax^2*(T8-2*T7+T6)$
7. $d(T8)/d(t) = 0.000262/deltax^2*(T9-2*T8+T7)$
8. $d(T9)/d(t) = 0.000262/deltax^2*(T9-2*T8+T7)$
9. $d(T10)/d(t) = 0.000262/deltax^2*(T11-2*T10+T9)$

$$(6.77)$$

Explicit Equations

1. $T1 = 11.98*exp\ (-(t-0.097)^2/(2*0.0576)^2)$
2. $T11 = (4*T10-T9)/3$
3. $deltax = .00167$

$$(6.78)$$

The output data of dimensionless temperatures are presented in Table 6A.1 in the appendix at the end of this chapter.

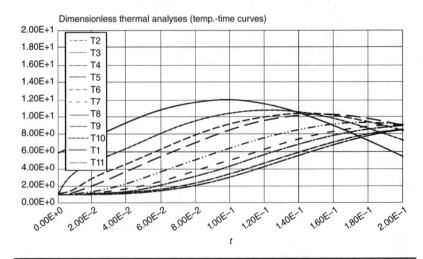

FIGURE 6.7 Dimensionless thermal analysis.

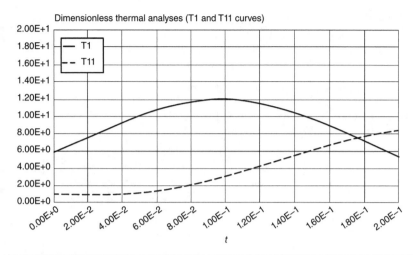

FIGURE 6.8 Dimensionless thermal analysis.

Model $T11 = a0 + a1*t + a2*t^2 + a3*t^3 + a4*t^4$

Variable	Value
a0	1.121439
a1	−14.90028
a2	280.2576
a3	1549.569
a4	−8349.566

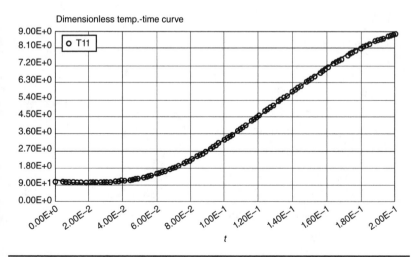

FIGURE 6.9 Dimensionless temperature-time curve.

Finally,

$$\theta_{11}(\tau) = 1.12 - 14.9\tau + 280.26\tau^2 + 1,549.6\tau^3 - 8,349.6\tau^4 \qquad (6.79)$$

Case 2: 882 K < T_{max} < 1,022 K, Fast Fire

Data $T_* = 600$ K; $\delta = 20$; $K_v = 0.05$; $\beta = 0.1$; $P = 0.157$; $a = 1.75 \times 10^{-4}$; $0 < \tau < 0.2$; $\alpha = 0.000387$

	Variable	Initial Value	Minimal Value	Maximal Value	Final Value
1	deltax	0.00167	0.00167	0.00167	0.00167
2	t	0	0	0.2	0.2
3	T1	3.40004	0.3005524	6.949824	0.3005524
4	T10	1	1	5.156834	4.034095
5	T11	1	0.9990051	5.164526	4.169509
6	T2	1	1	6.244502	1.033238
7	T3	1	1	5.989849	1.572796
8	T4	1	1	5.94678	1.857763
9	T5	1	1	5.94678	1.857763
10	T6	1	1	5.43254	2.745466
11	T7	1	1	5.253614	3.33503
12	T8	1	1	5.22322	3.627851
13	T9	1	1	5.22322	3.627851

TABLE 6.54   Calculated Values of DEQ Variables

Differential Equations

1. d(T2)/d(t) = 0.000387/deltax^2*(T3–2*T2+T1)
2. d(T3)/d(t) = 0.000387/deltax^2*(T4–2*T3+T2)
3. d(T4)/d(t) = 0.000387/deltax^2*(T5–2*T4+T3)
4. d(T5)/d(t) = 0.000387/deltax^2*(T5–2*T4+T3)
5. d(T6)/d(t) = 0.000387/deltax^2*(T7–2*T6+T5)
6. d(T7)/d(t) = 0.000387/deltax^2*(T8–2*T7+T6)
7. d(T8)/d(t) = 0.000387/deltax^2*(T9–2*T8+T7)
8. d(T9)/d(t) = 0.000387/deltax^2*(T9–2*T8+T7)
9. d(T10)/d(t) = 0.000387/deltax^2*(T11–2*T10+T9)

$$(6.80)$$

Explicit Equations

1. $T1 = 6.95{*}\exp\left(-(t-0.0646)^{\wedge}2/(2{*}0.0382)^{\wedge}2\right)$
2. $T11 = (4{*}T10-T9)/3$
3. $deltax = .00167$

(6.81)

The output data of dimensionless temperatures are presented in Table 6A.2 in the appendix at the end of this chapter.

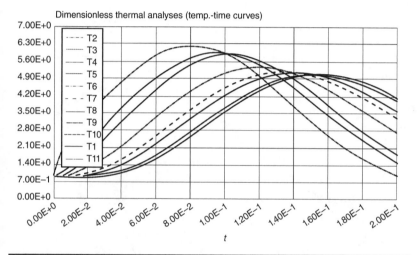

FIGURE 6.10 Dimensionless thermal analysis.

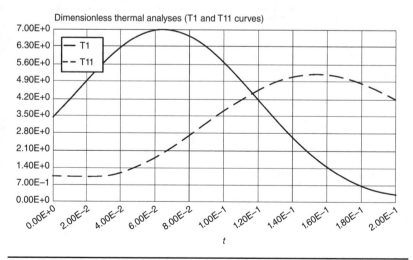

FIGURE 6.11 Dimensionless thermal analysis.

Model   T11 = a0 + a1*t + a2*t^2 + a3*t^3 + a4*t^4

Variable	Value
a0	1.262454
a1	−37.96157
a2	1105.821
a3	−5434.228
a4	6007.412

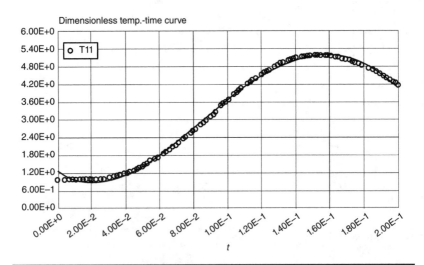

FIGURE 6.12   Dimensionless temperature-time curve.

Finally,

$$\theta_{11}(\tau) = 1.262 - 37.96\tau + 1{,}105.82\tau^2 - 5{,}434.23\tau^3 + 6{,}007.41\tau^4 \quad (6.82)$$

Case 3: 822 K < T_{max} < 882 K, Medium Fire

Data $T_* = 600$ K; $\delta = 20$; $K_v = 0.05$; $\beta = 0.1$; $P = 0.157$; $a = 1.38 \times 10^{-4}$; $0 < \tau < 0.2$; $\alpha = 0.000491$; $A = 4.55(1.219) = 5.55$ [increase (21.9 percent) owing to hydrodynamic (convection) effect (see Chap. 5)]; $a = 0.082$; $\sigma = 0.0598$

	Variable	Initial Value	Minimal Value	Maximal Value	Final Value
1	deltax	0.00167	0.00167	0.00167	0.00167
2	t	0	0	0.2	0.2
3	T1	3.540032	2.034908	5.549723	2.034908
4	T10	1	1	4.949241	4.391255
5	T11	1	0.9989602	4.951652	4.457427
6	T2	1	1	5.328492	2.637744
7	T3	1	1	5.257793	3.040648
8	T4	1	1	5.246016	3.242075
9	T5	1	1	5.246016	3.242075
10	T6	1	1	5.046699	3.731351
11	T7	1	1	4.982609	4.042221
12	T8	1	1	4.972566	4.192739
13	T9	1	1	4.972566	4.192739

TABLE **6.55** Calculated Values of DEQ Variables

Differential Equations

1. $d(T2)/d(t) = 0.000491/deltax^2*(T3-2*T2+T1)$
2. $d(T3)/d(t) = 0.000491/deltax^2*(T4-2*T3+T2)$
3. $d(T4)/d(t) = 0.000491/deltax^2*(T5-2*T4+T3)$
4. $d(T5)/d(t) = 0.000491/deltax^2*(T5-2*T4+T3)$
5. $d(T6)/d(t) = 0.000491/deltax^2*(T7-2*T6+T5)$
6. $d(T7)/d(t) = 0.000491/deltax^2*(T8-2*T7+T6)$
7. $d(T8)/d(t) = 0.000491/deltax^2*(T9-2*T8+T7)$
8. $d(T9)/d(t) = 0.000491/deltax^2*(T9-2*T8+T7)$
9. $d(T10)/d(t) = 0.000491/deltax^2*(T11-2*T10+T9)$

(6.83)

Explicit Equations

1. $T1 = 5.55*exp(-(t-0.0802)^2/(2*0.0598)^2)$
2. $T11 = (4*T10-T9)/3$
3. $deltax = .00167$

(6.84)

The output data of dimensionless temperatures are presented in Table 6A.3 in the appendix at the end of this chapter.

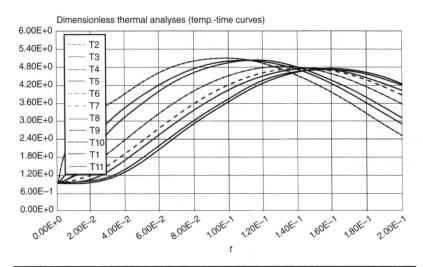

Figure 6.13 Dimensionless thermal analysis.

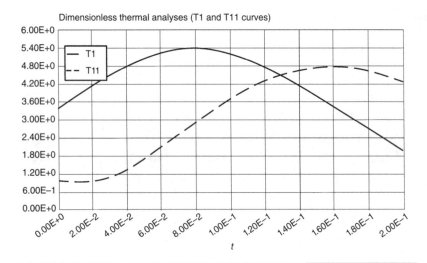

Figure 6.14 Dimensionless thermal analysis.

Model   T11 = a0 + a1*t + a2*t^2 + a3*t^3 + a4*t^4

Variable	Value
a0	1.097808
a1	–23.75761
a2	1062.501
a3	–6674.374
a4	1.189E+04

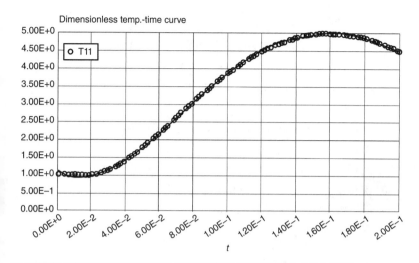

FIGURE 6.15   Dimensionless temperature-time curve.

Finally,

$$\theta_{11}(\tau) = 1.098 - 23.76\tau + 1{,}062.5\tau^2 - 6{,}674.4\tau^3 + 11{,}890\tau^4 \quad (6.85)$$

Case 4: 715 K < T_{max} < 822 K, Slow Fire

Data $T_* = 600$ K; $\delta = 20$; $K_v = 0.05$; $\beta = 0.1$; $P = 0.157$; $a = 1.24 \times 10^{-4}$; $0 < \tau < 0.2$; $\alpha = 0.000545$; $A = 3.73(1.321) = 4.92$ [increase (32.1 percent) owing to hydrodynamic (convection) effect (see Chap. 5)]; $a = 0.0893$; $\sigma = 0.0750$

	Variable	Initial Value	Minimal Value	Maximal Value	Final Value
1	deltax	0.00167	0.00167	0.00167	0.00167
2	t	0	0	0.2	0.2
3	T1	3.451769	2.853837	4.919993	2.853837
4	T10	1	1	4.603544	4.29427
5	T11	1	0.9990645	4.604955	4.331733
6	T2	1	1	4.80693	3.245411
7	T3	1	1	4.771624	3.501067
8	T4	1	1	4.766185	3.62714
9	T5	1	1	4.766185	3.62714
10	T6	1	1	4.655726	3.914792
11	T7	1	1	4.621554	4.095199
12	T8	1	1	4.616112	4.18188
13	T9	1	1	4.616112	4.18188

TABLE 6.56   Calculated Values of DEQ Variables

Differential Equations

1. $d(T2)/d(t) = 0.000545/deltax^2*(T3–2*T2+T1)$
2. $d(T3)/d(t) = 0.000545/deltax^2*(T4–2*T3+T2)$
3. $d(T4)/d(t) = 0.000545/deltax^2*(T5–2*T4+T3)$
4. $d(T5)/d(t) = 0.000545/deltax^2*(T5–2*T4+T3)$
5. $d(T6)/d(t) = 0.000545/deltax^2*(T7–2*T6+T5)$
6. $d(T7)/d(t) = 0.000545/deltax^2*(T8–2*T7+T6)$
7. $d(T8)/d(t) = 0.000545/deltax^2*(T9–2*T8+T7)$
8. $d(T9)/d(t) = 0.000545/deltax^2*(T9–2*T8+T7)$
9. $d(T10)/d(t) = 0.000545/deltax^2*(T11–2*T10+T9)$

$$(6.86)$$

Explicit Equations

1. $T1 = 4.92*exp\,(–(t–0.0893)^2/(2*0.0750)^2)$
2. $T11 = (4*T10–T9)/3$
3. $deltax = .00167$

$$(6.87)$$

The output data of dimensionless temperatures are presented in Table 6A.4 in the appendix at the end of this chapter.

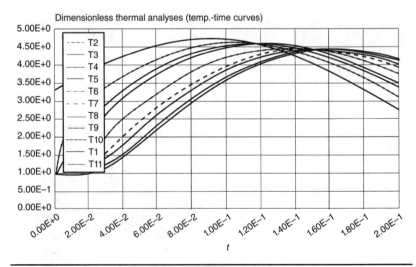

FIGURE 6.16   Dimensionless thermal analysis.

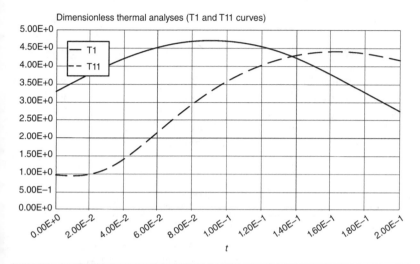

FIGURE 6.17   Dimensionless thermal analysis.

Model   T11 = a0 + a1*t + a2*t^2 + a3*t^3 + a4*t^4

Variable	Value
a0	1.008753
a1	−13.6981
a2	926.3878
a3	−6438.727
a4	1.286E+04

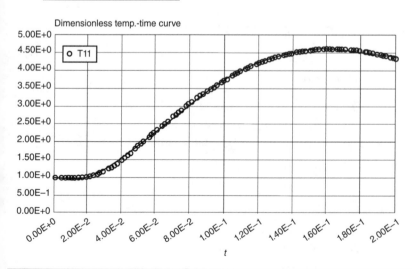

FIGURE 6.18   Dimensionless temperature-time curve.

Finally,

$$\theta_{11}(\tau) = 1.008 - 13.70\tau + 926.39\tau^2 - 6,438.73\tau^3 + 12,860\tau^4 \quad (6.88)$$

The reduction in maximum temperature is due to the quality of the insulation material. If the material is different (e.g., lightweight concrete with a specific heat $c_c = 837$ J/kg K, thermal conductivity $k_c = 0.61$ W/m K, and density $\rho_c = 1,760$ kg/m³, data at ambient temperature from ref. [15]), then the reduction in temperature is much smaller (see Case 1 below).

	Variable	Initial Value	Minimal Value	Maximal Value	Final Value
1	deltax	0.00167	0.00167	0.00167	0.00167
2	t	0	0	0.22	0.22
3	T1	5.895865	3.831428	11.97936	3.831428
4	T10	1	1	11.15332	8.382543
5	T11	1	0.998433	11.15678	8.540396
6	T2	1	1	11.68321	4.863079
7	T3	1	1	11.58777	5.561306
8	T4	1	1	11.57259	5.913252
9	T5	1	1	11.57259	5.913252
10	T6	1	1	11.29175	6.918435
11	T7	1	1	11.20169	7.580637
12	T8	1	1	11.18652	7.908982
13	T9	1	1	11.18652	7.908982

TABLE 6.57   Calculated Values of DEQ Variables

Differential Equations

1. $d(T2)/d(t) = 0.000638/\text{deltax}^2*(T3-2*T2+T1)$
2. $d(T3)/d(t) = 0.000638/\text{deltax}^2*(T4-2*T3+T2)$
3. $d(T4)/d(t) = 0.000638/\text{deltax}^2*(T5-2*T4+T3)$
4. $d(T5)/d(t) = 0.000638/\text{deltax}^2*(T5-2*T4+T3)$
5. $d(T6)/d(t) = 0.000638/\text{deltax}^2*(T7-2*T6+T5)$
6. $d(T7)/d(t) = 0.000638/\text{deltax}^2*(T8-2*T7+T6)$
7. $d(T8)/d(t) = 0.000638/\text{deltax}^2*(T9-2*T8+T7)$
8. $d(T9)/d(t) = 0.000638/\text{deltax}^2*(T9-2*T8+T7)$
9. $d(T10)/d(t) = 0.000638/\text{deltax}^2*(T11-2*T10+T9)$

$$(6.89)$$

Explicit Equations

1. $T1 = (4.12 + 7.5*(\log(102*t+1)))*0 + 1*11.98*\exp(-(t-0.097)^2/(2*0.0576)^2)$

2. $T11 = (4*T10 - T9)/3$

3. $\text{deltax} = .00167$

$$(6.90)$$

The output data of dimensionless temperatures are presented in Table 6A.5 in the appendix at the end of this chapter.

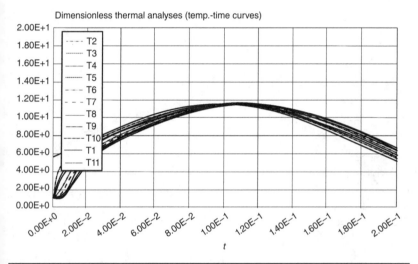

FIGURE 6.19 Dimensionless thermal analysis.

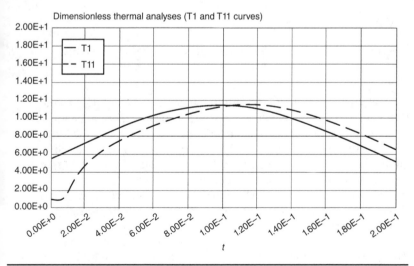

FIGURE 6.20 Dimensionless thermal analysis.

Model   T11 = a0+a1*t+a2*t^2+a3*t^3+a4*t^4+a5*t^5

Variable	Value
a0	1.099552
a1	−45.96392
a2	3000.926
a3	−2.464E+04
a4	7.876E+04
a5	−9.684E+04

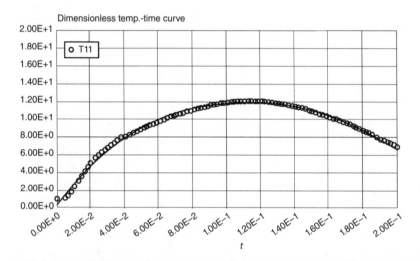

FIGURE 6.21   Dimensionless temperature-time curve.

Finally,

$$\theta_{11}(\tau) = 1.1 - 45.96\tau + 3{,}001\tau^2 - 24{,}640\tau^3 + 78{,}760\tau^4 - 96{,}840\tau^5$$

(6.91)

Let's now analyze the sensitivity of the thermal analysis to the thermal properties of insulating materials, such as thermal conductivity and specific heat (see Table 6.52), and to investigate the influence of different types of insulating materials used in fire protection engineering and their impact on the temperature-time curves of structural steel elements and systems during different fire exposures.

Building codes have been the most accepted solution to structural fire design. The combination of the test methods, listings, and prescriptive fire-resistance requirements of the building codes have resulted in very satisfactory overall fire performance of buildings. The goal of this work is not to alter this prescriptive-based system but rather to provide a partial basis for a complementary performance-based system for the provision of structural fire protection. The design approaches in the prescriptive methods are different from those in the performance-based methods, and they have different requirements and acceptance criteria. For example, the fire protection (fire rating) of a steel column has the main requirement that the average temperature of an unexposed side at any cross section does not exceed 1,000°F (538°C) [2,3]. Fire test standards impose an additional temperature limit of 1,200°F (649°C) at any one location along the member [4,5]. This 1,200°F (538°C) temperature is often referred to as the *critical temperature*, and typically it represents the temperature where a 50 percent strength loss occurs. These temperature limits can be used as the basis for a heat-transfer analysis, and they can represent the failure criteria for a test of an insulated column. The failure criteria obviously are completely different when a performance-based method is used: Ultimate strength capacity of a column must be greater than the ultimate design load, including the temperature-load effect and the stiffness reduction. In this case, the stiffness-reduction effect could be different for a short or slender column, an axially or eccentrically loaded column, a wind or gravity column, a pin or fixed-end column, a fully or partially loaded column, and so on—the whole range of purely structural engineering questions that should be answered before the ultimate strength capacity can be computed. Obviously, the single temperature limitation of 1,000°F (538°C) cannot fit all the different structural conditions just mentioned. However, this does not mean that one method is wrong and the other is correct. It means that these methods have different requirements and approaches to the same problem. Nevertheless, some of the properties of the "new" performance-based design method should be verified by comparisons with the existing prescriptive-based method. In this case, the fire rating of a column (the number of hours that fire exposure does not cause the average temperature to rise above 1,000°F during the standard fire test procedure) can be computed and compared with the heat-transfer analysis of an insulated steel column with the aid of the finite-differences method. There will always need to be a relationship between the standard fire test and performance-based real fire analysis because of the wealth of component knowledge available to designers from years of use of the test.

Proper fire safety design requires the appropriate selection of design fires against which the performance of a building is evaluated. Selection of the design fire directly affects all aspects of fire safety performance, including the structural fire. By using performance-based design methods, real fire effects are addressed based on credible design scenarios. This design method ensures that real fire behavior and its effect on structures are addressed and also provides a useful link back to the familiar benchmark of the standard fire test. Performance-based design methods sometimes permit increased design freedom over prescriptive code restrictions while maintaining overall fire safety requirements. Appropriate and cost-effective fire safety measures are derived. Practically, this can mean that the smaller thickness of an insulated material can be specified owing to the reduced fire resistance required. The typical fuel load is considered, and therefore, the temperatures from an agreed design fire rather than the standard furnace test can be used. The ventilation provision is calculated along with the volume of the compartment. This method is now part of the European building codes. Other factors also can be determined to take into account the consequence of an uncontrolled fire, such as the probability of fire occurrence and the benefits of sprinklers.

Performance-based design of structures for fire resistance requires the collaboration of fire protection engineers and structural engineers. Fire protection engineers typically analyze fire environments and the transfer of heat from the fire to elements of a structure. Structural engineers use the results of heat-transfer analysis to determine the structural response, considering thermally induced strains and the effects of changes in material properties at elevated temperatures. Typically, this collaboration is most effective when fire resistance is considered as early as the conceptual design stage, when flexibility in structural concepts allows the fire resistance considerations to influence the structural design.

Designing structural fire resistance on a performance basis generally includes the following steps:

1. *Predict the fire severity.* This is generally accomplished by computer modeling of the fire (zone method or parametric curves method, see Chaps. 2 and 3; FDS mathematical fire model, see Chap. 4; or approximate method, see Chap. 5).

2. *Conduct heat-transfer analyses.* Determine the thermal response of the structure (see Chap. 6). Computer fire modeling provides temperature boundary conditions for heat-transfer analyses.

3. *Evaluate the structural response.* This involves evaluation of the structural system response owing to structural design load combinations including the temperature-time load from the heat-transfer analyses (see Chap. 7).

This chapter provides general guidance on the approaches to and practical aspects of implementing a fire-resistant design (heat-transfer analyses only) for conventional applications. The boundary condition on the exposed side of an insulated material (temperature-time function) is taken from an approximate solution of conservation of energy, mass, and momentum equations (see Chap. 5), the standard fire test (temperature-time relationship), or the first method from the SFPE standard (2011).

Any live or environmental load (e.g., wind, snow, rain, etc.) traditionally is applied to structural elements as a uniformly distributed load, and this is done because the complexities involved in nonuniformly distributed application of loads to all structural members of a fire compartment would make structural analyses inefficient. The same principle should be applied to the structural elements subjected to the temperature-time load from the heat-transfer analyses; therefore, one-dimensional thermal analysis should be sufficient in most practical cases.

Dimensionless Eq. (6.72) allows us to analyze the whole range of insulating materials from steel to lightweight concrete and make a comparison of insulating capabilities based on one dimensionless thermal diffusivity parameter α_o. The solutions to Eq. (6.72) provide information about the fire rating of any particular material and the dimensionless temperature-time function T11 = θ_{11} (this is the most important part for further structural engineering analyses) that should be applied to structural members (or the structural system as a whole) as part of structural design load combinations [14]. These dimensionless functions will be obtained for each fire severity case (very fast, fast, medium, and slow fire scenarios), for the standard fire test temperature-time curve (transferred to dimensionless form), and for the first method from the SFPE standard (see Chap. 2). The dimensionless thermal diffusivity parameter α_o that represents the quality of different insulating materials will range from $\alpha_o = 0.000361$ for lightweight concrete (LWC) to $\alpha_o = 0.00963$ for steel. The results of computer analyses are presented and summarized graphically (see Figs. 6.22 through 6.28).

Case 1: 1,022 K < T_{max} < 1,305 K, Very Fast Fire.

Data $T_* = 600$ K; $K_v = 0.05$; $\beta = 0.1$; $0 < \tau < 0.22$; $h = 3.0$ m; $A = 11.98$; $a = 0.097$; $\sigma = 0.0576$

$\alpha_o = 0.000361$

Differential Equations

1. $d(T2)/d(t) = 0.000361/deltax^2*(T3–2*T2+T1)$
2. $d(T3)/d(t) = 0.000361/deltax^2*(T4–2*T3+T2)$
3. $d(T4)/d(t) = 0.000361/deltax^2*(T5–2*T4+T3)$
4. $d(T5)/d(t) = 0.000361/deltax^2*(T5–2*T4+T3)$
5. $d(T6)/d(t) = 0.000361/deltax^2*(T7–2*T6+T5)$
6. $d(T7)/d(t) = 0.000361/deltax^2*(T8–2*T7+T6)$
7. $d(T8)/d(t) = 0.000361/deltax^2*(T9–2*T8+T7)$
8. $d(T9)/d(t) = 0.000361/deltax^2*(T9–2*T8+T7)$
9. $d(T10)/d(t) = 0.000361/deltax^2*(T11–2*T10+T9)$

Explicit Equations

1. T1 = (4.12+7.5*(log(102*t+1)))*0+1*11.98*exp (–(t–0.097)^2/ (2*0.0576)^2)+15*0
2. T11 = (4*T10–T9)/3
3. deltax = .00167

Model   T11 = a0+a1*t+a2*t^2+a3*t^3+a4*t^4

Variable	Value
a0	1.287124
a1	–42.84519
a2	1258.746
a3	–4942.936
a4	4017.7

$$\theta 11 = 1.287–42.84*t+1258.7*(t^2) – 4942.9*(t^3)+4017.7*(t^4) \qquad (6.92)$$

$\alpha_o = 0.000387$

Model   T11 = a0+a1*t+a2*t^2+a3*t^3+a4*t^4

Variable	Value
a0	1.285156
a1	–45.86397
a2	1450.358
a3	–6378.498
a4	6880.415

$$\theta 11 = 1.285 – 45.86t+1450t^2 – 6379t^3+6880t^4 \qquad (6.93)$$

$\alpha_o = 0.000387$

Model   T11 = a0+a1*t+a2*t^2+a3*t^3+a4*t^4

Variable	Value
a0	1.065406
a1	–35.66941
a2	2128.95
a3	–1.3E+04
a4	2.185E+04

$$\theta 11 = 1.065–35.67t+2129t^2–13000t^3+21850t^4 \qquad (6.94)$$

$\alpha_o = 0.00126$

Model   T11 = a0+a1*t+a2*t^2+a3*t^3+a4*t^4

Variable	Value
a0	−0.0352148
a1	144.6155
a2	172.3332
a3	−6452.435
a4	1.492E+04

$$\theta 11 = -0.0352+144.6t+172.3t^2-6452t^3+14920t^4 \qquad (6.95)$$

$\alpha_o = 0.00963$

Model   T11 = a0+a1*t+a2*t^2+a3*t^3+a4*t^4

Variable	Value
a0	3.886105
a1	167.2151
a2	−837.7094
a3	−911.9151
a4	5783.558

$$\theta 11 = 3.886+167.2t-837.7t^2-911.9t^3+5783.6t^4 \qquad (6.96)$$

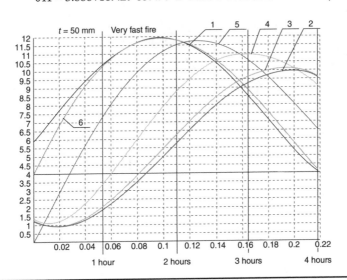

FIGURE 6.22   Temperature-time curves on unexposed side of insulating materials. 1 = temperature-time curve: very fast fire; 2 = LWC (lightweight concrete): α_o = 0.000361; 3 = SFRM1: α_o = 0.000387; 4 = SFRM2: α_o = 0.000560; 5 = ceramic blanket: α_o = 0.00126; 6 = steel: α_o = 0.00963.

Case 2: 882 K < T_{max} < 1,022 K, Fast Fire

Data $T_* = 600$ K; $K_v = 0.05$; $\beta = 0.1$; $0 < \tau < 0.22$; $h = 3.0$ m; $A = 6.95$; $a = 0.0646$; $\sigma = 0.0382$

$\alpha_o = 0.00963$

Model   T11 = a0+a1*t+a2*t^2+a3*t^3+a4*t^4

Variable	Value
a0	1.729763
a1	181.3106
a2	−1878.034
a3	4638.65
a4	145.5052

$$\theta 11 = 1.730+181.3t-1878t^2+4638.6t^3+145.5t^4 \tag{6.97}$$

Model   T1 = a0+a1*t+a2*t^2+a3*t^3+a4*t^4

Variable	Value
a0	2.790581
a1	148.2606
a2	−1559.157
a3	3213.739
a4	2710.243

$$\theta 1 = 2.79+148.26t-1559t^2+3213t^3+2710t^4 \tag{6.98}$$

$\alpha_o = 0.001265$

Model   T11 = a0+a1*t+a2*t^2+a3*t^3+a4*t^4

Variable	Value
a0	0.3260568
a1	74.68204
a2	754.9484
a3	−1.203E+04
a4	3.238E+04

$$\theta 11 = 0.326+74.68t+755t^2-12030t^3+32380t^4 \tag{6.99}$$

$\alpha_o = 0.000560$

Model   T11 = a0+a1*t+a2*t^2+a3*t^3+a4*t^4

Variable	Value
a0	1.279913
a1	−52.55424
a2	2092.03
a3	−1.468E+04
a4	2.886E+04

$$\theta 11 = 1.28-52.55t+2092t^2-14680t^3+28860t^4 \qquad (6.100)$$

$\alpha_o = 0.000387$

Model   T11 = a0+a1*t+a2*t^2+a3*t^3+a4*t^4

Variable	Value
a0	1.344398
a1	−46.93795
a2	1326.483
a3	−7284.365
a4	1.093E+04

$$\theta 11 = 1.344-46.94t+1326t^2-7284t^3+10930t^4 \qquad (6.101)$$

$\alpha_o = 0.000361$

Model   T11 = a0+a1*t+a2*t^2+a3*t^3+a4*t^4

Variable	Value
a0	1.314784
a1	−41.93564
a2	1133.073
a3	−5769.948
a4	7657.043

$$\theta 11 = 1.315-41.94t+1133t^2-5770t^3+7657t^4 \qquad (6.102)$$

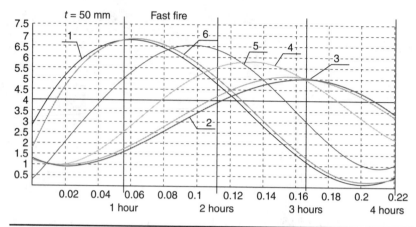

FIGURE 6.23 Temperature-time curves on unexposed side of insulating materials.
1 = temperature-time curve: fast fire; 2 = LWC (lightweight concrete): $\alpha_o = 0.000361$;
3 = SFRM1: $\alpha_o = 0.000387$; 4 = SFRM2: $\alpha_o = 0.000560$; 5 = ceramic blanket: $\alpha_o = 0.00126$; 6 = steel: $\alpha_o = 0.00963$.

Case 3: 822 K< T_{max} < 882 K, Medium Fire

Data $T_* = 600$ K; $K_v = 0.05$; $\beta = 0.1$; $0 < \tau < 0.22$; $h = 3.0$ m; $A = 5.55$;
$a = 0.082$; $\sigma = 0.0598$.

$\alpha_o = 0.000361$

Model   T1 = a0+a1*t+a2*t^2+a3*t^3+a4*t^4

Variable	Value
a0	3.489564
a1	44.89825
a2	−121.2822
a3	−2066.824
a4	6824.361

$$\theta1 = 3.49+44.90t-121.28t^2-2067t^3+6824t^4 \qquad (6.103)$$

Model   T11 = a0+a1*t+a2*t^2+a3*t^3+a4*t^4

Variable	Value
a0	1.150003
a1	−23.01009
a2	699.2014
a3	−3335.397
a4	4196.365

$$\theta11 = 1.15-23t+699t^2-3335t^3+4196t^4 \qquad (6.104)$$

$\alpha_o = 0.000387$

Model T11 = a0+a1*t+a2*t^2+a3*t^3+a4*t^4

Variable	Value
a0	1.147483
a1	−24.23644
a2	790.9997
a3	−4073.303
a4	5755.977

$$\theta 11 = 1.147{-}24.24t{+}791t^2{-}4073t^3{+}5756t^4 \qquad (6.105)$$

$\alpha_o = 0.000560$

Model T11 = a0+a1*t+a2*t^2+a3*t^3+a4*t^4

Variable	Value
a0	1.005138
a1	−15.56684
a2	1059.889
a3	−7148.904
a4	1.325E+04

$$\theta 11 = 1.0{-}15.57t{+}1060t^2{-}7149t^3{+}13250t^4 \qquad (6.106)$$

$\alpha_o = 0.001265$

Model T11 = a0+a1*t+a2*t^2+a3*t^3+a4*t^4

Variable	Value
a0	0.3999889
a1	84.40776
a2	−203.7752
a3	−1982.904
a4	6189.187

$$\theta 11 = 0.4{+}84.4t{-}203.77t^2{-}1982.9t^3{+}6189t^4 \qquad (6.107)$$

$\alpha_o = 0.00963$

Model   T11 = a0+a1*t+a2*t^2+a3*t^3+a4*t^4

Variable	Value
a0	2.464071
a1	89.18897
a2	−762.6042
a3	1759.079
a4	−1045.094

$$\theta11 = 2.464+89.19-762.6t^2+1759t^3-1045t^4 \qquad (6.108)$$

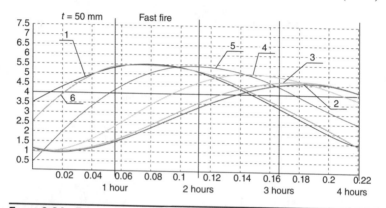

Figure 6.24   Temperature-time curves on unexposed side of insulating materials. 1 = temperature-time curve: medium fire; 2 = LWC (lightweight concrete): $\alpha_o = 0.000361$; 3 = SFRM1: $\alpha_o = 0.000387$; 4 = SFRM2: $\alpha_o = 0.000560$; 5 = ceramic blanket: $\alpha_o = 0.00126$; 6 = steel: $\alpha_o = 0.00963$.

Case 4: 715 K < T_{max} < 822 K, Slow Fire

Data $T_* = 600$ K; $K_v = 0.05$; $\beta = 0.1$; $0 < \tau < 0.2$; $h = 3.0$ m; $A = 3.73$; $a = 0.0893$; $\sigma = 0.0750$

$\alpha_o = 0.000361$

Model   T1 = a0+a1*t+a2*t^2+a3*t^3+a4*t^4

Variable	Value
a0	2.610917
a1	21.26155
a2	−35.71044
a3	−934.5785
a4	2624.658

$$\theta1 = 2.61+21.26t-35.71t^2-934.6t^3+2624.6t^4 \qquad (6.109)$$

Model T11 = a0+a1*t+a2*t^2+a3*t^3+a4*t^4

Variable	Value
a0	1.078639
a1	−12.77657
a2	404.2378
a3	−1951.279
a4	2616.182

$$\theta 11 = 1.08 - 12.77t + 404.2t^2 - 1951.3t^3 + 2616t^4 \qquad (6.110)$$

$\alpha_o = \mathbf{0.000387}$

Model T11 = a0+a1*t+a2*t^2+a3*t^3+a4*t^4

Variable	Value
a0	1.074352
a1	−13.1964
a2	452.3959
a3	−2345.768
a4	3458.22

$$\theta 11 = 1.07 - 13.2t + 452.4t^2 - 2345.8t^3 + 3458.2t^4 \qquad (6.111)$$

$\alpha_o = \mathbf{0.000560}$

Model T11 = a0+a1*t+a2*t^2+a3*t^3+a4*t^4

Variable	Value
a0	0.9714077
a1	−5.770396
a2	560.76
a3	−3788.002
a4	7066.753

$$\theta 11 = 0.97 - 5.77t + 560.8t^2 - 3788t^3 + 7066.7t^4 \qquad (6.112)$$

$\alpha_o = 0.001265$

Model   T11 = a0+a1*t+a2*t^2+a3*t^3+a4*t^4

Variable	Value
a0	0.6118335
a1	56.17791
a2	−274.6358
a3	87.28526
a4	767.3997

$$\theta11 = 0.612+56.18t-274.6t^2+87.28t^3+767.4t^4 \qquad (6.113)$$

$\alpha_o = 0.00963$

Model   T11 = a0+a1*t+a2*t^2+a3*t^3+a4*t^4

Variable	Value
a0	1.961021
a1	50.76648
a2	−476.9701
a3	1714.559
a4	−2796.058

$$\theta11 = 1.96+50.77t-477t^2+1714.6t^3-2796t^4 \qquad (6.114)$$

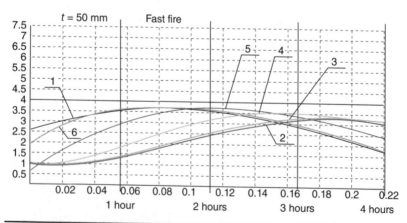

FIGURE 6.25 Temperature-time curves on unexposed side of insulating materials. 1 = temperature-time curve: slow fire; 2 = LWC (lightweight concrete): α_o = 0.000361; 3 = SFRM1: α_o = 0.000387; 4 = SFRM2: α_o = 0.000560; 5 = ceramic blanket: α_o = 0.00126; 6 = steel: α_o = 0.00963.

Case 5: Standard Temperature-Time Curve from 11

$\alpha_o = 0.000361$

Model   T1 = a0+a1*t+a2*t^2+a3*t^3+a4*t^4

Variable	Value
a0	4.791537
a1	166.732
a2	−1566.867
a3	7550.473
a4	−1.357E+04

$$\theta1 = 4.79+167.0t-1567t^2+7550t^3-13570t^4 \qquad (6.115)$$

Model   T11 = a0+a1*t+a2*t^2+a3*t^3+a4*t^4

Variable	Value
a0	1.316742
a1	−47.19386
a2	1373.4
a3	−6325.138
a4	9225.814

$$\theta11 = 1.317-47.2t+1373t^2-6325t^3+9226t^4 \qquad (6.116)$$

$\alpha_o = 0.000387$

Model   T11 = a0+a1*t+a2*t^2+a3*t^3+a4*t^4

Variable	Value
a0	1.312023
a1	−49.99155
a2	1558.515
a3	−7747.927
a4	1.228E+04

$$\theta11 = 1.31-50t+1558t^2-7748t^3+12280t^4 \qquad (6.117)$$

$\alpha_o = \mathbf{0.00056}$

Model $T11 = a0+a1*t+a2*t\wedge2+a3*t\wedge3+a4*t\wedge4$

Variable	Value
a0	1.016799
a1	−33.72727
a2	2092.89
a3	−1.344E+04
a4	2.624E+04

$$\theta11 = 1.02-33.7t+2093t\wedge2-13440t\wedge3+26240t\wedge4 \qquad (6.118)$$

$\alpha_o = \mathbf{0.001265}$

Model $T11 = a0+a1*t+a2*t\wedge2+a3*t\wedge3+a4*t\wedge4$

Variable	Value
a0	−0.459196
a1	179.3144
a2	−658.9157
a3	−551.6951
a4	5521.045

$$\theta11 = -0.46+179t-659t\wedge2-551t\wedge3+5521t\wedge4 \qquad (6.119)$$

$\alpha_o = \mathbf{0.00963}$

Model $T11 = a0+a1*t+a2*t\wedge2+a3*t\wedge3+a4*t\wedge4$

Variable	Value
a0	2.93415
a1	240.3297
a2	−2608.86
a3	1.349E+04
a4	−2.527E+04

$$\theta11 = 2.934+240.3t-2609t\wedge2+13490t\wedge3-25270t\wedge4 \qquad (6.120)$$

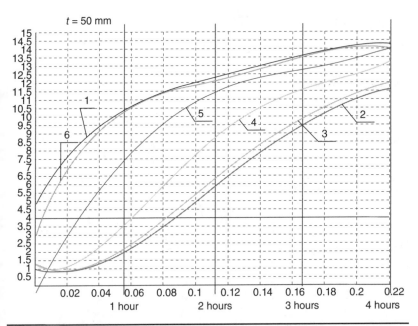

FIGURE 6.26 Temperature-time curves on unexposed side of insulating materials.
1 = temperature-time curve: standard temperature-time curve; 2 = LWC (lightweight concrete): α_o = 0.000361; 3 = SFRM1: α_o = 0.000387; 4 = SFRM2: α_o = 0.000560; 5 = ceramic blanket: α_o = 0.00126; 6 = steel: α_o = 0.00963.

Case 6: SFPE Parametric Temperature-Time Curve $T = 1200°C$ [15]

$\alpha_o = 0.000361$

Model $T11 = a0+a1*t+a2*t^2+a3*t^3+a4*t^4$

Variable	Value
a0	1.372095
a1	−76.28298
a2	2842.297
a3	−1.7E+04
a4	3.131E+04

$$\theta11 = 1.372–76.28t+2842t^2–17000t^3+31310t^4 \qquad (6.121)$$

$\alpha_o = 0.000387$

Model T11 = a0+a1*t+a2*t^2+a3*t^3+a4*t^4

Variable	Value
a0	1.273991
a1	−71.67631
a2	3024.854
a3	−1.896E+04
a4	3.617E+04

$$\theta11 = 1.274–71.68t+3025t^2–18960t^3+36170t^4 \qquad (6.122)$$

$\alpha_o = 0.00056$

Model T11 = a0+a1*t+a2*t^2+a3*t^3+a4*t^4

Variable	Value
a0	0.1634749
a1	32.30412
a2	2551.548
a3	−2.12E+04
a4	4.721E+04

$$\theta11 = 0.1635+32.3t+2551.6t^2–21200t^3+47210t^4 \qquad (6.123)$$

$\alpha_o = 0.001265$

Model T11 = a0+a1*t+a2*t^2+a3*t^3+a4*t^4

Variable	Value
a0	−2.240143
a1	516.19
a2	−5489.403
a3	2.479E+04
a4	−4.04E+04

$$\theta11 = -2.24+516.2t–5489.4t^2+24790t^3–40400t^4 \qquad (6.124)$$

$\alpha_o = 0.00963$

Model $T11 = a0+a1*t+a2*t^2+a3*t^3+a4*t^4$

Variable	Value
a0	9.871585
a1	267.5125
a2	−4150.495
a3	2.473E+04
a4	−4.991E+04

$$\theta11 = 9.87+267.5t-4150t^2+24730t^3-49910t^4 \qquad (6.125)$$

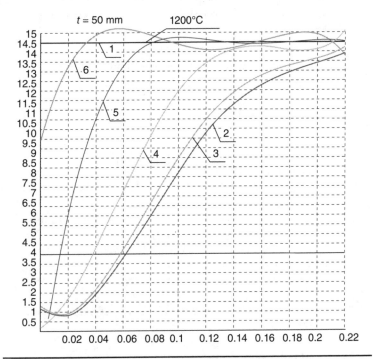

FIGURE 6.27 Temperature-time curves on unexposed side of insulating materials. 1 = temperature-time curve: SFPE temperature-time curve; 2 = LWC (lightweight concrete): $\alpha_o = 0.000361$; 3 = SFRM1: $\alpha_o = 0.000387$; 4 = SFRM2: $\alpha_o = 0.000560$; 5 = ceramic blanket: $\alpha_o = 0.00126$; 6 = steel: $\alpha_o = 0.00963$.

Finally, the following examples illustrate the step-by-step procedures of practical computations of column fire rating, time-equivalence method (Law-Pettersson method; see Chap. 2), comparison with the

Eurocode parametric method, SFL calculations for a three-story hotel, use of vermiculate material for insulating steel structural elements, and one investigating case of study.

Example 6.1

Let's compare the fire rating of a given insulated steel column based on the IBC equation (see below) and Eq. (6.116) or Fig. 6.26. The graphical solution based on Fig. 6.26 could be very useful during the preliminary stage of fire protection and structural design.

Data

Insulating material: Lightweight concrete; concrete density: 110 pcf.

Column width: W12 × 72 m; weight: 72 plf.

Ambient temperature specific heat of concrete: 0.2 Btu/ft °F

Ambient temperature thermal conductivity of concrete: 0.35 Btu/h ft °F.

Thickness of concrete cover: 2 inches.

Actual moisture content of concrete by volume: 4 percent.

Fire endurance rating at the actual moisture condition (minutes): R.

IBC Item 720.5.1.4 lists the equation for fire endurance at zero moisture as

$$R_o = 10\left(\frac{W}{D}\right)^{0.7} + 17\left(\frac{h^{1.6}}{k_c^{0.2}}\right)\left\{1 + 26\left[\frac{H}{\rho_c(c_c)h(L+h)}\right]^{0.8}\right\} \qquad (6.126)$$

where R_o = fire endurance rating at zero moisture content (min.)
W = weight of steel column (lb/ft)
D = inside perimeter of the fire protection (in.)
h = thickness of concrete cover (in.)
k_c = ambient temperature thermal conductivity of concrete (Btu/h ft °F)
H = ambient temperature thermal capacity of the steel column = 0.11 W (Btu/ft °F)
ρ_c = concrete density (pcf)
c_c = ambient temperature specific heat of concrete
L = Interior dimension of one side of a square concrete box protection (in.)

$$L = \frac{12 + 12.25}{2} = 12.125 \text{ in.}$$

$$h = 2 \text{ in}$$

$$R = R_o(1 + 0.03 \text{ m})$$

After substituting data into Eq. (6.125), we have

$$R_o = 10\left(\frac{72}{48.5}\right)^{0.7} + 17\left(\frac{2^{1.6}}{0.35^{0.2}}\right)\left\{1 + 26\left[\frac{7.92}{110(0.2)2(12.125 + 2)}\right]^{0.8}\right\}$$

$$= 93.1 \text{ minutes}$$

Critical dimensionless temperature for the column (unexposed side at any cross section does not exceed 1,000°F): $\theta 11_{col.cr} = (540 - 300)/60 = 4$.

From Fig. 6.26 and Eq. (6.116), we have for the standard temperature-time curve:

$$\theta 11 = 4 = 1.317 - 47.2t + 1373t^2 - 6325t^3 + 9226t^4$$

$t = \tau = 0.851$ and $R_o = 0.0851(4)/0.22 = 1.547 \text{ hours} =$
92.8 minutes   *Okay.*

The intersection of this horizontal line ($\theta 11 = 4$) with the corresponding curve of $\theta 11$ gives the dimensionless time (fire rating) for a given material. This gives you an approximate graphical solution for the fire-rating calculations.

Example 6.2
Let's compare Law's and Pettersson's time-equivalence concepts [see Chap. 2, Eqs. (2.5) and (2.6)] based on data from real (natural) fire compartment experiments with the theoretical data from the preceding.

Data

Fire compartment dimensions: $H = 3$ m, height; $B = 6$ m; $L = 12$ m
Total area: $A_t = 72(2) + 2(18) + 2(36) = 252 \text{ m}^2$
Floor area: $A_f = 6(12) = 72 \text{ m}^2$
Openings area: $A_v = 0.05(72) = 3.6 \text{ m}^2$
Height of each opening: $h = 1$ m
Fire load: $L = 100 \text{ kg/m}^2$ medium fire

$$\tau_e = \frac{L}{\sqrt{A_v(A_t - A_v)}} = \frac{(100)}{\sqrt{(3.6)(252 - 3.6)}} = 3.35 \text{ hours} \qquad \text{by Law}$$

$$\tau_e = \frac{1.21L}{\sqrt{A_t A_v \sqrt{h}}} = \frac{1.21(100)}{\sqrt{252(3.6)1}} = 4.0 \text{ hours} \qquad \text{by Pettersson}$$

If the fire load $L = 200 \text{ kg/m}^2$, very fast fire, then the time-equivalence time $\tau_e = 7.7$ hours (by Law) and $\tau_e = 8.0$ hours (by Pettersson).

Finally, from Ingberg's fuel-load fire severity relationship (see Table 2.2 in Chap. 2):

Fire load $L = 100$ kg/m^2, medium fire: $\tau_e = 2.0$ hours $<< 4.0$ hours

Fire load $L = 200$ kg/m^2, very fast fire: $\tau_e = 4.0$ hours $<< 8.0$ hours

From Fig. 6.28, based on Law's concept of time equivalence, we have

$\tau_e = 0.097$ hour (graphical solution)   medium fire

$\tau_e = 0.097(4)/0.22 = 1.76$ hours   medium fire

$\tau_e = 0.18$ hour (graphical solution)   very fast fire

$\tau_e = 0.18(4)/0.22 = 3.27$ hours   very fast fire

As was stated in Chap. 2, the Law-Pettersson method provides very approximate quantitative results and reasonably acceptable qualitative results with respect to real fire behavior. At the same time, this example had shown that the theoretical approach provides good quantitative and qualitative results regarding real fire duration and the dependency of fire severity category (see Fig. 6.28).

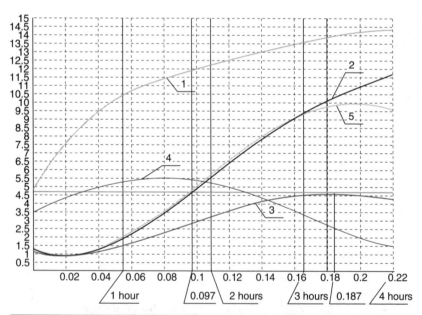

Figure 6.28 Temperature-time curves on exposed and unexposed sides of LWC material. 1 = temperature-time curve: standard temperature-time curve, exposed side; 2 = LWC (lightweight concrete): $\alpha_o = 0.000361$, standard test curve, unexposed side; 3 = LWC (lightweight concrete): $\alpha_o = 0.000361$, medium fire, unexposed side; 4 = LWC (lightweight concrete): $\alpha_o = 0.000361$, medium fire, exposed side; 5 = LWC (lightweight concrete): $\alpha_o = 0.000361$, very fast fire, unexposed side.

It is important also to underline here that the standard temperature-time curve (unexposed side, curve 2 in Fig. 6.28) is very close to the very fast fire temperature-time curve (unexposed side, curve 5 in Fig. 6.28). Therefore, the standard fire test (ASTM E 119) provides very conservative results and is not always economically justifiable.

Example 6.3: Comparison with Eurocode Parametric Method

Fire Compartment Data

$$H = 3 \text{ m}; B = 6 \text{ m}; L = 12 \text{ m}$$
$$A_f = (6)(12) = 72 \text{ m}^2; A_t = 2(72) + 2(3)(12 + 6) = 252 \text{ m}^2$$
Opening factor (assume $h = 1.0$ m): $F_v = 0.02$
Fuel load: 700 MJ/m^2—very fast fire

The step-by-step comparison procedure is as follows:

1. Determine the opening area (based on Eurocode):
$$A_v = 0.02(252) = 5.04 \text{ m}^2$$

2. Determine the opening factor (based on dimensionless analysis):
$$K_v = 5.04/72 = 0.07$$

3. Use the look-up method (Eurocode): From Table 6.50,
$$T_{max} = 841°C \quad \text{or} \quad T_{max} = 841 + 273 = 1{,}114°K$$
$$t^*_{max} = 0.5 \text{ hour}$$

4. Use the look-up method (dimensionless analysis): From Table 6.44,
$$\theta_{max} = 10.446 \quad (K_v = 0.06) \quad \text{and} \quad \theta_{max} = 8.592 \quad (K_v = 0.08)$$
Therefore,
$$\theta_{max} = (10.446 + 8.592)/2 = 9.519 \quad (K_v = 0.07 \text{ by linear interpolation})$$
Finally [see Eq. (6.65)],
$$T_{max} = 60(9.519) + 600 = 1{,}171 \text{ K} \quad (5 \text{ percent difference})$$

5. Use the computational method (Eurocode):
   a. Calculate the following parameters [see Eqs. (6.61) through (6.64)]:
$$\Gamma = \frac{(F_v/0.04)^2}{(b/1{,}160)^2} = \frac{(0.02/0.4)^2}{1} = 0.25 \quad b = 1{,}160$$

$$\Gamma/F_v = 0.25/0.02 = 12.5$$

$$q_{t,d} = q_{f,d}A_f/A_t = 700(72)/252 = 200$$

$$t^* = 0.20(10^{-3})q_{t,d}\Gamma/F_v = 0.20(10^{-3})(200)12.5 = 0.5 \text{ hour}$$

$$C = (1 - 0.324e^{-0.2(0.5)} - 0.204e^{-1.7(0.5)} - 0.472e^{-19(0.5)}) = 0.62$$

b. Finally,

$$T_t = 20 + 1{,}325C = 20 + 1{,}325(0.62) = 841°C + 273 = 1{,}114 \text{ K}$$

6. Use the computational method [dimensionless analysis; see Eq. (6.55)]:

   a. $\theta_{\max 1} = 31.67 - 701.2(0.07) + 7{,}924.8(0.07)^2 - 41{,}170(0.07)^3 + 78{,}660(0.07)^4 = 9.1848$

   b. Finally [see Eq. (6.65)]:

$$T_{\max} = 60(9.1848) + 600 = 1{,}151 \text{ K} \quad \text{(3 percent difference)}$$

   c. From Table 5.3, find the value of τ that gives the maximum value of $\theta_{\max 1}$: $\tau = 0.0554$.

   d. Find t^* based on Eq. (6.71):

$$\tau = \frac{a_2}{h^2}(3{,}600)t^*$$

$$0.0554 = \frac{2.58(10^{-4})}{9}t^*$$

$$t^* = 0.537(\text{hour}) \approx 0.5(\text{hour}) \quad \text{(7.4 percent difference)}$$

Note Thermal diffusivity $a_2 = 2.58 \times 10^{-4}$ is obtained from Eq. (6.73).

In order to simplify the computations the fire compartment size and geometric dimensions will remain the same for all the following examples.

Example 6.4

Preliminary building design (for a three-story hotel, construction type IIA) had shown that owing to mechanical and architectural requirements, the minimum opening area of the fire compartment is 7.2 m². The design fire load is 30 kg/m². Obtain approximate data for the following items:

Maximum gas temperature in the compartment owing to the fire

The duration of the fire t^*

Final expression for the temperature-time curve (SFL)

1. Opening factor K_v = 7.2/72 = 0.1, and the fuel load FL = 30(16.7) = 501 MJ/m².

   From Table 5.2, define the fire severity case: Case 1, ultrafast.

2. From Table 6.6, ($K_v = 0.1$), read the values of maximum dimensionless temperature. *Note:* For fuel load $FL = 500$ MJ/m², read the value of y0 but not y:

$$\theta_{max} = y0 = 5.29$$

3. Calculate the real maximum temperature based on Eq. (6.65):

$$T_{max} = 60(5.29) + 600 = 917.4 \text{ K}$$

4. From Table 5.3, find the value of τ that gives the maximum value of θ_{max1}: $\tau = 0.0442$.

5. Find t^* based on Eq. (6.71):

$$\tau = \frac{a_2}{h^2} (3,600)t^*$$

$$0.0422 = \frac{2.58 \times 10^{-4} (3,600)}{9} t^*$$

$$t^* = 0.409 \text{ (hour)}$$

6. Modify Eq. (5.5) as follows:

$$\theta = A \exp[-(\tau/a - 1)^2 / 2(\sigma/a)^2]$$

In this case, $A = 5.29$, $\sigma/a = 0.0576/0.097 = 0.6$, and $\tau_m = a = 0.0422$—dimensionless time at the maximum value of temperature.

7. Finally: SFL is as follows:

$$T = 60(5.29)\exp\left(-\frac{\left(\frac{t}{0.409} - 1\right)^2}{2(0.36)}\right) + 600$$

$$= 317.4\exp\left(-\frac{\left(\frac{t}{0.409} - 1\right)^2}{2(0.36)}\right) + 600$$

where t is time in hours and T is gas temperature in kelvins.

Example 6.5

The structural engineer has established the fact that the structural elements and the system as a whole will not require any reinforcement or modifications if the maximum temperature in the case of fire (in a given compartment) on the unexposed side of fire-protected steel elements will not exceed 400°C (fire severity case: Case 3, medium). Vermiculite material will be used for insulation of the structural steel.

It is required to estimate for the preliminary design the following parameters of a postflashover fire development:

Maximum gas temperature in the fire compartment

Vertical openings area required in this case

The duration of the estimated fire t^*

Analytical expression for the temperature-time curve

1. It has been established in this chapter (see above) that for vermiculite material, the reduction in maximum temperature is approximately 30 percent. Therefore, the maximum temperature inside the fire compartment can be estimated as follows:

$$T_{max} = 1.3(400 + 273) = 875 \text{ K}$$

2. Obtain the dimensionless temperature θ_{max} from Eq. (6.65):

$$\theta_{max} = (875 - 600)/60 = 4.58$$

3. Obtain the opening factor K_v from Eq. (6.57):

$$\theta_{max3} = 6.178 - 43.92K_v + 366.6K_v^2 - 1,685.1K_v^3 + 3,027.8K_v^4$$

However, since we are just estimating the solution, for the first approximation, let's use the graphical method. From Fig. 6.29, graphically find the corresponding value of factor K_v.

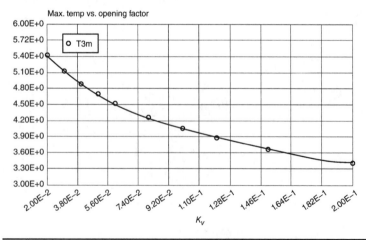

FIGURE 6.29   Maximum temperatures versus opening factor.

In this case, $K_v = 0.056$. In order to check the result, let's plug in this number back into Eq. (6.57). The result is okay.

4. The openings area required is: $A_v = 0.056(72) = 4.03 \text{ m}^2$.
5. From Table 5.5, find the value of τ that gives the maximum value of θ_{max}: $\tau = 0.0467$.
6. Find t^* based on Eq. (6.71):

$$\tau = \frac{a_2}{h^2}(3{,}600)t^*$$

$$0.0467 = \frac{1.38 \times 10^{-4} \ (3{,}600)}{9} t^*$$

$$t^* = 0.847 \ (\text{hour})$$

Note Thermal diffusivity $a_2 = 1.38 \times 10^{-4}$ is obtained from Eq. (6.73).

7. Equation (5.11) in this case is as follows:

$$\theta = A \ \exp[-(\tau/a - 1)^2/2(\sigma/a)^2]$$

In this case, $A = 4.58$, $\sigma/a = 0.0598/0.0802 = 0.7456$, and $\tau_m = a = 0.0467$—dimensionless time at the maximum value of temperature.

8. Finally, estimate SFL as follows:

$$T = 60(4.58) \ \exp\left(-\frac{\left(\frac{t}{0.847} - 1\right)^2}{2(0.556)}\right) + 600$$

$$= 274.8 \ \exp\left(-\frac{\left(\frac{t}{0.847} - 1\right)^2}{2(0.556)}\right) + 600$$

where t is the time in hours and T is the gas temperature in kelvins.

Example 6.6

During the fire investigation of a building, the following information was revealed: The thickness of the fire protection material (the vermiculite-type material) at the structural member was 40 mm instead of the 50 mm specified on the design drawings (fire severity: Case 1, very fast fire). The questions in this case are as follows:

1. What is the difference in maximum temperatures at the face of the structural steel member (actual versus design temperatures)?

2. What is the difference in the duration of fire? *Note:* The term *duration of fire* is defined here as the time when the maximum temperature occurs at the face of a structural steel element.

   a. Since the thickness of insulation material had been changed, the original system of differential [Eqs. (6.52) and (6.77)] has to be changed and rerun. (Change parameter: deltax = 40/(10)(1000)3 = 0.001333.)

   b. Below is the new computer output (t_{ins} = 40 mm).

	Variable	Initial Value	Minimal Value	Maximal Value	Final Value
1	deltax	0.00133	0.00133	0.00133	0.00133
2	t	0	0	0.2	0.2
3	T1	5.895865	5.386139	11.97982	5.386139
4	T10	1	1	10.27653	10.10623
5	T11	1	0.9979613	10.28364	10.1747
6	T2	1	1	11.34714	6.903462
7	T3	1	1	11.13702	7.886978
8	T4	1	1	11.10186	8.368901
9	T5	1	1	11.10186	8.368901
10	T6	1	1	10.55303	9.204653
11	T7	1	1	10.37297	9.684741
12	T8	1	1	10.34315	9.90081
13	T9	1	1	10.34315	9.90081

TABLE 6.58   Calculated Values of DEQ Variables

Differential Equations

1. $d(T2)/d(t) = 0.000262/deltax^2*(T3-2*T2+T1)$
2. $d(T3)/d(t) = 0.000262/deltax^2*(T4-2*T3+T2)$
3. $d(T4)/d(t) = 0.000262/deltax^2*(T5-2*T4+T3)$
4. $d(T5)/d(t) = 0.000262/deltax^2*(T5-2*T4+T3)$
5. $d(T6)/d(t) = 0.000262/deltax^2*(T7-2*T6+T5)$

6. $d(T7)/d(t) = 0.000262/deltax^2*(T8-2*T7+T6)$

7. $d(T8)/d(t) = 0.000262/deltax^2*(T9-2*T8+T7)$

8. $d(T9)/d(t) = 0.000262/deltax^2*(T9-2*T8+T7)$

9. $d(T10)/d(t) = 0.000262/deltax^2*(T11-2*T10+T9)$

$$(6.127)$$

Explicit Equations

1. $T1 = 11.98*exp(-(t-0.097)^2/(2*0.0576)^2)$

2. $T11 = (4*T10-T9)/3$

3. $deltax = .00133$

$$(6.128)$$

The output data of dimensionless temperatures are presented in Table 6A.6 in the appendix at the end of this chapter.

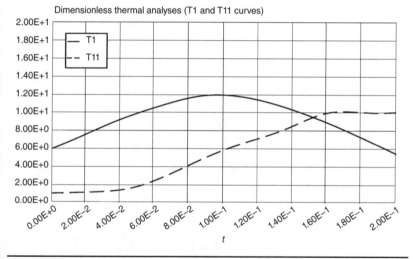

Figure 6.30   Dimensionless thermal analysis.

Model $T11 = a0+a1*t+a2*t^2+a3*t^3+a4*t^4$

Variable	Value
a0	1.249333
a1	−44.54751
a2	1542.122
a3	−7062.975
a4	7879.802

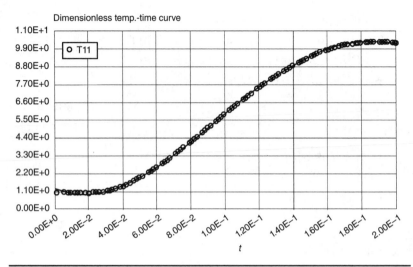

FIGURE 6.31   Dimensionless temperature-time curve.

Finally,

$$\theta_{11}(\tau) = 1.249 - 44.547\tau + 1,542\tau^2 - 7,063\tau^3 + 7,880\tau^4 \quad (6.129)$$

3. From Table 6.53, find the values of θ_{max1} and θ_{max11}:

$$\theta_{max1} = 11.98 \quad \text{and} \quad \theta_{max11} = 8.458$$

Therefore, maximum gas temperature (original design):

$$T_{max} = 60(11.98) + 600 = 1,318.8 \text{ K} = 1,045°C$$

Maximum steel temperature (original design):

$$T_{max} = 60(8.458) + 600 = 1,107.5 \text{ K} = 834.5°C$$

4. From Table 6.63, find the value of θ_{max1} and θ_{max11}:

$$\theta_{max1} = 11.98 \quad \text{and} \quad \theta_{max11} = 10.28$$

Therefore, maximum gas temperature (actual case):

$$T_{max} = 60(11.98) + 600 = 1,318.8 \text{ K} = 1,045°C$$

Maximum steel temperature (original design):

$$T_{max} = 60(10.28) + 600 = 1,217 \text{ K} = 944°C$$

Finally, the temperature difference $944 - 834.5 = 109.4°C$.

5. From Table 6.54 (original design), the dimensionless time τ value (at the maximum temperature value) is $\tau = 0.233$.

6. From Table 6.64 (actual case), the dimensionless time τ value (at the maximum temperature) is $\tau = 0.186$. The difference is $\Delta\tau = 0.233 - 0.186 = 0.047$. Finally, based on Eq. (6.71),

$$\Delta\tau = \frac{a_2}{h^2}(3,600)\Delta t$$

$$0.047 = \frac{2.58 \times 10^{-4}\,(3,600)}{9}\Delta t$$

$$\Delta t = 0.455 \text{ (hour)}$$

Note: Thermal diffusivity $a_2 = 2.58 \times 10^{-4}$ is obtained from Eq. (6.73).

Conclusions

1. Fire rating of structural elements based on the standard temperature-time curve is very conservative and is practically equivalent to the very fast fire scenario. Any other fire scenarios (fast, medium, and slow) provide larger fire ratings. This may result in smaller thicknesses of insulating materials when the performance-based design method is used.

2. The analytical formulas for temperature-time curves on the unexposed side of an insulated material are provided. This is very important for the structural engineering analyses and design that follow.

3. The real fire scenarios (very fast, fast, medium, and slow) have both stages of fire development: fire duration and decay period. In many cases in structural engineering analyses, the decay period is as important as the duration of the fire itself (e.g., in the case where the structural element or system has large inelastic deformations owing to structural fire load).

4. Dimensionless thermal analysis in this case provides solutions in graphical and simple analytical forms. These can be used for fast preliminary design purposes in the early stages of structural design, and they will allow structural and fire protection engineers to root out the less important fire scenarios.

5. The results of dimensionless thermal analysis are tied to the benchmarks of the standard fire test with respect to fire rating of structural elements and different insulating materials, and this is illustrated in Example 6.1, which shows very good correlation between the two methods with respect to fire rating of an insulated steel column.

References

1. Issen, L. A., *Single-Family Residential Fire and Live Loads Survey* (NBSIR 80-2155). National Bureau of Standards, Gaithersburg, MD, 1980.
2. Culver, C. G., *Survey Results for Fire Loads and Live Loads in Office Buildings* (NBS BSS 085). National Bureau of Standards, Gaithersburg, MD, 1976.
3. England, J. P., Young, S. A., Hui, M. C., and Kurban, N., *Guide for the Design of Fire Resistant Barriers and Structures*. Warrington Fire Research Party, Ltd., and Building Control Commission, Melborne, Australia, 2000.
4. Ingberg, S. H., "Tests of the Severity of Building Fires." *NFPA Quarterly* 22(1):43–61, 1928.
5. Law, M., "Review of Formula for *T*-Equivalent," in *Fire Safety Science: Proceedings of the Fifth International Symposium*, pp. 985–996. Intl. Assoc. for Fire Safety Science, Melbourne, Australia, 1997.
6. Pettersson, O., *Fire Engineering Design of Steel Structures* (Publication 50). Swedish Institute of Steel Construction, Stockholm, Sweden, 1976, pp. 33–41.
7. Heaney, A. C., "A Reliability-Based Study Concerning Live Loads and Codified Structural Design." Ph.D. Thesis presented to the University of Waterloo, Waterloo, Ontario, Canada, 1971.
8. IFEG, *International Fire Engineering Guidelines*, DBH, NZ; ABCB, Australia; NRC, Canada; International Code Council (ICC), USA, 2005.
9. Heskestad, G., and Delichatsios, M. A., "The Initial Convective Flow in Fire," *17th Symposium on Combustion*. Combustion Institute, Philadelphia, 1978.
10. Drysdale, Dougal, *An Introduction to Fire Dynamics*, 2nd ed., Wiley, West Sussex, England, 1999.
11. National Institute of Standards and Technology (NIST). "Best Practice Guidelines for Structural Fire Resistance Design of Concrete and Steel Buildings" (NIST 7563). NIST, Gaithersburg, MD, 2009.
12. Schiesser, W. E., *The Numerical Method of Lines*. Academic Press, San Diego, 1991.
13. International Standard Organization (ISO). "*Plastics—Determination of Thermal Conductivity and Thermal Diffusivity, Part 2: Transient Plane Heat Source (Hot Disc) Method*" (ISO 22007-2). ISO, Brussels, 2008.
14. *SFPE Handbook of Fire Protection Engineering*, 2nd ed., SFPE, NFPA, 1995.
15. American Institute of Steel Construction (AISC). "Fire Resistance of Structural Steel Framing" (Design Guide 19). AISC, New York, 2003.

Appendix 6A

τ	θ_2	θ_3	θ_4	θ_5
0	1.	1.	1.	1.
0.0045699	2.511636	1.275636	1.03977	1.03977
0.0065584	2.938987	1.472476	1.097016	1.097016
0.0080611	3.211224	1.628287	1.156918	1.156918
0.0106633	3.613064	1.899735	1.289816	1.289816
0.0127016	3.885637	2.109101	1.414991	1.414991
0.0150716	4.171802	2.347119	1.578272	1.578272
0.0164185	4.323802	2.479812	1.677716	1.677716
0.019411	4.642308	2.768913	1.911519	1.911519
0.021011	4.804588	2.920783	2.042058	2.042058
0.022611	4.962785	3.07115	2.175463	2.175463
0.024211	5.117722	3.220262	2.311143	2.311143
0.027411	5.420161	3.515518	2.587511	2.587511
0.029011	5.568486	3.661964	2.727506	2.727506
0.030611	5.715261	3.807769	2.868354	2.868354
0.032211	5.860672	3.953008	3.009852	3.009852
0.035411	6.147884	4.241978	3.294167	3.294167
0.037011	6.289823	4.38576	3.436732	3.436732
0.038611	6.43069	4.529081	3.579434	3.579434
0.040211	6.570485	4.671934	3.722189	3.722189
0.043411	6.846775	4.956163	4.007571	4.007571
0.045011	6.983195	5.097482	4.15007	4.15007
0.046611	7.118397	5.238222	4.292364	4.292364
0.048211	7.252321	5.378342	4.434396	4.434396
0.051411	7.516066	5.656527	4.717466	4.717466
0.053011	7.645742	5.794489	4.858398	4.858398
0.054611	7.773851	5.931624	4.998857	4.998857
0.056211	7.900314	6.067872	5.138789	5.138789
0.059411	8.147972	6.33746	5.416853	5.416853
0.061011	8.269001	6.470673	5.554874	5.554874
0.062611	8.388052	6.602744	5.692142	5.692142
0.064211	8.505041	6.733605	5.828599	5.828599
0.067411	8.732496	6.991421	6.098837	6.098837

TABLE **6A.1**   Calculated Values of DEQ Variables

τ	θ₂	θ₃	θ₄	θ₅
0.069011	8.842796	7.118237	6.232492	6.232492
0.070611	8.950701	7.243562	6.365088	6.365088
0.072211	9.05613	7.367326	6.496558	6.496558
0.075411	9.259244	7.609881	6.755857	6.755857
0.077011	9.356773	7.728527	6.883551	6.883551
0.078611	9.451516	7.845323	7.009851	7.009851
0.080211	9.543401	7.960197	7.134688	7.134688
0.083411	9.718311	8.183893	7.379696	7.379696
0.085011	9.801201	8.292573	7.499727	7.499727
0.086611	9.880961	8.399049	7.618018	7.618018
0.088211	9.957531	8.50325	7.734498	7.734498
0.091411	10.10086	8.704562	7.961748	7.961748
0.093011	10.16752	8.80154	8.07238	8.07238
0.094611	10.23076	8.89598	8.180925	8.180925
0.096211	10.29055	8.987821	8.287317	8.287317
0.099411	10.39958	9.163461	8.493374	8.493374
0.101011	10.44875	9.247143	8.592909	8.592909
0.102611	10.49431	9.327993	8.69003	8.69003
0.104211	10.53622	9.405958	8.784674	8.784674
0.107411	10.60902	9.553027	8.96629	8.96629
0.109011	10.63986	9.622037	9.053146	9.053146
0.110611	10.66697	9.68797	9.137292	9.137292
0.112211	10.69034	9.750786	9.218674	9.218674
0.115411	10.72584	9.86691	9.372937	9.372937
0.117011	10.73796	9.920148	9.445721	9.445721
0.118611	10.74633	9.970129	9.515545	9.515545
0.120211	10.75095	10.01682	9.582364	9.582364
0.123411	10.74901	10.10026	9.706829	9.706829
0.125011	10.74247	10.13696	9.7644	9.7644
0.126611	10.73226	10.17029	9.818816	9.818816
0.128211	10.71838	10.20023	9.870049	9.870049
0.131411	10.67978	10.24995	9.962848	9.962848
0.133011	10.65513	10.2697	10.00437	10.00437
0.134611	10.62694	10.28605	10.04261	10.04261

TABLE 6A.1 Calculated Values of DEQ Variables (*Continued*)

τ	θ_2	θ_3	θ_4	θ_5
0.136211	10.59528	10.299	10.07755	10.07755
0.139411	10.5217	10.31471	10.13748	10.13748
0.141011	10.47988	10.3175	10.16245	10.16245
0.142611	10.43478	10.31692	10.18409	10.18409
0.144211	10.38646	10.313	10.20238	10.20238
0.147411	10.28039	10.29521	10.22897	10.22897
0.149011	10.22276	10.2814	10.23726	10.23726
0.150611	10.16217	10.26433	10.24224	10.24224
0.152211	10.09868	10.24405	10.24391	10.24391
0.155411	9.963301	10.194	10.23739	10.23739
0.157011	9.891558	10.1643	10.22925	10.22925
0.158611	9.817219	10.13155	10.21788	10.21788
0.160211	9.740362	10.09578	10.20331	10.20331
0.163411	9.579423	10.01539	10.16468	10.16468
0.165011	9.495507	9.970879	10.1407	10.1407
0.166611	9.409407	9.923555	10.11364	10.11364
0.168211	9.321209	9.873479	10.08357	10.08357
0.171411	9.13887	9.765302	10.0145	10.0145
0.173011	9.044905	9.707324	9.975605	9.975605
0.174611	8.949195	9.646838	9.933866	9.933866
0.176211	8.85183	9.583908	9.889336	9.889336
0.179411	8.652492	9.450985	9.792114	9.792114
0.181011	8.5507	9.381131	9.739536	9.739536
0.182611	8.447611	9.309108	9.68439	9.68439
0.184211	8.343315	9.234989	9.626739	9.626739
0.187411	8.131461	9.080752	9.504168	9.504168
0.189011	8.024079	9.000782	9.439378	9.439378
0.190611	7.915843	8.919011	9.372339	9.372339
0.192211	7.806841	8.835514	9.303121	9.303121
0.195411	7.586878	8.663649	9.158419	9.158419
0.197011	7.476086	8.575433	9.083076	9.083076
0.198611	7.364864	8.485796	9.005835	9.005835
0.2	7.268025	8.406891	8.937297	8.937297

TABLE 6A.1 Calculated Values of DEQ Variables (*Continued*)

θ_6	θ_7	θ_8	θ_9	θ_{10}
1.	1.	1.	1.	1.
1.003974	1.000325	1.000024	1.000024	1.000001
1.013469	1.001546	1.000163	1.000163	1.00001
1.026118	1.003621	1.000472	1.000472	1.000036
1.061078	1.010864	1.001875	1.001875	1.000192
1.100642	1.020811	1.004275	1.004275	1.000526
1.159861	1.038131	1.009258	1.009258	1.001364
1.199508	1.051016	1.013449	1.013449	1.002167
1.301761	1.088052	1.027166	1.027166	1.00521
1.363684	1.112734	1.037389	1.037389	1.007782
1.430116	1.140838	1.049874	1.049874	1.011192
1.50066	1.172338	1.064783	1.064783	1.015583
1.652641	1.245311	1.102373	1.102373	1.027884
1.733441	1.286639	1.12524	1.12524	1.036073
1.817074	1.33108	1.150908	1.150908	1.045797
1.903301	1.378544	1.179413	1.179413	1.057178
2.0827	1.482168	1.244996	1.244996	1.085358
2.175513	1.538141	1.282068	1.282068	1.102354
2.270194	1.596764	1.321968	1.321968	1.121401
2.366608	1.657948	1.364665	1.364665	1.142572
2.564159	1.787651	1.458288	1.458288	1.191527
2.665085	1.855999	1.509116	1.509116	1.219406
2.767318	1.926571	1.562549	1.562549	1.249599
2.870771	1.999287	1.618528	1.618528	1.282132
3.081017	2.15084	1.737876	1.737876	1.354277
3.187659	2.229528	1.801115	1.801115	1.393899
3.295218	2.31006	1.866643	1.866643	1.435886
3.403626	2.392363	1.934393	1.934393	1.480225
3.622719	2.562003	2.076284	2.076284	1.575893
3.733274	2.649201	2.15029	2.15029	1.627177
3.844413	2.737892	2.226244	2.226244	1.680722
3.956074	2.828009	2.304079	2.304079	1.736495
4.180702	3.012251	2.465116	2.465116	1.854578

TABLE 6A.1 Calculated Values of DEQ Variables (*Continued*)

θ_6	θ_7	θ_8	θ_9	θ_{10}
4.293541	3.106242	2.548183	2.548183	1.916806
4.406643	3.201391	2.632858	2.632858	1.981099
4.519944	3.297631	2.719073	2.719073	2.047411
4.746882	3.493117	2.895855	2.895855	2.185894
4.860388	3.592231	2.986288	2.986288	2.257961
4.97383	3.69217	3.077992	3.077992	2.331842
5.087142	3.792868	3.1709	3.1709	2.407481
5.313107	3.996272	3.360062	3.360062	2.563809
5.425625	4.098844	3.456181	3.456181	2.644383
5.537744	4.201909	3.553237	3.553237	2.726485
5.649396	4.305397	3.651163	3.651163	2.810055
5.871025	4.513379	3.849355	3.849355	2.98136
5.980867	4.617738	3.949488	3.949488	3.068971
6.089969	4.722252	4.050222	4.050222	3.157806
6.198265	4.826853	4.151491	4.151491	3.247803
6.412162	5.036051	4.355364	4.355364	3.431026
6.51763	5.140512	4.457833	4.457833	3.524125
6.622021	5.244791	4.560568	4.560568	3.618132
6.72527	5.348822	4.663502	4.663502	3.712981
6.928076	5.555868	4.869696	4.869696	3.904947
7.027503	5.65875	4.972822	4.972822	4.001935
7.125529	5.761117	5.075878	5.075878	4.099504
7.222089	5.862902	5.178796	5.178796	4.197591
7.410568	6.064463	5.383956	5.383956	4.395053
7.502365	6.164109	5.486063	5.486063	4.494297
7.592454	6.262914	5.587768	5.587768	4.593795
7.680779	6.360813	5.689004	5.689004	4.693481
7.851909	6.553641	5.889808	5.889808	4.893154
7.934605	6.648446	5.989246	5.989246	4.99301
8.015319	6.742097	6.087955	6.087955	5.092791
8.094	6.834534	6.185873	6.185873	5.192433
8.245067	7.01553	6.37908	6.37908	5.391034
8.31736	7.103973	6.474245	6.474245	5.489865

TABLE 6A.1 Calculated Values of DEQ Variables (*Continued*)

θ_6	θ_7	θ_8	θ_9	θ_{10}
8.387433	7.190971	6.568369	6.568369	5.588296
8.455244	7.276469	6.661391	6.661391	5.686263
8.583921	7.442749	6.843896	6.843896	5.880553
8.644713	7.523428	6.933262	6.933262	5.97675
8.703092	7.602398	7.021293	7.021293	6.072231
8.759028	7.679611	7.107935	7.107935	6.166934
8.86345	7.828578	7.276832	7.276832	6.353766
8.911881	7.900242	7.358982	7.358982	6.445775
8.957759	7.969969	7.439529	7.439529	6.536766
9.001064	8.037718	7.518426	7.518426	6.626682
9.079877	8.167123	7.671073	7.671073	6.803059
9.115352	8.228706	7.744731	7.744731	6.889408
9.148189	8.288163	7.816552	7.816552	6.974459
9.178378	8.345461	7.886493	7.886493	7.058157
9.230779	8.45346	8.020573	8.020573	7.22129
9.252981	8.504104	8.084635	8.084635	7.300622
9.272516	8.552478	8.146663	8.146663	7.3784
9.289384	8.598558	8.206621	8.206621	7.454577
9.315134	8.683754	8.320199	8.320199	7.60194
9.324029	8.722832	8.373759	8.373759	7.673039
9.330282	8.759542	8.425128	8.425128	7.74236
9.333907	8.793872	8.474282	8.474282	7.809862
9.333325	8.855344	8.565848	8.565848	7.939255
9.329152	8.882469	8.608218	8.608218	8.001072
9.322419	8.907179	8.648288	8.648288	8.060924
9.313146	8.92947	8.686041	8.686041	8.118777
9.287081	8.96679	8.754541	8.754541	8.228366
9.270343	8.981821	8.785266	8.785266	8.280045
9.251172	8.994438	8.813627	8.813627	8.329611
9.2296	9.004647	8.839619	8.839619	8.377041
9.179386	9.017873	8.884477	8.884477	8.465401
9.150815	9.020912	8.903338	8.903338	8.506293
9.119984	9.021585	8.919822	8.919822	8.544969
9.091417	9.020262	8.932206	8.932206	8.576736

TABLE 6A.1   Calculated Values of DEQ Variables (*Continued*)

θ_1	θ_{11}	θ_1	θ_{11}
5.895865	1.	11.56656	1.94924
6.293273	0.9999934	11.62469	2.015186
6.4681	0.999959	11.6786	2.083126
6.600827	0.9998909	11.72823	2.153008
6.831625	0.9996311	11.81446	2.298392
7.013037	0.9992767	11.85095	2.373784
7.224344	0.9987323	11.88296	2.450901
7.344478	0.9984062	11.91047	2.529686
7.61116	0.9978919	11.95184	2.692028
7.7534	0.9979132	11.96564	2.775466
7.895252	0.9982973	11.97485	2.860334
8.036597	0.9991827	11.97944	2.946573
8.317293	1.003055	11.97475	3.122913
8.456402	1.00635	11.96549	3.212889
8.59452	1.01076	11.95161	3.303986
8.731525	1.016433	11.93315	3.39614
9.001697	1.032145	11.88255	3.583364
9.134614	1.042449	11.85047	3.678306
9.265918	1.054545	11.81392	3.774047
9.395485	1.068542	11.77294	3.870523
9.648903	1.102607	11.67788	4.065419
9.772507	1.122835	11.62391	4.163708
9.893877	1.145282	11.56573	4.26247
10.01289	1.17	11.5034	4.36164
10.24337	1.22641	11.36658	4.560936
10.35459	1.258161	11.29224	4.660932
10.46299	1.2923	11.21407	4.76107
10.56844	1.328835	11.13214	4.861286
10.77007	1.409096	10.95738	5.061686
10.86604	1.452806	10.86474	5.161738
10.95863	1.498881	10.76873	5.261605
11.04775	1.547301	10.66945	5.361221
11.21517	1.651066	10.46151	5.559439
11.29329	1.706347	10.35308	5.657912
11.36758	1.763847	10.24181	5.755876
11.43793	1.823524	10.12784	5.853267

TABLE 6A.1 Calculated Values of DEQ Variables (*Continued*)

θ_1	θ_{11}
9.89222	6.046078
9.770818	6.141373
9.647184	6.235845
9.52144	6.329433
9.26412	6.51372
9.132792	6.604301
8.999854	6.693761
8.86543	6.782046
8.592623	6.954862
8.45449	7.039284
8.315369	7.122313
8.175383	7.203895
7.8933	7.362521
7.751442	7.439466

θ_1	θ_{11}
7.609197	7.514771
7.466682	7.588389
7.181292	7.730391
7.03864	7.79869
6.89616	7.865136
6.753958	7.929689
6.470798	8.052975
6.330036	8.111638
6.189949	8.168272
6.050626	8.222848
5.774631	8.32571
5.638127	8.373945
5.502727	8.420018
5.386139	8.458245

TABLE **6A.1** Calculated Values of DEQ Variables (*Continued*)

τ	θ$_2$	θ$_3$	θ$_4$	θ$_5$
0	1.	1.	1.	1.
0.0041884	1.925929	1.214842	1.041605	1.041605
0.00652	2.230935	1.393826	1.115419	1.115419
0.0086943	2.461235	1.561263	1.21089	1.21089
0.0104468	2.625494	1.693688	1.301446	1.301446
0.012475	2.801299	1.844132	1.417034	1.417034
0.0148104	2.992134	2.01466	1.560181	1.560181
0.0161076	3.094693	2.108572	1.643024	1.643024
0.018968	3.315335	2.314551	1.831541	1.831541
0.0205167	3.432548	2.425744	1.936081	1.936081
0.0221167	3.552398	2.540518	2.045424	2.045424
0.0253167	3.788807	2.76987	2.267107	2.267107
0.0269167	3.905415	2.884416	2.379035	2.379035
0.0285167	4.020897	2.998806	2.49147	2.49147
0.0301167	4.135156	3.112962	2.604271	2.604271
0.0332529	4.355097	3.335676	2.825951	2.825951
0.0347933	4.460907	3.444356	2.934889	2.934889
0.0363194	4.564108	3.551422	3.042708	3.042708
0.0393372	4.762835	3.760912	3.255181	3.255181
0.040832	4.858362	3.863362	3.359864	3.359864
0.0423193	4.951298	3.964266	3.463501	3.463501
0.0452749	5.129214	4.161298	3.667524	3.667524
0.0467451	5.214082	4.257334	3.767832	3.767832
0.0482112	5.296137	4.351643	3.866939	3.866939
0.0511338	5.451535	4.534824	4.061314	4.061314
0.0525915	5.524731	4.623557	4.156453	4.156453
0.0540474	5.594827	4.710287	4.250133	4.250133
0.0569564	5.725413	4.877426	4.432815	4.432815
0.0584103	5.785753	4.957675	4.521658	4.521658
0.0613145	5.895985	5.110877	4.69366	4.69366
0.0627651	5.945734	5.183647	4.776625	4.776625
0.0642153	5.991832	5.253766	4.857461	4.857461
0.067116	6.072863	5.385758	5.012426	5.012426
0.0685675	6.107697	5.447485	5.086393	5.086393
0.0700201	6.138682	5.506269	5.157906	5.157906
0.0729309	6.188933	5.614723	5.293245	5.293245

TABLE 6A.2 Calculated Values of DEQ Variables

τ	θ_2	θ_3	θ_4	θ_5
0.0743899	6.20812	5.664253	5.356905	5.356905
0.0773171	6.234445	5.753576	5.475797	5.475797
0.0787863	6.24152	5.793234	5.530865	5.530865
0.0802599	6.244502	5.82947	5.582905	5.582905
0.0832223	6.238087	5.891419	5.677577	5.677577
0.0847122	6.228649	5.917006	5.720047	5.720047
0.0862087	6.215035	5.93892	5.759165	5.759165
0.0892241	6.175213	5.971484	5.827023	5.827023
0.0907445	6.148977	5.982013	5.8556	5.8556
0.0922745	6.118506	5.988626	5.880499	5.880499
0.0953669	6.044801	5.989849	5.918924	5.918924
0.0969315	6.001531	5.984329	5.932274	5.932274
0.09851	5.953953	5.974625	5.941592	5.941592
0.100104	5.902036	5.960662	5.94678	5.94678
0.103304	5.786608	5.92023	5.944462	5.944462
0.104904	5.723573	5.893945	5.936989	5.936989
0.106504	5.657194	5.863719	5.925363	5.925363
0.108104	5.587629	5.829642	5.909635	5.909635
0.111304	5.439595	5.750321	5.866118	5.866118
0.112904	5.361462	5.705288	5.838471	5.838471
0.114504	5.280813	5.656824	5.807007	5.807007
0.116104	5.197822	5.60505	5.771814	5.771814
0.119304	5.025521	5.492073	5.690634	5.690634
0.120904	4.936562	5.431134	5.644856	5.644856
0.122504	4.845967	5.367409	5.59577	5.59577
0.124104	4.753911	5.30104	5.543494	5.543494
0.127304	4.56611	5.16094	5.429869	5.429869
0.128904	4.470707	5.087502	5.368778	5.368778
0.130504	4.374526	5.012004	5.305014	5.305014
0.132104	4.27773	4.934593	5.238714	5.238714
0.135304	4.082934	4.774635	5.099066	5.099066
0.136904	3.985241	4.692385	5.026003	5.026003
0.138504	3.887549	4.608821	4.950973	4.950973
0.140104	3.79	4.524087	4.87412	4.87412
0.143304	3.595873	4.351694	4.715529	4.715529
0.144904	3.499553	4.264318	4.634079	4.634079

TABLE 6A.2 Calculated Values of DEQ Variables (*Continued*)

τ	θ_2	θ_3	θ_4	θ_5
0.1464891	3.404778	4.177162	4.552161	4.552161
0.1480534	3.311986	4.0907	4.470255	4.470255
0.1511283	3.132117	3.919912	4.306683	4.306683
0.1526421	3.044973	3.83565	4.225145	4.225145
0.1541418	2.959652	3.752187	4.143854	4.143854
0.1571034	2.794394	3.587782	3.982248	3.982248
0.1585673	2.714417	3.506904	3.902045	3.902045
0.1600208	2.636185	3.426947	3.822313	3.822313
0.1629001	2.484884	3.269908	3.664456	3.664456
0.1643271	2.411781	3.192874	3.586424	3.586424
0.1671584	2.270578	3.041891	3.43236	3.43236
0.1685638	2.202443	2.967979	3.356405	3.356405
0.1713563	2.07101	2.8234	3.206823	3.206823
0.1727441	2.007673	2.75276	3.133258	3.133258
0.1741269	1.945894	2.683236	3.060548	3.060548
0.1768787	1.826928	2.547571	2.917794	2.917794
0.1782484	1.769697	2.481445	2.847794	2.847794
0.1809767	1.659625	2.352633	2.710649	2.710649
0.182336	1.606738	2.289951	2.643536	2.643536
0.1850461	1.505144	2.168045	2.512301	2.512301
0.1863974	1.456388	2.108818	2.448202	2.448202
0.1890938	1.362837	1.993804	2.323089	2.323089
0.1904394	1.317993	1.938009	2.262089	2.262089
0.1931266	1.23204	1.829815	2.143226	2.143226
0.1944685	1.190881	1.777403	2.08537	2.08537
0.1971503	1.112075	1.675907	1.972815	1.972815
0.1984906	1.074378	1.626805	1.918116	1.918116
0.2	1.033238	1.572796	1.857763	1.857763

TABLE **6A.2**   Calculated Values of DEQ Variables (*Continued*)

θ_6	θ_7	θ_8	θ_9	θ_{10}
1.	1.	1.	1.	1.
1.005476	1.000595	1.000059	1.000059	1.000003
1.02231	1.003623	1.000564	1.000564	1.000052
1.051447	1.010718	1.002222	1.002222	1.000276
1.084525	1.020505	1.005084	1.005084	1.000766
1.132701	1.037079	1.010875	1.010875	1.001973
1.199777	1.063609	1.021812	1.021812	1.004721
1.241716	1.081846	1.03021	1.03021	1.007121
1.344216	1.130765	1.055347	1.055347	1.015357
1.404702	1.162101	1.073015	1.073015	1.02187
1.470377	1.197909	1.094392	1.094392	1.030376
1.610166	1.279477	1.146828	1.146828	1.053552
1.683756	1.324973	1.177885	1.177885	1.068538
1.759521	1.373437	1.212126	1.212126	1.085959
1.837278	1.424741	1.249493	1.249493	1.105916
1.994858	1.533077	1.331534	1.331534	1.152663
2.074496	1.589837	1.375942	1.375942	1.17941
2.154692	1.648239	1.4225	1.4225	1.208396
2.316641	1.769685	1.521737	1.521737	1.27301
2.398317	1.832587	1.574254	1.574254	1.30858
2.480398	1.896849	1.628598	1.628598	1.34628
2.645569	2.029176	1.742463	1.742463	1.427909
2.728548	2.097104	1.801835	1.801835	1.471751
2.811711	2.166118	1.862737	1.862737	1.517548
2.978336	2.307133	1.988852	1.988852	1.614815
3.061669	2.378999	2.053925	2.053925	1.666182
3.144925	2.451681	2.120252	2.120252	1.719303
3.310925	2.599218	2.256394	2.256394	1.830591
3.393523	2.673935	2.326074	2.326074	1.88865
3.557276	2.824671	2.468083	2.468083	2.009118
3.638246	2.900513	2.54024	2.54024	2.071379
3.718512	2.976575	2.613071	2.613071	2.134909
3.876621	3.129071	2.760475	2.760475	2.265529
3.95431	3.205357	2.834907	2.834907	2.332494
4.030982	3.281569	2.909729	2.909729	2.400478
4.18096	3.433473	3.060261	3.060261	2.539248

TABLE 6A.2 Calculated Values of DEQ Variables (*Continued*)

θ_6	θ_7	θ_8	θ_9	θ_{10}
4.254105	3.509014	3.135824	3.135824	2.609906
4.396296	3.658885	3.287177	3.287177	2.753444
4.465179	3.733061	3.362818	3.362818	2.826192
4.532476	3.806625	3.438335	3.438335	2.899504
4.661979	3.951599	3.58869	3.58869	3.047551
4.724017	4.022847	3.663372	3.663372	3.12215
4.784133	4.093161	3.73762	3.73762	3.19704
4.898254	4.230653	3.884495	3.884495	3.347418
4.952087	4.297663	3.956959	3.956959	3.422765
5.00365	4.363403	4.028667	4.028667	3.498123
5.099607	4.490726	4.16948	4.16948	3.64859
5.143816	4.552133	4.238417	4.238417	3.723556
5.185386	4.611916	4.306261	4.306261	3.798252
5.224217	4.669985	4.372927	4.372927	3.872606
5.29244	4.779248	4.500773	4.500773	4.018168
5.321602	4.830064	4.561507	4.561507	4.088872
5.347426	4.87824	4.619995	4.619995	4.158047
5.369887	4.92371	4.676152	4.676152	4.225589
5.40467	5.006302	4.781153	4.781153	4.355366
5.416982	5.043318	4.82985	4.82985	4.417409
5.42591	5.077422	4.875921	4.875921	4.477432
5.431466	5.108576	4.919306	4.919306	4.535346
5.43254	5.161908	4.997797	4.997797	4.64452
5.428112	5.18404	5.032807	5.032807	4.695625
5.420419	5.203128	5.064939	5.064939	4.744313
5.409504	5.219163	5.094159	5.094159	4.790519
5.378201	5.242064	5.143753	5.143753	4.875242
5.357925	5.248941	5.164088	5.164088	4.913652
5.334648	5.252785	5.181429	5.181429	4.949364
5.308439	5.253614	5.195772	5.195772	4.982337
5.247513	5.246332	5.215463	5.215463	5.039929
5.212954	5.238283	5.220826	5.220826	5.064492
5.175774	5.227346	5.22322	5.22322	5.086203
5.136061	5.213566	5.222666	5.222666	5.105049
5.0494	5.177668	5.212819	5.212819	5.134109
5.00264	5.15566	5.203593	5.203593	5.144321

TABLE 6A.2   Calculated Values of DEQ Variables (*Continued*)

θ_6	θ_7	θ_8	θ_9	θ_{10}
4.954189	5.131267	5.191674	5.191674	5.151603
4.904391	5.104726	5.177243	5.177243	5.156036
4.801131	5.045658	5.141329	5.141329	5.156834
4.74786	5.013344	5.120068	5.120068	5.153412
4.693611	4.97931	5.09674	5.09674	5.14757
4.582513	4.906445	5.044256	5.044256	5.12898
4.52582	4.867785	5.015275	5.015275	5.116394
4.468461	4.827748	4.984579	4.984579	5.101716
4.352024	4.743847	4.918351	4.918351	5.066376
4.293075	4.700128	4.882967	4.882967	5.045851
4.174024	4.609493	4.808013	4.808013	4.999423
4.114036	4.562708	4.768576	4.768576	4.973646
3.993401	4.466509	4.686106	4.686106	4.91727
3.932856	4.417213	4.643195	4.643195	4.886788
3.872222	4.367194	4.599246	4.599246	4.854855
3.750864	4.265202	4.50846	4.50846	4.786853
3.690226	4.213332	4.461733	4.461733	4.750893
3.569224	4.108088	4.365877	4.365877	4.675317
3.508935	4.054808	4.316848	4.316848	4.635804
3.388948	3.947154	4.216834	4.216834	4.553577
3.329315	3.892864	4.165942	4.165942	4.510959
3.210913	3.783562	4.062613	4.062613	4.422951
3.152199	3.728626	4.010263	4.010263	4.377653
3.03587	3.618365	3.904396	3.904396	4.284689
2.978302	3.563109	3.850959	3.850959	4.237111
2.864461	3.452512	3.743269	3.743269	4.13997
2.808228	3.397232	3.689089	3.689089	4.090489
2.745466	3.33503	3.627851	3.627851	4.034095

TABLE 6A.2   Calculated Values of DEQ Variables (*Continued*)

θ_1	θ_{11}	θ_1	θ_{11}
3.40004	1.	6.836815	2.434601
3.719134	0.9999848	6.760079	2.575533
3.899404	0.9998812	6.714455	2.647317
4.068535	0.9996278	6.664052	2.719894
4.205228	0.9993274	6.549107	2.867172
4.363423	0.9990051	6.484675	2.941742
4.545026	0.9990244	6.415682	3.016847
4.645393	0.9994253	6.264264	3.168392
4.864684	1.002028	6.181969	3.2447
4.981866	1.004821	6.095374	3.321275
5.101496	1.009038	5.909541	3.47496
5.335384	1.022461	5.810431	3.551936
5.449144	1.032089	5.707266	3.628915
5.560451	1.043904	5.600097	3.705833
5.669056	1.058057	5.376839	3.8573
5.873163	1.093039	5.261643	3.931328
5.968718	1.113899	5.144401	4.004065
6.060059	1.137028	5.025361	4.075401
6.23016	1.1901	4.782878	4.213438
6.308885	1.220022	4.659924	4.279929
6.383343	1.252174	4.536151	4.344602
6.519255	1.323058	4.411794	4.407359
6.580593	1.361723	4.162246	4.526761
6.637437	1.402485	4.037501	4.583231
6.737402	1.490136	3.91306	4.637438
6.7804	1.536935	3.78913	4.689305
6.818666	1.585654	3.543582	4.785738
6.880781	1.688657	3.422336	4.830173
6.904531	1.742841	3.30234	4.872009
6.937159	1.856129	3.183757	4.911193
6.945992	1.915092	2.951437	4.981418
6.949824	1.975522	2.837977	5.01238
6.942466	2.100547	2.726486	5.040531
6.931283	2.165023	2.617078	5.065843
6.915108	2.230727	2.404921	5.107873
6.867851	2.365577	2.30235	5.124563

TABLE 6A.2 Calculated Values of DEQ Variables (*Continued*)

θ_1	θ_{11}	θ_1	θ_{11}
2.203143	5.138247	0.8900626	4.940058
2.10765	5.148967	0.8016906	4.87965
1.927122	5.162002	0.7602966	4.847279
1.841821	5.164526	0.6827911	4.778464
1.759673	5.164514	0.6465631	4.742123
1.604414	5.157221	0.5788736	4.665824
1.531112	5.1501	0.5473027	4.625965
1.460576	5.140762	0.488444	4.543064
1.327465	5.115718	0.4610539	4.500116
1.264731	5.100146	0.4101056	4.411454
1.146526	5.063226	0.3864521	4.365828
1.090909	5.042002	0.3425576	4.272203
0.9862958	4.994324	0.3222285	4.224289
0.9371646	4.967986	0.3005524	4.169509

TABLE 6A.2   Calculated Values of DEQ Variables (*Continued*)

τ	θ_2	θ_3	θ_4	θ_5
0	1.	1.	1.	1.
0.0041013	2.068799	1.297118	1.071509	1.071509
0.0061119	2.325717	1.48696	1.168964	1.168964
0.0080246	2.510316	1.656088	1.28571	1.28571
0.0103675	2.693706	1.847299	1.445871	1.445871
0.0122076	2.818638	1.987473	1.577576	1.577576
0.0143118	2.949027	2.139408	1.72964	1.72964
0.0167096	3.08706	2.303958	1.901301	1.901301
0.0180226	3.159246	2.390833	1.993752	1.993752
0.0208621	3.309189	2.572018	2.188848	2.188848
0.0223715	3.386043	2.664977	2.289633	2.289633
0.0255059	3.540286	2.851321	2.492257	2.492257
0.0271052	3.61643	2.943136	2.592203	2.592203
0.0287052	3.690981	3.032912	2.689932	2.689932
0.0303052	3.763945	3.120672	2.785448	2.785448
0.0335052	3.905211	3.290378	2.970068	2.970068
0.035099	3.973278	3.372105	3.058941	3.058941
0.036686	4.039546	3.451684	3.145467	3.145467
0.0382672	4.10407	3.529215	3.229764	3.229764
0.0414136	4.228	3.678386	3.392001	3.392001
0.0429782	4.287399	3.750074	3.470015	3.470015
0.0445374	4.345105	3.819882	3.546035	3.546035
0.0460914	4.401132	3.887856	3.620118	3.620118
0.0491852	4.508209	4.018448	3.76268	3.76268
0.0507256	4.559281	4.081132	3.831248	3.831248
0.0522622	4.608718	4.142111	3.898059	3.898059
0.055325	4.702713	4.259044	4.026537	4.026537
0.0568516	4.747277	4.315034	4.088258	4.088258
0.0583755	4.790221	4.369393	4.148331	4.148331
0.0614157	4.871247	4.473261	4.263604	4.263604
0.0629326	4.909327	4.522788	4.318832	4.318832
0.0644475	4.945782	4.570718	4.37247	4.37247
0.0674727	5.013804	4.661802	4.475008	4.475008
0.0689832	5.045363	4.704961	4.52392	4.52392
0.0704927	5.075282	4.746531	4.571265	4.571265
0.0720013	5.103556	4.786512	4.617045	4.617045

TABLE 6A.3   Calculated Values of DEQ Variables

τ	θ₂	θ₃	θ₄	θ₅
0.0750165	5.155151	4.861698	4.703912	4.703912
0.0765236	5.178462	4.896897	4.744996	4.744996
0.0780305	5.200109	4.930496	4.78451	4.78451
0.0810445	5.23839	4.992873	4.858816	4.858816
0.082552	5.255016	5.021642	4.893598	4.893598
0.08406	5.269958	5.04879	4.926792	4.926792
0.0870785	5.29478	5.0982	4.98839	4.98839
0.0885893	5.304651	5.12045	5.016782	5.016782
0.0901014	5.312824	5.141056	5.043559	5.043559
0.0931301	5.324068	5.177307	5.092237	5.092237
0.0946472	5.327134	5.19294	5.114123	5.114123
0.0961663	5.328492	5.206904	5.134364	5.134364
0.0992114	5.326085	5.229798	5.169873	5.169873
0.1007379	5.322317	5.238716	5.185126	5.185126
0.1022672	5.316839	5.245941	5.1987	5.1987
0.1053354	5.300755	5.255287	5.220779	5.220779
0.1068747	5.290151	5.257398	5.229267	5.229267
0.1084179	5.27784	5.257793	5.236042	5.236042
0.1115168	5.248105	5.253419	5.244425	5.244425
0.1130732	5.230685	5.248639	5.246016	5.246016
0.1146345	5.211567	5.242124	5.245862	5.245862
0.116201	5.190752	5.233869	5.243955	5.243955
0.1193515	5.144045	5.21212	5.234847	5.234847
0.1209361	5.118158	5.198617	5.227629	5.227629
0.1225276	5.090585	5.183353	5.218622	5.218622
0.1241263	5.061327	5.166322	5.207817	5.207817
0.1273263	4.998203	5.127219	5.180974	5.180974
0.1289263	4.964427	5.105207	5.164982	5.164982
0.1305263	4.929223	5.081591	5.147305	5.147305
0.1321263	4.892627	5.056398	5.127968	5.127968
0.1353263	4.81541	5.001397	5.084407	5.084407
0.1369263	4.774868	4.971652	5.060237	5.060237
0.1385263	4.73309	4.940451	5.03451	5.03451
0.1401263	4.690117	4.90783	5.007256	5.007256
0.1433263	4.600755	4.838464	4.948291	4.948291

TABLE 6A.3 Calculated Values of DEQ Variables (*Continued*)

τ	θ₂	θ₃	θ₄	θ₅
0.1449263	4.554452	4.80179	4.916643	4.916643
0.1465263	4.507125	4.763838	4.883596	4.883596
0.1481263	4.458818	4.724646	4.849186	4.849186
0.1513263	4.359444	4.642697	4.776413	4.776413
0.1529263	4.308466	4.60002	4.738126	4.738126
0.1545263	4.256689	4.556262	4.698621	4.698621
0.1561263	4.204156	4.511463	4.657937	4.657937
0.1593263	4.097011	4.418913	4.573192	4.573192
0.1609263	4.042488	4.371245	4.529211	4.529211
0.1625263	3.987394	4.322707	4.484212	4.484212
0.1641263	3.931773	4.27334	4.438237	4.438237
0.1673263	3.81913	4.172296	4.343526	4.343526
0.1689263	3.762198	4.120705	4.294875	4.294875
0.1705263	3.704918	4.068459	4.245417	4.245417
0.1721263	3.647334	4.015602	4.195195	4.195195
0.1753263	3.531428	3.90823	4.092633	4.092633
0.1769263	3.47319	3.8538	4.040379	4.040379
0.1785263	3.414819	3.798932	3.987533	3.987533
0.1801263	3.356356	3.743668	3.934139	3.934139
0.1833263	3.239312	3.632122	3.825878	3.825878
0.1849263	3.18081	3.575923	3.771096	3.771096
0.1865263	3.122374	3.519495	3.715936	3.715936
0.1881263	3.064039	3.462878	3.66044	3.66044
0.1913263	2.947819	3.349238	3.548603	3.548603
0.1929263	2.890005	3.292293	3.492344	3.492344
0.1945263	2.832432	3.235315	3.435912	3.435912
0.1961263	2.775135	3.178342	3.379346	3.379346
0.1993263	2.66149	3.064556	3.265966	3.265966
0.2	2.637744	3.040648	3.242075	3.242075

TABLE 6A.3 Calculated Values of DEQ Variables (*Continued*)

θ_6	θ_7	θ_8	θ_9	θ_{10}
1.	1.	1.	1.	1.
1.011467	1.001527	1.00019	1.00019	1.000014
1.037988	1.007223	1.001344	1.001344	1.000149
1.079552	1.01906	1.004641	1.004641	1.000685
1.149427	1.044006	1.013682	1.013682	1.002641
1.215862	1.072035	1.025992	1.025992	1.005942
1.301	1.112825	1.046717	1.046717	1.012555
1.406478	1.169686	1.079661	1.079661	1.024958
1.46695	1.205031	1.101981	1.101981	1.03437
1.602118	1.290349	1.160325	1.160325	1.061941
1.675662	1.340061	1.19667	1.19667	1.080903
1.830576	1.451411	1.282899	1.282899	1.130265
1.910273	1.511853	1.331968	1.331968	1.160671
1.990206	1.574414	1.384127	1.384127	1.194562
2.07019	1.638818	1.439082	1.439082	1.231832
2.229842	1.772268	1.556335	1.556335	1.31599
2.309009	1.840646	1.6179	1.6179	1.362385
2.387499	1.909756	1.680995	1.680995	1.411314
2.465291	1.979467	1.745426	1.745426	1.462596
2.61864	2.12016	1.877557	1.877557	1.571459
2.694066	2.190843	1.944863	1.944863	1.628617
2.768613	2.261595	2.012783	2.012783	1.687344
2.842257	2.332325	2.081184	2.081184	1.747469
2.986755	2.473383	2.21895	2.21895	1.87128
3.057573	2.543563	2.288103	2.288103	1.934673
3.127415	2.613423	2.357308	2.357308	1.998877
3.264105	2.751945	2.495543	2.495543	2.129221
3.330923	2.820501	2.564421	2.564421	2.195134
3.396703	2.888522	2.633047	2.633047	2.261402
3.525089	3.022781	2.769301	2.769301	2.394623
3.587667	3.088938	2.836817	2.836817	2.461402
3.649149	3.154397	2.903857	2.903857	2.528185
3.768766	3.283082	3.036321	3.036321	2.661475
3.826873	3.346242	3.101659	3.101659	2.727847
3.883826	3.408574	3.166348	3.166348	2.793954
3.93961	3.470047	3.23035	3.23035	2.859739

TABLE 6A.3   Calculated Values of DEQ Variables (*Continued*)

θ_6	θ_7	θ_8	θ_9	θ_{10}
4.047617	3.590311	3.356151	3.356151	2.990125
4.099809	3.649048	3.417885	3.417885	3.054628
4.150775	3.706822	3.478798	3.478798	3.118607
4.248967	3.819382	3.598044	3.598044	3.244825
4.296165	3.874122	3.656321	3.656321	3.306982
4.342076	3.927805	3.713664	3.713664	3.368453
4.42998	4.03191	3.825443	3.825443	3.489194
4.471944	4.082289	3.87983	3.87983	3.548395
4.512561	4.131525	3.933181	3.933181	3.606773
4.589699	4.226483	4.036683	4.036683	3.720934
4.626188	4.272165	4.086787	4.086787	3.776658
4.661272	4.316621	4.135763	4.135763	3.831438
4.72716	4.401778	4.23024	4.23024	3.938056
4.757935	4.442439	4.275696	4.275696	3.989839
4.787244	4.481796	4.319935	4.319935	4.040569
4.841403	4.55652	4.404676	4.404676	4.138767
4.866223	4.591847	4.445133	4.445133	4.186183
4.889516	4.625793	4.484287	4.484287	4.232443
4.931464	4.68946	4.558597	4.558597	4.321396
4.950088	4.719142	4.59371	4.59371	4.364038
4.967125	4.747365	4.627433	4.627433	4.405425
4.982559	4.774107	4.659744	4.659744	4.445531
5.008558	4.823073	4.720041	4.720041	4.521799
5.01909	4.845255	4.747983	4.747983	4.557912
5.027956	4.865875	4.774422	4.774422	4.592642
5.035138	4.884913	4.799337	4.799337	4.625964
5.044344	4.917956	4.844218	4.844218	4.687888
5.046374	4.931941	4.864152	4.864152	4.716437
5.046699	4.944234	4.882407	4.882407	4.743363
5.04533	4.954836	4.898982	4.898982	4.768657
5.037553	4.970979	4.927081	4.927081	4.814312
5.031172	4.976527	4.938605	4.938605	4.834661
5.023148	4.980402	4.948447	4.948447	4.853351
5.013497	4.982609	4.956609	4.956609	4.870377
4.989386	4.982059	4.967913	4.967913	4.899431
4.974964	4.979321	4.971068	4.971068	4.911459

TABLE 6A.3   Calculated Values of DEQ Variables (*Continued*)

θ_6	θ_7	θ_8	θ_9	θ_{10}
4.958991	4.974958	4.972566	4.972566	4.92182
4.941491	4.968983	4.972418	4.972418	4.930519
4.901998	4.952256	4.967226	4.967226	4.942944
4.880056	4.941536	4.962205	4.962205	4.946681
4.856684	4.92927	4.955586	4.955586	4.948777
4.831909	4.915475	4.947384	4.947384	4.949241
4.778267	4.883386	4.926297	4.926297	4.945313
4.749458	4.865135	4.913447	4.913447	4.940945
4.719365	4.845443	4.899085	4.899085	4.93499
4.688018	4.824335	4.883233	4.883233	4.927464
4.621697	4.777972	4.847142	4.847142	4.907762
4.586789	4.752772	4.82695	4.82695	4.895621
4.550762	4.726262	4.80536	4.80536	4.881977
4.513649	4.698471	4.782397	4.782397	4.866852
4.436313	4.639167	4.732458	4.732458	4.832239
4.396162	4.607715	4.705539	4.705539	4.812797
4.35507	4.575105	4.677357	4.677357	4.791962
4.313075	4.54137	4.647943	4.647943	4.769761
4.226526	4.470654	4.585539	4.585539	4.72136
4.182048	4.433741	4.552613	4.552613	4.695215
4.136817	4.395837	4.51858	4.51858	4.667811
4.090872	4.356977	4.483474	4.483474	4.639176
3.996994	4.276531	4.410173	4.410173	4.578336
3.949136	4.235015	4.372047	4.372047	4.546192
3.900718	4.192686	4.332984	4.332984	4.512941
3.851776	4.14958	4.293018	4.293018	4.478614
3.752476	4.061181	4.210522	4.210522	4.406866
3.731351	4.042221	4.192739	4.192739	4.391255

TABLE 6A.3   Calculated Values of DEQ Variables (*Continued*)

θ_1	θ_{11}	θ_1	θ_{11}
3.540032	1.	5.539585	2.868117
3.702282	0.9999551	5.544758	2.933542
3.781271	0.9997505	5.548174	2.998544
3.855953	0.9993659	5.549723	3.127086
3.94669	0.9989602	5.547854	3.190536
4.017292	0.9992582	5.544222	3.253383
4.097193	1.001167	5.531673	3.377111
4.187021	1.006723	5.52276	3.437917
4.2356	1.011832	5.512092	3.49797
4.339012	1.029146	5.485509	3.615685
4.392991	1.042314	5.469604	3.673281
4.502655	1.079387	5.451967	3.729996
4.557244	1.103572	5.411521	3.840661
4.610872	1.131374	5.38873	3.894553
4.663462	1.162749	5.36424	3.947447
4.765327	1.235876	5.310202	4.05013
4.814319	1.277213	5.280677	4.099866
4.861885	1.32142	5.249497	4.148495
4.908024	1.36832	5.182224	4.242329
4.995946	1.469426	5.146158	4.287481
5.037656	1.523202	5.108489	4.331422
5.077836	1.578864	5.069234	4.374126
5.116471	1.63623	4.986017	4.455719
5.189049	1.755391	4.942086	4.494554
5.222962	1.816864	4.896624	4.532048
5.255274	1.8794	4.849646	4.568173
5.315037	2.007114	4.751861	4.635778
5.342463	2.072039	4.701185	4.667199
5.368235	2.137521	4.649385	4.697015
5.41477	2.26973	4.596511	4.725215
5.43551	2.336263	4.487737	4.776723
5.454552	2.402961	4.431939	4.800013
5.487505	2.536526	4.375269	4.821652
5.501398	2.603243	4.317778	4.841633
5.513558	2.669823	4.200538	4.876604
5.52398	2.736201	4.140893	4.891589

TABLE 6A.3 Calculated Values of DEQ Variables (*Continued*)

θ_1	θ_{11}
4.080635	4.904905
4.019814	4.916552
3.896692	4.93485
3.834493	4.941506
3.771937	4.946507
3.709073	4.94986
3.582623	4.951652
3.519133	4.950111
3.455532	4.946959
3.391866	4.942208
3.264523	4.927969
3.200937	4.918511
3.137466	4.907516
3.074154	4.895003

θ_1	θ_{11}
2.948168	4.865499
2.885575	4.848549
2.8233	4.830164
2.76138	4.810367
2.638751	4.766634
2.578109	4.742749
2.517959	4.717554
2.458333	4.691077
2.340767	4.634391
2.282884	4.604241
2.225636	4.572926
2.169046	4.540479
2.057936	4.472315
2.034908	4.457427

TABLE 6A.3   Calculated Values of DEQ Variables (*Continued*)

τ	θ_2	θ_3	θ_4	θ_5
0	1.	1.	1.	1.
0.0041566	2.086665	1.330806	1.089216	1.089216
0.0061063	2.315835	1.519774	1.198002	1.198002
0.0086748	2.530454	1.743147	1.373664	1.373664
0.0102319	2.634986	1.866014	1.487711	1.487711
0.0120118	2.740827	1.997366	1.619466	1.619466
0.014049	2.85025	2.138137	1.76847	1.76847
0.0163758	2.965019	2.288786	1.933426	1.933426
0.0190093	3.085772	2.448501	2.1116	2.1116
0.0204361	3.148039	2.530854	2.204119	2.204119
0.0234674	3.274236	2.697135	2.391392	2.391392
0.0250459	3.337032	2.779423	2.484086	2.484086
0.0266459	3.398822	2.860052	2.574835	2.574835
0.0282459	3.458836	2.938025	2.662489	2.662489
0.0314459	3.573843	3.086512	2.829052	2.829052
0.0330459	3.628959	3.157253	2.908224	2.908224
0.0346459	3.682552	3.225792	2.984818	2.984818
0.0362459	3.734665	3.292221	3.05895	3.05895
0.0394459	3.834598	3.419073	3.200244	3.200244
0.0410459	3.882484	3.479645	3.267597	3.267597
0.0426459	3.929021	3.538402	3.332872	3.332872
0.044242	3.974123	3.595268	3.395995	3.395995
0.0474224	4.060156	3.703594	3.516134	3.516134
0.0490073	4.101151	3.755186	3.57332	3.57332
0.0505887	4.140837	3.805144	3.628684	3.628684
0.0521669	4.17924	3.853521	3.682296	3.682296
0.0553145	4.252288	3.945719	3.784508	3.784508
0.0568842	4.286972	3.989627	3.833219	3.833219
0.0584514	4.320454	4.032124	3.880399	3.880399
0.0600162	4.352749	4.073244	3.926094	3.926094
0.0631393	4.413831	4.151478	4.013182	4.013182
0.0646979	4.442643	4.188644	4.054648	4.054648
0.0662548	4.470315	4.224542	4.09477	4.09477
0.0693637	4.522276	4.292618	4.171094	4.171094
0.070916	4.54658	4.324832	4.207345	4.207345
0.0724671	4.569776	4.355852	4.242351	4.242351

TABLE 6A.4 Calculated Values of DEQ Variables

τ	θ₂	θ₃	θ₄	θ₅
0.0740171	4.591868	4.385692	4.276132	4.276132
0.0771142	4.632763	4.441883	4.340085	4.340085
0.0786616	4.651574	4.468256	4.370288	4.370288
0.0802084	4.669299	4.493493	4.399326	4.399326
0.0833006	4.701501	4.540593	4.45395	4.45395
0.0848462	4.715984	4.562471	4.479557	4.479557
0.0863918	4.72939	4.583242	4.504039	4.504039
0.0894829	4.752984	4.621482	4.549652	4.549652
0.0910288	4.763174	4.638961	4.570797	4.570797
0.0925749	4.772295	4.65535	4.59084	4.59084
0.0941216	4.780348	4.670653	4.609787	4.609787
0.0972167	4.793257	4.698011	4.644404	4.644404
0.0987655	4.798114	4.71007	4.660081	4.660081
0.1003152	4.801909	4.721052	4.674673	4.674673
0.1034179	4.806316	4.739791	4.70061	4.70061
0.1049712	4.80693	4.747551	4.711958	4.711958
0.1065259	4.806487	4.754238	4.722227	4.722227
0.1080822	4.804988	4.759855	4.731419	4.731419
0.1112001	4.798829	4.767877	4.746571	4.746571
0.112762	4.794172	4.770285	4.752531	4.752531
0.114326	4.788466	4.771624	4.757415	4.757415
0.1174611	4.773914	4.7711	4.763954	4.763954
0.1190324	4.765072	4.769237	4.765608	4.765608
0.1206065	4.75519	4.766308	4.766185	4.766185
0.1221835	4.744269	4.762314	4.765684	4.765684
0.125347	4.719323	4.75113	4.761449	4.761449
0.1269339	4.705303	4.743942	4.757713	4.757713
0.1285244	4.690255	4.73569	4.752898	4.752898
0.1301187	4.674182	4.726375	4.747003	4.747003
0.1333172	4.639005	4.704577	4.731985	4.731985
0.1349172	4.619964	4.692135	4.722891	4.722891
0.1365172	4.59998	4.678682	4.712756	4.712756
0.1381172	4.579065	4.66423	4.701588	4.701588
0.1413172	4.534508	4.632376	4.676198	4.676198
0.1429172	4.510896	4.614998	4.661998	4.661998

TABLE 6A.4 Calculated Values of DEQ Variables (*Continued*)

τ	θ_2	θ_3	θ_4	θ_5
0.1445172	4.486416	4.596672	4.646811	4.646811
0.1461172	4.461086	4.57741	4.630649	4.630649
0.1493172	4.40794	4.536137	4.595449	4.595449
0.1509172	4.38016	4.514155	4.576438	4.576438
0.1525172	4.3516	4.491298	4.556506	4.556506
0.1541172	4.322277	4.46758	4.535666	4.535666
0.1573172	4.261421	4.417627	4.491325	4.491325
0.1589172	4.229926	4.391427	4.467854	4.467854
0.1605172	4.197746	4.364433	4.443538	4.443538
0.1621172	4.1649	4.336664	4.418393	4.418393
0.1653172	4.097294	4.278871	4.365683	4.365683
0.1669172	4.062574	4.248884	4.338154	4.338154
0.1685172	4.027272	4.218195	4.309865	4.309865
0.1701172	3.991407	4.186824	4.280835	4.280835
0.1733172	3.918075	4.122111	4.220626	4.220626
0.1749172	3.88065	4.088809	4.189484	4.189484
0.1765172	3.842749	4.054904	4.157677	4.157677
0.1781172	3.804392	4.020415	4.125223	4.125223
0.1813172	3.726398	3.94977	4.058457	4.058457
0.1829172	3.686805	3.913654	4.024184	4.024184
0.1845172	3.646843	3.877038	3.989345	3.989345
0.1861172	3.606533	3.839942	3.953961	3.953961
0.1893172	3.524959	3.764393	3.881638	3.881638
0.1909172	3.483737	3.725984	3.844741	3.844741
0.1925172	3.442254	3.687178	3.807382	3.807382
0.1941172	3.400532	3.647999	3.769581	3.769581
0.1973172	3.316453	3.5686	3.692738	3.692738
0.1989172	3.274138	3.528424	3.653738	3.653738
0.2	3.245411	3.501067	3.62714	3.62714

TABLE 6A.4 Calculated Values of DEQ Variables (*Continued*)

θ_6	θ_7	θ_8	θ_9	θ_{10}
1.	1.	1.	1.	1.
1.015872	1.002354	1.00033	1.00033	1.000027
1.048488	1.010095	1.002085	1.002085	1.000258
1.119332	1.033237	1.009661	1.009661	1.001729
1.174595	1.05534	1.018759	1.018759	1.003988
1.245991	1.087941	1.034385	1.034385	1.008622
1.335299	1.134103	1.059812	1.059812	1.017562
1.443581	1.196921	1.098996	1.098996	1.033758
1.570549	1.278831	1.155928	1.155928	1.061133
1.64029	1.32719	1.19194	1.19194	1.080314
1.788991	1.437366	1.279096	1.279096	1.131515
1.866196	1.498014	1.329523	1.329523	1.163744
1.944031	1.561317	1.383667	1.383667	1.200162
2.021292	1.626149	1.440493	1.440493	1.240182
2.17365	1.759302	1.560813	1.560813	1.330173
2.24858	1.827141	1.623676	1.623676	1.379664
2.32259	1.895543	1.687958	1.687958	1.431813
2.395631	1.964321	1.753404	1.753404	1.486362
2.538644	2.10236	1.886865	1.886865	1.601643
2.60855	2.171335	1.954472	1.954472	1.661883
2.677356	2.240115	2.022422	2.022422	1.723546
2.744874	2.308423	2.090385	2.090385	1.786255
2.876027	2.443261	2.225803	2.225803	1.913986
2.939668	2.509652	2.293029	2.293029	1.978645
3.002025	2.575277	2.359804	2.359804	2.043615
3.063105	2.640087	2.426046	2.426046	2.108752
3.181467	2.7671	2.556645	2.556645	2.239012
3.238766	2.829233	2.620883	2.620883	2.303908
3.294823	2.890414	2.684346	2.684346	2.368515
3.349649	2.950618	2.74699	2.74699	2.432746
3.455641	3.068022	2.86968	2.86968	2.559772
3.506826	3.125191	2.929665	2.929665	2.622433
3.556818	3.181322	2.988709	2.988709	2.68445
3.653255	3.290435	3.103898	3.103898	2.806354
3.699718	3.343402	3.160009	3.160009	2.866158
3.745021	3.395303	3.215113	3.215113	2.925148

TABLE **6A.4** Calculated Values of DEQ Variables (*Continued*)

θ_6	θ_7	θ_8	θ_9	θ_{10}
3.789172	3.446134	3.2692	3.2692	2.983293
3.87405	3.544576	3.374284	3.374284	3.096944
3.91479	3.592184	3.425269	3.425269	3.152405
3.954405	3.638715	3.475208	3.475208	3.20693
4.030287	3.728546	3.571933	3.571933	3.313112
4.066563	3.771846	3.618713	3.618713	3.364743
4.101736	3.814071	3.664435	3.664435	3.415385
4.168789	3.895297	3.752701	3.752701	3.51367
4.200676	3.934299	3.795243	3.795243	3.561297
4.231475	3.972228	3.836722	3.836722	3.607906
4.261187	4.009086	3.87714	3.87714	3.653491
4.317365	4.07959	3.954788	3.954788	3.741573
4.343834	4.113237	3.992017	3.992017	3.784063
4.369225	4.145816	4.028184	4.028184	3.825514
4.416778	4.207767	4.097327	4.097327	3.90529
4.438941	4.237139	4.130303	4.130303	3.943609
4.46003	4.265444	4.162214	4.162214	3.980881
4.480045	4.292679	4.19306	4.19306	4.017102
4.516851	4.343941	4.251554	4.251554	4.086385
4.533641	4.367966	4.2792	4.2792	4.119443
4.549356	4.39092	4.305776	4.305776	4.151442
4.577557	4.433606	4.355716	4.355716	4.212256
4.59004	4.453336	4.379076	4.379076	4.241066
4.601444	4.471988	4.40136	4.40136	4.268808
4.611767	4.48956	4.422566	4.422566	4.29548
4.629163	4.521457	4.461733	4.461733	4.345599
4.636233	4.535776	4.479689	4.479689	4.369041
4.642215	4.549006	4.496556	4.496556	4.391399
4.647107	4.561143	4.512331	4.512331	4.41267
4.653608	4.582111	4.540567	4.540567	4.451905
4.655212	4.590916	4.55299	4.55299	4.469814
4.655726	4.598603	4.564284	4.564284	4.486581
4.655156	4.605178	4.574452	4.574452	4.502209
4.650795	4.61501	4.591429	4.591429	4.530058
4.64702	4.618279	4.598247	4.598247	4.542286
4.642192	4.620458	4.603958	4.603958	4.553389

TABLE 6A.4 Calculated Values of DEQ Variables (*Continued*)

θ_6	θ_7	θ_8	θ_9	θ_{10}
4.63632	4.621554	4.608567	4.608567	4.563369
4.621483	4.620524	4.614505	4.614505	4.579983
4.612537	4.618412	4.615846	4.615846	4.586626
4.602587	4.615248	4.616112	4.616112	4.592167
4.591642	4.611038	4.615309	4.615309	4.596611
4.566818	4.599519	4.610526	4.610526	4.602233
4.552961	4.592229	4.606564	4.606564	4.603424
4.538157	4.583931	4.601565	4.601565	4.603544
4.52242	4.574636	4.59554	4.59554	4.602601
4.488197	4.553097	4.580446	4.580446	4.597557
4.469739	4.540876	4.571398	4.571398	4.593472
4.450402	4.527703	4.561363	4.561363	4.588357
4.4302	4.51359	4.550353	4.550353	4.582222
4.387265	4.482595	4.525451	4.525451	4.566926
4.364563	4.46574	4.511583	4.511583	4.557787
4.341058	4.447997	4.496788	4.496788	4.547668
4.316767	4.429382	4.481077	4.481077	4.536579
4.265894	4.389588	4.446965	4.446965	4.511542
4.239345	4.368441	4.42859	4.42859	4.497616
4.212079	4.346479	4.409356	4.409356	4.48277
4.184113	4.32372	4.389276	4.389276	4.467015
4.126151	4.275871	4.346641	4.346641	4.432835
4.096193	4.250814	4.324116	4.324116	4.414437
4.065606	4.225025	4.300808	4.300808	4.395186
4.034412	4.19852	4.276732	4.276732	4.375096
3.970272	4.143433	4.226343	4.226343	4.332462
3.937365	4.114886	4.200063	4.200063	4.309948
3.914792	4.095199	4.18188	4.18188	4.29427

TABLE 6A.4   Calculated Values of DEQ Variables (*Continued*)

θ_1	θ_{11}	θ_1	θ_{11}
3.451769	1.	4.869191	2.887991
3.564818	0.9999262	4.887636	3.004497
3.6172	0.9996486	4.895314	3.06145
3.685481	0.9990846	4.901959	3.117504
3.726437	0.9990645	4.912136	3.226839
3.772813	1.000035	4.915664	3.280086
3.825283	1.003479	4.918151	3.332369
3.88435	1.012012	4.919993	3.433993
3.95001	1.029535	4.919347	3.483315
3.985022	1.043105	4.917655	3.531634
4.057997	1.082321	4.914919	3.578941
4.0952	1.108485	4.906314	3.670502
4.132325	1.138994	4.900447	3.714745
4.168838	1.173411	4.89354	3.757957
4.239938	1.253293	4.876609	3.841277
4.274482	1.298326	4.866591	3.881378
4.308328	1.346431	4.855541	3.920437
4.341453	1.397348	4.843462	3.958449
4.405461	1.506569	4.816234	4.031329
4.436304	1.564354	4.801092	4.066191
4.466346	1.62392	4.784937	4.099998
4.495497	1.684878	4.749608	4.164436
4.551082	1.810047	4.730444	4.195063
4.577499	1.87385	4.710286	4.224625
4.602988	1.938218	4.689142	4.253118
4.627541	2.002987	4.643916	4.306888
4.67381	2.133134	4.619847	4.332158
4.695512	2.19825	4.594815	4.356347
4.716248	2.263239	4.568827	4.37945
4.736013	2.327998	4.51406	4.422351
4.772602	2.456469	4.485379	4.442089
4.789414	2.520023	4.455866	4.46068
4.80523	2.58303	4.42554	4.478127
4.833852	2.707173	4.362526	4.509601
4.846649	2.768208	4.329879	4.523633
4.85843	2.828493	4.296498	4.536532

TABLE 6A.4 Calculated Values of DEQ Variables (*Continued*)

θ_1	θ_{11}	θ_1	θ_{11}
4.262404	4.548303	3.552023	4.573189
4.192163	4.568476	3.508635	4.564628
4.156059	4.576886	3.464987	4.55508
4.119329	4.584185	3.377009	4.533067
4.081994	4.590378	3.332723	4.520625
4.005601	4.599468	3.288269	4.507241
3.966588	4.602377	3.243671	4.492928
3.927061	4.604204	3.154126	4.461566
3.887044	4.604955	3.109223	4.444544
3.805629	4.603261	3.064263	4.426645
3.764279	4.60083	3.019265	4.407885
3.72253	4.597355	2.929242	4.367835
3.680407	4.592845	2.884257	4.346577
3.595131	4.580752	2.853837	4.331733

TABLE 6A.4 Calculated Values of DEQ Variables (*Continued*)

τ	θ₂	θ₃	θ₄	θ₅
0	1.	1.	1.	1.
0.0042383	4.780979	3.642092	3.027371	3.027371
0.0060699	5.255596	4.36068	3.875976	3.875976
0.0083599	5.731016	5.054708	4.690881	4.690881
0.0101903	6.046988	5.494018	5.198659	5.198659
0.0125179	6.391986	5.94976	5.715899	5.715899
0.0142175	6.615014	6.230209	6.028233	6.028233
0.0160506	6.835889	6.497116	6.320697	6.320697
0.0180111	7.05542	6.752494	6.595966	6.595966
0.0200089	7.266174	6.989543	6.84759	6.84759
0.0225962	7.525278	7.272135	7.143174	7.143174
0.024187	7.67888	7.436058	7.312787	7.312787
0.0264616	7.892867	7.661101	7.543894	7.543894
0.028643	8.093134	7.86914	7.756157	7.756157
0.0300583	8.220915	8.00095	7.890135	7.890135
0.0321347	8.405686	8.19065	8.082459	8.082459
0.034165	8.583483	8.372514	8.266454	8.266454
0.036157	8.755293	8.547884	8.443661	8.443661
0.0381166	8.921769	8.717644	8.61509	8.61509
0.0400482	9.083351	8.882386	8.781421	8.781421
0.0425863	9.291682	9.094911	8.996035	8.996035
0.0444654	9.44284	9.249286	9.152	9.152
0.0463259	9.589769	9.399538	9.303891	9.303891
0.0481699	9.732538	9.545766	9.451821	9.451821
0.0506054	9.916489	9.734568	9.64301	9.64301
0.0524166	10.04966	9.871573	9.781899	9.781899
0.0542156	10.17871	10.00464	9.916932	9.916932
0.0560037	10.30363	10.13374	10.04809	10.04809
0.0583724	10.4637	10.29966	10.2169	10.2169
0.0601382	10.57884	10.41939	10.33889	10.33889
0.06248	10.72572	10.57266	10.49531	10.49531
0.0642276	10.83084	10.68277	10.60789	10.60789
0.0665473	10.96418	10.82302	10.75155	10.75155
0.0682799	11.05901	10.92322	10.8544	10.8544
0.0700069	11.14935	11.01908	10.95299	10.95299
0.0723017	11.26272	11.14003	11.0777	11.0777

TABLE 6A.5 Calculated Values of DEQ Variables

τ	θ₂	θ₃	θ₄	θ₅
0.0740174	11.34238	11.22553	11.1661	11.1661
0.0762986	11.44132	11.33248	11.27702	11.27702
0.0780052	11.51002	11.40733	11.35492	11.35492
0.0802756	11.5942	11.49989	11.45164	11.45164
0.082541	11.6698	11.58406	11.54007	11.54007
0.0842371	11.72081	11.64162	11.60088	11.60088
0.0864954	11.78117	11.71085	11.67453	11.67453
0.0881871	11.82067	11.7571	11.72413	11.72413
0.0904405	11.86555	11.8111	11.78267	11.78267
0.0921293	11.89336	11.84582	11.82084	11.82084
0.0943798	11.92259	11.88435	11.86401	11.86401
0.0960673	11.93859	11.90738	11.89055	11.89055
0.098317	11.952	11.93024	11.91812	11.91812
0.1000045	11.95612	11.94147	11.9329	11.9329
0.1022553	11.95364	11.94851	11.94469	11.94469
0.1045076	11.94208	11.94648	11.94743	11.94743
0.1061983	11.92743	11.93898	11.94351	11.94351
0.1084548	11.89996	11.92103	11.93032	11.93032
0.1101494	11.8734	11.90159	11.91444	11.91444
0.1124122	11.83007	11.86771	11.88529	11.88529
0.1141123	11.79166	11.83634	11.85745	11.85745
0.1163836	11.73259	11.78658	11.81237	11.81237
0.1180909	11.68243	11.74334	11.7726	11.7726
0.120373	11.60778	11.67782	11.71166	11.71166
0.1220894	11.546	11.6228	11.66003	11.66003
0.124385	11.45597	11.54164	11.58335	11.58335
0.1261125	11.38274	11.47497	11.51997	11.51997
0.1284246	11.27756	11.37837	11.42769	11.42769
0.1301656	11.19307	11.30019	11.35268	11.35268
0.1324975	11.073	11.18835	11.24499	11.24499
0.1342548	10.97745	11.09881	11.15849	11.15849
0.1360202	10.87721	11.00447	11.06711	11.06711
0.1383875	10.73635	10.87123	10.93773	10.93773
0.1401741	10.62533	10.76576	10.83506	10.83506
0.1425725	10.4702	10.61778	10.6907	10.6907
0.1443849	10.34856	10.50131	10.57685	10.57685

TABLE 6A.5 Calculated Values of DEQ Variables (*Continued*)

τ	θ₂	θ₃	θ₄	θ₅
0.1462099	10.2224	10.38015	10.45822	10.45822
0.1480489	10.09173	10.2543	10.33481	10.33481
0.150525	9.910483	10.07918	10.16281	10.16281
0.1524022	9.769251	9.942325	10.02818	10.02818
0.1542988	9.623434	9.800683	9.888665	9.888665
0.1562172	9.47296	9.654178	9.744184	9.744184
0.15816	9.317716	9.502691	9.594616	9.594616
0.1601307	9.157543	9.346056	9.439791	9.439791
0.1621336	8.992206	9.184031	9.279464	9.279464
0.1641741	8.821366	9.016272	9.113288	9.113288
0.1662597	8.644526	8.842272	8.940752	8.940752
0.1684007	8.460934	8.661273	8.761095	8.761095
0.1706124	8.269397	8.472073	8.573109	8.573109
0.1721379	8.136341	8.340426	8.4422	8.4422
0.1745334	7.926117	8.132088	8.234854	8.234854
0.1762369	7.775856	7.982926	8.086277	8.086277
0.1780928	7.611634	7.819675	7.923548	7.923548
0.1803864	7.40816	7.617081	7.721439	7.721439
0.182631	7.208794	7.418239	7.522903	7.522903
0.1844559	7.04675	7.25638	7.361171	7.361171
0.1861085	6.900167	7.109783	7.214598	7.214598
0.1884274	6.694988	6.904299	7.009003	7.009003
0.1906236	6.501453	6.710177	6.814626	6.814626
0.192038	6.377348	6.585544	6.689754	6.689754
0.1941005	6.197271	6.404494	6.508252	6.508252
0.1961044	6.023497	6.229552	6.332758	6.332758
0.1980595	5.855257	6.059967	6.162531	6.162531
0.2	5.689688	5.89287	5.9947	5.9947

TABLE 6A.5   Calculated Values of DEQ Variables (*Continued*)

θ_6	θ_7	θ_8	θ_9	θ_{10}
1.	1.	1.	1.	1.
2.031586	1.496671	1.272658	1.272658	1.104013
2.735287	2.035088	1.708019	1.708019	1.368262
3.540206	2.778228	2.401573	2.401573	1.913481
4.108564	3.363962	2.987819	2.987819	2.44853
4.738334	4.055869	3.705956	3.705956	3.164247
5.139171	4.513101	4.190204	4.190204	3.673089
5.523336	4.959124	4.667123	4.667123	4.188588
5.887285	5.385099	5.124852	5.124852	4.692457
6.217013	5.770809	5.539675	5.539675	5.153482
6.595124	6.209183	6.009757	6.009757	5.677071
6.805897	6.450379	6.267082	6.267082	5.962673
7.084951	6.765145	6.600867	6.600867	6.330607
7.333117	7.040209	6.89031	6.89031	6.646385
7.486066	7.207412	7.065145	7.065145	6.835356
7.701248	7.439775	7.306714	7.306714	7.09409
7.903013	7.654871	7.52895	7.52895	7.329681
8.094275	7.856628	7.736314	7.736314	7.547515
8.27706	8.047825	7.931988	7.931988	7.751484
8.452798	8.230456	8.118265	8.118265	7.94443
8.677869	8.463068	8.354833	8.354833	8.188073
8.840634	8.630647	8.524911	8.524911	8.362492
8.998754	8.793107	8.689603	8.689603	8.530954
9.152557	8.950939	8.84949	8.84949	8.694211
9.351292	9.154775	9.055902	9.055902	8.904716
9.495776	9.302995	9.205995	9.205995	9.057701
9.636442	9.447393	9.352254	9.352254	9.206778
9.773332	9.588058	9.494795	9.494795	9.352116
9.950003	9.769889	9.679182	9.679182	9.540265
10.07811	9.90199	9.813259	9.813259	9.677225
10.243	10.07241	9.986408	9.986408	9.854339
10.36218	10.1959	10.11203	10.11203	9.983038
10.515	10.35471	10.27379	10.27379	10.14907
10.625	10.46937	10.39075	10.39075	10.26936
10.73096	10.58015	10.50391	10.50391	10.38596
10.86586	10.72171	10.64877	10.64877	10.53559

TABLE 6A.5 Calculated Values of DEQ Variables (*Continued*)

θ_6	θ_7	θ_8	θ_9	θ_{10}
10.96217	10.8232	10.75282	10.75282	10.64335
11.08398	10.95215	10.8853	10.8853	10.78096
11.17029	11.04399	10.97987	10.97987	10.87951
11.27855	11.15982	11.09946	11.09946	11.00457
11.37887	11.26796	11.21148	11.21148	11.12224
11.44882	11.34391	11.29041	11.29041	11.20555
11.53491	11.43821	11.38878	11.38878	11.30988
11.59404	11.50362	11.45731	11.45731	11.383
11.66552	11.58365	11.54158	11.54158	11.47352
11.71355	11.6382	11.59938	11.59938	11.53609
11.77008	11.70356	11.66914	11.66914	11.61234
11.80678	11.74698	11.71591	11.71591	11.66406
11.84804	11.79733	11.77077	11.77077	11.72562
11.8732	11.82937	11.80624	11.80624	11.76619
11.89896	11.86439	11.84589	11.84589	11.8127
11.91576	11.89055	11.87671	11.87671	11.85047
11.92245	11.9043	11.89399	11.89399	11.873
11.92346	11.91476	11.90917	11.90917	11.89523
11.91826	11.91667	11.91464	11.91464	11.90602
11.90336	11.91127	11.91398	11.91398	11.91248
11.88619	11.90123	11.90751	11.90751	11.91136
11.85532	11.87984	11.89087	11.89087	11.90186
11.82616	11.85778	11.87237	11.87237	11.8887
11.77929	11.82033	11.83964	11.83964	11.86309
11.73815	11.78622	11.80904	11.80904	11.83781
11.67532	11.73268	11.76016	11.76016	11.79598
11.62223	11.68649	11.71745	11.71745	11.75851
11.54351	11.61688	11.6524	11.6524	11.70041
11.47853	11.55865	11.59756	11.59756	11.65072
11.38401	11.47299	11.51636	11.51636	11.57631
11.30722	11.40274	11.44941	11.44941	11.51438
11.2254	11.32736	11.37727	11.37727	11.4472
11.10849	11.21887	11.27302	11.27302	11.34945
11.01496	11.13153	11.18879	11.18879	11.27
10.88249	11.0071	11.06842	11.06842	11.15587
10.77731	10.9078	10.97209	10.97209	11.06411

TABLE 6A.5   Calculated Values of DEQ Variables (*Continued*)

θ_6	θ_7	θ_8	θ_9	θ_{10}
10.66714	10.80336	10.87055	10.87055	10.96705
10.55195	10.69377	10.76378	10.76378	10.86465
10.39054	10.53956	10.61323	10.61323	10.71976
10.26356	10.4178	10.4941	10.4941	10.60475
10.13143	10.29071	10.36957	10.36957	10.48421
9.994063	10.15821	10.23954	10.23954	10.35804
9.851325	10.02015	10.10386	10.10386	10.22609
9.703037	9.876349	9.96234	9.96234	10.08818
9.548945	9.726548	9.814725	9.814725	9.94402
9.388691	9.570382	9.660644	9.660644	9.793253
9.221757	9.407327	9.499571	9.499571	9.635345
9.047373	9.236606	9.330727	9.330727	9.469513
8.864325	9.057003	9.152893	9.152893	9.294539
8.736518	8.931371	9.02838	9.02838	9.171848
8.533548	8.731484	8.830085	8.830085	8.976165
8.387713	8.587596	8.687205	8.687205	8.834958
8.227625	8.429397	8.529989	8.529989	8.67939
8.028286	8.23206	8.333701	8.333701	8.484888
7.831934	8.037317	8.139807	8.139807	8.292476
7.671602	7.878039	7.981094	7.981094	8.134775
7.526008	7.733207	7.836675	7.836675	7.991126
7.321335	7.529299	7.633194	7.633194	7.788492
7.127352	7.335717	7.439856	7.439856	7.59571
7.002492	7.210954	7.315167	7.315167	7.471255
6.820679	7.029058	7.133268	7.133268	7.289523
6.644515	6.852566	6.956648	6.956648	7.112876
6.473296	6.680798	6.784638	6.784638	6.940663
6.304159	6.510899	6.614392	6.614392	6.770045

TABLE 6A.5 Calculated Values of DEQ Variables (*Continued*)

θ_1	θ_{11}	θ_1	θ_{11}
5.895865	1.	11.51255	10.60687
6.26422	1.047798	11.59932	10.74618
6.42506	1.25501	11.65869	10.84605
6.627268	1.750783	11.73015	10.97294
6.789592	2.268767	11.79275	11.0925
6.996666	2.983677	11.83385	11.17726
7.148167	3.500718	11.8808	11.28358
7.31167	4.029076	11.91009	11.35823
7.486475	4.548326	11.94122	11.45083
7.664347	5.024751	11.9586	11.515
7.893943	5.566176	11.9738	11.59341
8.034483	5.861203	11.97921	11.64678
8.234333	6.24052	11.97843	11.71057
8.424489	6.565076	11.97185	11.75283
8.546927	6.758759	11.95509	11.80164
8.725018	7.023216	11.92923	11.84172
8.897146	7.263258	11.90387	11.866
9.063858	7.484581	11.86214	11.89059
9.22552	7.691316	11.82493	11.90314
9.382383	7.886486	11.76748	11.91198
9.584358	8.132487	11.71855	11.91264
9.730598	8.308352	11.64558	11.90552
9.87242	8.478071	11.5851	11.89414
10.00986	8.642452	11.49686	11.8709
10.18633	8.85432	11.42503	11.84739
10.31359	9.00827	11.32179	11.80792
10.43647	9.158286	11.23883	11.7722
10.55495	9.304557	11.12092	11.71641
10.70602	9.49396	11.02708	11.66844
10.8141	9.631881	10.89485	11.59629
10.95118	9.810315	10.79039	11.53604
11.04865	9.940041	10.68146	11.47051
11.17143	10.10749	10.52935	11.37492
11.25807	10.22889	10.41016	11.29707
11.33999	10.34664	10.24452	11.18502
11.4418	10.49787	10.11529	11.09479

TABLE 6A.5   Calculated Values of DEQ Variables (*Continued*)

θ_1	θ_{11}	θ_1	θ_{11}
9.981812	10.99922	7.616108	9.024859
9.844093	10.89828	7.464372	8.884209
9.653886	10.75527	7.29888	8.729191
9.50628	10.64163	7.094314	8.535284
9.354393	10.52242	6.89438	8.343366
9.198162	10.39754	6.73223	8.186002
9.037486	10.26684	6.585822	8.04261
8.872213	10.13012	6.381311	7.840258
8.702118	9.987118	6.188849	7.647661
8.526873	9.837455	6.065655	7.523284
8.345992	9.680602	5.887207	7.341609
8.158736	9.515775	5.715347	7.164952
7.963926	9.341754	5.549274	6.992671
7.828913	9.219671	5.386139	6.821929

TABLE 6A.5   Calculated Values of DEQ Variables (*Continued*)

τ	θ₂	θ₃	θ₄	θ₅
0	1.	1.	1.	1.
0.0042183	2.910064	1.474006	1.099089	1.099089
0.0062402	3.403391	1.793282	1.239537	1.239537
0.0081174	3.76064	2.080816	1.410721	1.410721
0.0104697	4.13035	2.421689	1.661429	1.661429
0.0123444	4.387738	2.679225	1.878653	1.878653
0.0144945	4.658204	2.962302	2.1379	2.1379
0.0169749	4.949579	3.276361	2.442893	2.442893
0.0183429	5.103957	3.445055	2.611802	2.611802
0.021355	5.433221	3.807506	2.982085	2.982085
0.022955	5.603701	3.995805	3.176941	3.176941
0.024555	5.771807	4.181563	3.370131	3.370131
0.026155	5.937867	4.364997	3.561494	3.561494
0.029355	6.264632	4.725489	3.938442	3.938442
0.030955	6.425555	4.90276	4.123979	4.123979
0.032555	6.58491	5.078147	4.307564	4.307564
0.034155	6.74271	5.251699	4.489217	4.489217
0.037355	7.053604	5.593437	4.84683	4.84683
0.038955	7.206642	5.761661	5.022839	5.022839
0.040555	7.358019	5.928133	5.197014	5.197014
0.042155	7.507686	6.092852	5.369371	5.369371
0.045355	7.801663	6.416989	5.708678	5.708678
0.046955	7.94585	6.576375	5.875639	5.875639
0.048555	8.088079	6.73394	6.040802	6.040802
0.050155	8.228282	6.889656	6.20416	6.20416
0.053355	8.502319	7.195402	6.525403	6.525403
0.054955	8.636003	7.345353	6.683249	6.683249
0.056555	8.767363	7.493297	6.839209	6.839209
0.058155	8.896321	7.639186	6.993254	6.993254
0.061355	9.146718	7.924594	7.295452	7.295452
0.062955	9.267999	8.064004	7.443525	7.443525
0.064555	9.386562	8.201141	7.589523	7.589523
0.066155	9.50233	8.335947	7.733396	7.733396
0.069355	9.725163	8.598314	8.014565	8.014565
0.070955	9.832074	8.725749	8.151753	8.151753

TABLE 6A.6   Calculated Values of DEQ Variables

τ	θ_2	θ_3	θ_4	θ_5
0.072555	9.935881	8.850599	8.286601	8.286601
0.074155	10.03651	8.972798	8.41905	8.41905
0.077355	10.22793	9.208981	8.676508	8.676508
0.078955	10.31859	9.322833	8.801393	8.801393
0.080555	10.40578	9.433769	8.923632	8.923632
0.082155	10.48945	9.541725	9.04316	9.04316
0.085355	10.64596	9.748436	9.273829	9.273829
0.086955	10.71868	9.847063	9.384841	9.384841
0.088555	10.78764	9.942455	9.492886	9.492886
0.090155	10.85278	10.03455	9.597901	9.597901
0.093355	10.97143	10.20862	9.798591	9.798591
0.094955	11.02485	10.29048	9.894144	9.894144
0.096555	11.07427	10.36882	9.986422	9.986422
0.098155	11.11967	10.44358	10.07537	10.07537
0.101355	11.19827	10.5822	10.24304	10.24304
0.102955	11.23142	10.64596	10.32166	10.32166
0.104555	11.26043	10.70596	10.39673	10.39673
0.106155	11.2853	10.76218	10.46821	10.46821
0.109355	11.32254	10.86309	10.6002	10.6002
0.110955	11.33491	10.90772	10.66063	10.66063
0.112555	11.34311	10.94844	10.7173	10.7173
0.114155	11.34714	10.98522	10.77017	10.77017
0.117355	11.34272	11.04689	10.8644	10.8644
0.118955	11.33431	11.07176	10.9057	10.9057
0.120555	11.32179	11.09263	10.9431	10.9431
0.122155	11.30518	11.1095	10.97657	10.97657
0.125355	11.25982	11.13124	11.03168	11.03168
0.126955	11.23114	11.13612	11.05329	11.05329
0.128555	11.19851	11.13702	11.07093	11.07093
0.130155	11.16197	11.13395	11.08461	11.08461
0.133355	11.07739	11.11597	11.10007	11.10007
0.134955	11.02945	11.10111	11.10186	11.10186
0.136555	10.97782	11.08237	11.09972	11.09972
0.138155	10.92256	11.05979	11.09365	11.09365
0.141355	10.80142	11.00322	11.06982	11.06982

TABLE 6A.6   Calculated Values of DEQ Variables (*Continued*)

τ	θ₂	θ₃	θ₄	θ₅
0.142955	10.73568	10.96932	11.05212	11.05212
0.144555	10.6666	10.93174	11.03059	11.03059
0.146155	10.59425	10.89052	11.00528	11.00528
0.149355	10.44006	10.7974	10.94344	10.94344
0.150955	10.35839	10.7456	10.90701	10.90701
0.152555	10.27379	10.6904	10.86696	10.86696
0.154155	10.18635	10.63185	10.82334	10.82334
0.157355	10.00331	10.505	10.72564	10.72564
0.158955	9.907896	10.43684	10.67166	10.67166
0.160555	9.81002	10.36562	10.61436	10.61436
0.162155	9.709778	10.29142	10.55379	10.55379
0.165355	9.502594	10.1344	10.42314	10.42314
0.166955	9.395854	10.05175	10.3532	10.3532
0.168555	9.287153	9.966439	10.28029	10.28029
0.170155	9.176593	9.878569	10.20447	10.20447
0.173355	8.950312	9.695494	10.04448	10.04448
0.174955	8.834799	9.600471	9.960465	9.960465
0.176555	8.717843	9.503247	9.873884	9.873884
0.178155	8.599547	9.403916	9.784825	9.784825
0.181355	8.359353	9.199312	9.599627	9.599627
0.182955	8.237661	9.094229	9.50367	9.50367
0.184555	8.115042	8.987421	9.405597	9.405597
0.186155	7.991597	8.878984	9.3055	9.3055
0.189355	7.742629	8.657612	9.099613	9.099613
0.190955	7.617305	8.54487	8.994012	8.994012
0.192555	7.491549	8.430887	8.886766	8.886766
0.194155	7.365457	8.315759	8.777972	8.777972
0.197355	7.11264	8.08245	8.556119	8.556119
0.198955	6.986098	7.964459	8.443252	8.443252
0.2	6.903462	7.886978	8.368901	8.368901

TABLE 6A.6   Calculated Values of DEQ Variables (*Continued*)

θ_6	θ_7	θ_8	θ_9	θ_{10}
1.	1.	1.	1.	1.
1.013964	1.001626	1.000174	1.000174	1.000011
1.047409	1.007877	1.001257	1.001257	1.000118
1.100704	1.021032	1.004364	1.004364	1.000543
1.196764	1.050772	1.013505	1.013505	1.002195
1.293689	1.086367	1.026837	1.026837	1.00518
1.423022	1.140571	1.050533	1.050533	1.011514
1.591213	1.220349	1.09077	1.09077	1.024276
1.690839	1.27182	1.119338	1.119338	1.034474
1.923265	1.402319	1.198763	1.198763	1.066542
2.052505	1.480459	1.25018	1.25018	1.089643
2.184891	1.564168	1.307853	1.307853	1.117354
2.319915	1.653047	1.371594	1.371594	1.149902
2.596298	1.844839	1.516325	1.516325	1.230097
2.737044	1.947077	1.596782	1.596782	1.277898
2.87917	2.053132	1.682258	1.682258	1.33085
3.02248	2.16272	1.772469	1.772469	1.388915
3.311996	2.391461	1.965968	1.965968	1.520043
3.45792	2.510137	2.068713	2.068713	1.59287
3.604459	2.631388	2.175104	2.175104	1.670346
3.7515	2.75501	2.284895	2.284895	1.752305
4.04668	3.008602	2.513729	2.513729	1.928961
4.194627	3.138214	2.632328	2.632328	2.023282
4.34269	3.26948	2.753435	2.753435	2.121341
4.490781	3.40224	2.876852	2.876852	2.222944
4.786706	3.67165	3.12987	3.12987	2.436006
4.934372	3.808014	3.25912	3.25912	2.54708
5.081732	3.945305	3.389974	3.389974	2.660933
5.228703	4.083392	3.522277	3.522277	2.777381
5.521157	4.361465	3.790626	3.790626	3.01735
5.666479	4.501211	3.926388	3.926388	3.140527
5.811093	4.641279	4.063029	4.063029	3.26561
5.954917	4.781556	4.200417	4.200417	3.392441
6.23988	5.062313	4.476941	4.476941	3.650724
6.38086	5.202583	4.615837	4.615837	3.78188
6.520734	5.342646	4.755002	4.755002	3.914189

TABLE 6A.6 Calculated Values of DEQ Variables (*Continued*)

θ_6	θ_7	θ_8	θ_9	θ_{10}
6.659422	5.482401	4.894324	4.894324	4.047513
6.932923	5.760604	5.173009	5.173009	4.316676
7.067579	5.898861	5.312161	5.312161	4.45226
7.200734	6.036431	5.451051	5.451051	4.588346
7.332309	6.173221	5.589579	5.589579	4.724816
7.590408	6.444105	5.865157	5.865157	4.998445
7.716777	6.578021	6.002017	6.002017	5.135379
7.841257	6.710802	6.138133	6.138133	5.272248
7.963771	6.842364	6.273413	6.273413	5.408945
8.202604	7.101487	6.541105	6.541105	5.681411
8.318773	7.228883	6.67334	6.67334	5.816978
8.432681	7.354724	6.804385	6.804385	5.95197
8.544255	7.47893	6.934154	6.934154	6.08629
8.760122	7.722118	7.189527	7.189527	6.35254
8.864277	7.840944	7.314966	7.314966	6.484284
8.965824	7.957822	7.438797	7.438797	6.614987
9.064699	8.072678	7.56094	7.56094	6.74456
9.254176	8.296025	7.79985	7.79985	6.999969
9.344657	8.404374	7.916463	7.916463	7.125633
9.432223	8.510412	8.031079	8.031079	7.249826
9.516816	8.614071	8.143626	8.143626	7.372465
9.676874	8.81399	8.362222	8.362222	7.612759
9.752238	8.910121	8.468131	8.468131	7.730257
9.824428	9.003617	8.571688	8.571688	7.845886
9.8934	9.094419	8.672829	8.672829	7.95957
10.02152	9.267711	8.8676	8.8676	8.18081
10.0806	9.350093	8.961106	8.961106	8.288222
10.1363	9.429562	9.051947	9.051947	8.393403
10.1886	9.50607	9.140065	9.140065	8.496285
10.28289	9.650017	9.30791	9.30791	8.694888
10.32482	9.717369	9.387534	9.387534	8.790482
10.36326	9.781586	9.464225	9.464225	8.883523
10.39818	9.842632	9.537937	9.537937	8.973952
10.45742	9.955073	9.676249	9.676249	9.146747
10.48173	10.00641	9.740766	9.740766	9.229005
10.50248	10.05445	9.802141	9.802141	9.308436

TABLE **6A.6** Calculated Values of DEQ Variables (*Continued*)

θ_6	θ_7	θ_8	θ_9	θ_{10}
10.51968	10.09917	9.860339	9.860339	9.384989
10.54343	10.17858	9.967079	9.967079	9.529285
10.55	10.21323	10.01556	10.01556	9.59694
10.55303	10.2445	10.06076	10.06076	9.661548
10.55256	10.27238	10.10264	10.10264	9.72307
10.54114	10.31792	10.17641	10.17641	9.836726
10.53024	10.33558	10.20827	10.20827	9.888798
10.51592	10.34984	10.23675	10.23675	9.937662
10.49819	10.3607	10.26186	10.26186	9.983296
10.45269	10.37225	10.30195	10.30195	10.06479
10.42499	10.37297	10.31692	10.31692	10.10061
10.39403	10.37034	10.32852	10.32852	10.13313
10.35986	10.36438	10.33675	10.33675	10.16234
10.2821	10.34256	10.34315	10.34315	10.21081
10.23859	10.32674	10.34135	10.34135	10.23005
10.19207	10.30771	10.33623	10.33623	10.24597
10.1426	10.28547	10.32783	10.32783	10.25857
10.03503	10.23157	10.30126	10.30126	10.27384
9.977043	10.19997	10.28314	10.28314	10.27653
9.916346	10.16534	10.26185	10.26185	10.27594
9.853001	10.12771	10.23741	10.23741	10.27208
9.718639	10.04368	10.17926	10.17926	10.25467
9.647759	9.997365	10.14562	10.14562	10.24117
9.57451	9.948262	10.109	10.109	10.22449
9.498963	9.896424	10.06945	10.06945	10.20469
9.341274	9.784772	9.981735	9.981735	10.1558
9.259284	9.725079	9.933675	9.933675	10.12679
9.204653	9.684741	9.90081	9.90081	10.10623

TABLE 6A.6 Calculated Values of DEQ Variables (*Continued*)

θ_1	θ_{11}	θ_1	θ_{11}
5.895865	1.	11.51802	3.765243
6.262468	0.9999564	11.63663	4.031232
6.440065	0.9997391	11.68963	4.165626
6.605801	0.9992687	11.73834	4.300778
6.814416	0.9984244	11.78271	4.436562
6.981218	0.9979613	11.85821	4.709541
7.172866	0.9985078	11.88926	4.8465
7.394102	1.002111	11.91579	4.983619
7.516047	1.006187	11.93778	5.120789
7.783933	1.022468	11.96801	5.394846
7.925686	1.036131	11.97623	5.531524
8.066908	1.053855	11.97982	5.667831
8.207478	1.076004	11.9788	5.803669
8.486183	1.134688	11.96289	6.073544
8.624073	1.171603	11.94803	6.20739
8.760822	1.213715	11.92859	6.340384
8.896307	1.261063	11.90458	6.472434
9.162983	1.371401	11.84299	6.733342
9.293925	1.434256	11.80549	6.862023
9.423101	1.502093	11.76356	6.989408
9.550387	1.574775	11.71726	7.115411
9.798793	1.734038	11.61176	7.362938
9.919667	1.820266	11.55268	7.484299
10.03816	1.910643	11.48946	7.603951
10.15415	2.004975	11.42219	7.721817
10.37814	2.204717	11.27576	7.95188
10.48591	2.309733	11.19677	8.063928
10.59072	2.417919	11.11404	8.173889
10.69244	2.529082	11.02767	8.281692
10.88623	2.759591	10.84438	8.490548
10.97808	2.878573	10.74766	8.591465
11.06644	2.999804	10.64769	8.689956
11.15121	3.123115	10.54458	8.785957
11.30959	3.375319	10.32939	8.970246
11.38304	3.503895	10.21754	9.058418
11.45254	3.633919	10.10299	9.143867

TABLE 6A.6   Calculated Values of DEQ Variables (*Continued*)

θ_1	θ_{11}		θ_1	θ_{11}
9.985878	9.226539		7.578577	10.19296
9.744423	9.383353		7.436018	10.21589
9.620325	9.457399		7.293327	10.23549
9.494143	9.528477		7.00799	10.2647
9.366003	9.596546		6.865561	10.27432
9.104344	9.723497		6.723433	10.28063
8.971075	9.782308		6.581707	10.28364
8.836348	9.837966		6.299859	10.27981
8.700286	9.89044		6.159928	10.27302
8.42466	9.985733		6.020782	10.26299
8.285343	10.02851		5.882509	10.24976
8.145186	10.068		5.608922	10.21382
8.004312	10.10421		5.473769	10.19116
7.720889	10.16669		5.386139	10.1747

TABLE **6A.6**   Calculated Values of DEQ Variables (*Continued*)

CHAPTER 7

Structural Analysis and Design

Notation

y	Total displacement of a one degree of freedom (ODOF) system
y_d	Dynamic portion of a total displacement y
Δ_{1t}	Static portion of total displacement owing to temperature load
α	Coefficient of linear expansion
α_o	Coefficient of linear expansion for steel
L	Linear dimension of a structural element
ω	Natural frequency (vertical or horizontal vibrations) of a structural system (or element)
$K_d = \dfrac{y_d}{\Delta_{1t}}$	Dynamic coefficient
$y01 = y_d/L$	Dimensionless displacement
$T(t)$	Temperature-time functions defined by Eqs. (6.78), (6.81), (6.84), and (6.87)
Time	$t = \dfrac{h^2}{a}\tau$ (s)
Temperature	$T = \dfrac{RT_*^2}{E}\theta + T_*$ (K), where $T_* = 600$ K is the baseline temperature
θ	Dimensionless temperature
τ	Dimensionless time
$K_v = A_o h/V$	Dimensionless opening factor

A_o	Total area of vertical and horizontal openings
σ	Stress value
ε	Strain value
E	Hook's modulus of elastisity (short-term modulus of elastisity)
H	Long-term modulus of elastisity
$n = H/E$	Relaxation time
$K(t - \tau)$	Kernel of integral equation (7.26)
g	Gravitational acceleration
W	Total gravity load
w_u	Ultimate design load (klf)
M_u	Ultimate bending moment (kip-ft)
V_u	Ultimate shear (kip)
N_u	Ultimate axial force (kip)
δ_{11}	Deformation from unit force

7.1   Introduction

There are many factors affecting structural behavior in fire, such as material degradation at elevated temperatures, restrained thermal expansion, thermal bowing, and the degree of redundancy available when the structure acts as a whole. Each factor is addressed separately, but in an integrated structure exposed to fire, they all will interact to generate more complex structural behavior. Traditionally, steel fire design has been based on fire-resistance testing, although fire resistance by calculation also has been implemented for many years.

The Eurocodes are a collection of the most recent methodologies for structural design, such as *Eurocode 3 (EC3): Design of Steel Structures*, Part 1.2. Structural Fire Design, and *Eurocode 4 (EC4): Design of Steel and Composite Structures*, Part 1.2. Each Eurocode is supplemented by a national application document (NAD) appropriate to the country. All Eurocodes are presented in a limit-state format, where partial safety factors are used to modify loads and material strengths. *EC3* and *EC4* are very similar to BS 5950, Part 8, although some of the terminology differs. *EC3* and *EC4*, Part 1.2, and BS 5950, Part 8, are concerned only with calculating the fire resistance of steel or composite sections. Three levels of calculation are described in *EC3* and *EC4*, tabular methods, simple calculation models, and advanced calculation models.

Tabular methods are look-up tables for direct design based on parameters such as loading, geometry, and reinforcement. They relate to most common designs. *Simple calculations* are based on principles such as plastic analysis, taking into account reduction in material strength with temperature. These are more accurate than the tabular methods. *Advanced calculation methods* relate to computer analyses and are not used in general design.

The structural engineering community needs structural fire load (SFL) information that is understandable by a reasonably intelligent professional. The main objective of this chapter is to obtain the approximate analytical solution of a structural system subjected to SFL, but in such a simple form that it can be used in ordinary structural engineering practice. The methodology of finding the solution in this case is similar to that for any other environmental structural design load (e.g., wind load, seismic load, etc.). The main idea here is based on substituting a very complex analysis of corresponding dynamic structural system with an equivalent system with one degree of freedom (ODOF) that has a simple form and is easy to use. In the case of wind load, for example, most international codes and standards use the so-called *gust-loading factor* approach for assessing the dynamic along wind loads and their effects on the structure. Variations of these models have been adopted by major international codes and standards. Although a similar theoretical basis is used in the SFL case by introducing a dynamic coefficient, there is a considerable difference in application of such a method based on fire severity classification.

7.2 Fire Severity and Dynamic Coefficient

One can expect a significant dynamic effect of a temperature-time load on structural analysis and design if the fire growth constant is large. Similar to the wind load, any thermal load has two components: static and dynamic (the second derivative of the temperature-time function multiplied by the mass and the coefficient of linear expansion produce the dynamic force acting on a structural system). If the temperature acceleration is high, so is the dynamic force. In order to calculate the dynamic effect (dynamic coefficient), an ODOF structural system is analyzed below:

$$\ddot{y}_d + \omega^2 y_d = -\ddot{\Delta}_{1t} \qquad (7.1)$$

where y_d is the dynamic portion of a total displacement y, that is,

$$y = \Delta_{1t} + y_d \qquad (7.2)$$

and Δ_{1t} is the static portion of total displacement owing to temperature:

$$\ddot{\Delta}_{1t} = \alpha \ddot{T}(t)L \qquad (7.3)$$

$$\Delta_{1t} = \alpha T(t)L \qquad (7.4)$$

where α is the coefficient of linear expansion, L is the linear dimension of a structural element, and ω is the natural frequency (vertical or horizontal vibrations) of the structural system (or element), which can be calculated using any classical method or approximate formula from ASCE-7-05 [1] for horizontal vibrations only.

The dynamic coefficient now is defined as

$$K_d = \frac{y_d}{\Delta_{1t}} \qquad (7.5)$$

In order to obtain the dynamic coefficient, let's now substitute Eq. (7.3) into Eq. (7.1). Equation (7.1) then can be rewritten after introducing the dimensionless displacement $y01 = y_d/L$ and real time t from Eq. (6.70) as follows:

$$\ddot{y}01 + \omega^2 y01 = -A\ddot{T}(t) \qquad (7.6)$$

where $\quad A = \alpha_o T_{max}$

α_o = coefficient of linear expansion for steel
T_{max} = maximum gas temperature for each fire severity case
$T(t)$ = temperature-time functions defined by Eqs. (6.78), (6.81), (6.84), and (6.87), where dimensionless time τ is substituted by real time t [see Eq. (6.70)]

All these differential equations (and solutions) are presented below for each fire severity case (the natural frequencies ω range from 1 to 5 Hz). The dynamic coefficient for all four fire severity cases is zero if the natural frequency of a structural system is more then 5 Hz.

The results are summarized and presented below (see Table 7.14). The most significant dynamic effect the thermal load has occurs in Cases 1, 2, and 3 (very fast, fast, and medium fire growth category). It is also shown that the most vulnerable are the flexible structures ($\omega < 1.0$ Hz) [1]. In the case of very fast, and fast fires, some type of beating vibrations can be induced (see Figs. 7.1 and 7.2). [Compare with "normal" oscillations in the case of more rigid structures ($\omega > 1.0$ Hz); see Fig. 7.3.]

For detailed information regarding all practical applications of SFL in steel and reinforced-concrete design, see Razdolsky [2,3]. For material on the temperature effect on steel and reinforced-concrete materials using general creep theory, see Razdolsky [3]. The dynamic coefficient values for each fire severity case are presented below (the structural system is substituted here by an ODOF system with corresponding natural frequency ω).

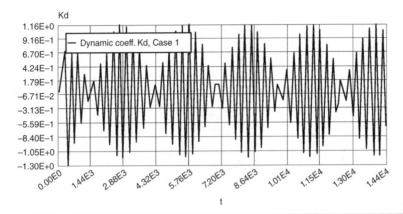

Figure 7.1 Beating vibrations owing to the dynamic portion of the temperature-time curve. $\omega = 0.5$ Hz, very fast fire.

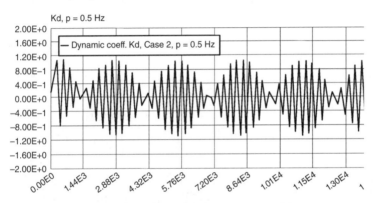

Figure 7.2 Beating vibrations owing to the dynamic portion of the temperature-time curve. $\omega = 0.5$ Hz, fast fire.

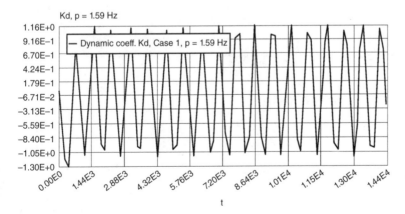

Figure 7.3 Normal vibrations owing to the dynamic portion of the temperature-time curve. $\omega = 1.59$ Hz, very fast fire.

Case 1: Ultrafast Fire

$K_v = 0.05$ $\omega = 0.5$ Hz

	Variable	Initial Value	Minimal Value	Maximal Value	Final Value
1	t	0	0	3600.	3600.
2	y01	0	−0.0355972	1.62933	0.1325948
3	y1	0	−2.170322	2.169438	−1.743039

TABLE 7.1   Calculated Values of DEQ Variables

Differential Equations

1. $d(y01)/d(t) = y1$
2. $d(y1)/d(t) = -(10)*y01+(1)*0.01225*(10^0)*(560+0.27*t- 8.41*(10^-5)*t^2)$

$$(7.7)$$

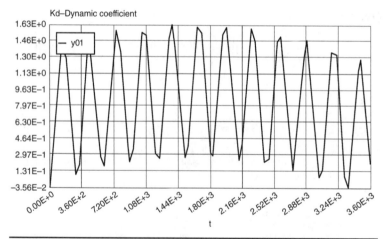

FIGURE 7.4   Dynamic coefficient.

$K_v = 0.05$ $\omega = 1.0$ Hz

	Variable	Initial Value	Minimal Value	Maximal Value	Final Value
1	t	0	0	3600.	3600.
2	y01	0	−0.0357968	0.4135937	−0.0357968
3	y1	0	−1.092049	1.090919	−0.1105263

TABLE 7.2   Calculated Values of DEQ Variables

Differential Equations

1. $d(y01)/d(t) = y1$
2. $d(y1)/d(t) = -(39.5)*y01+(1)*0.01225*(10^0)*(560+0.27*t-8.41*(10^-5)*t^2)$

$$(7.8)$$

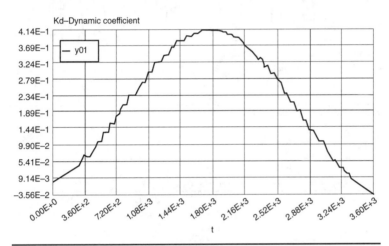

FIGURE 7.5 Dynamic coefficient.

$K_v = 0.05$ $\omega = 1.59$ Hz

	Variable	Initial Value	Minimal Value	Maximal Value	Final Value
1	t	0	0	3600.	3600.
2	y01	0	−0.0115896	0.163518	−0.1147551
3	y1	0	−0.6865162	0.6859081	−0.3229965

TABLE 7.3 Calculated Values of DEQ Variables

Differential Equations

1. $d(y01)/d(t) = y1$
2. $d(y1)/d(t) = -(100)*y01+(1)*0.01225*(10^0)*(560+0.27*t-8.41*(10^-5)*t^2)$

$$(7.9)$$

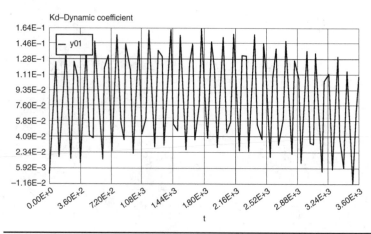

Figure 7.6 Dynamic coefficient.

$K_v = 0.05$ $\omega = 5.0$ Hz

	Variable	Initial Value	Minimal Value	Maximal Value	Final Value
1	t	0	0	3600.	3600.
2	y01	0	−0.0009372	0.0163063	0.0122615
3	y1	0	−0.2173595	0.2169672	−0.0222569

Table 7.4 Calculated Values of DEQ Variables

Differential Equations

1. $d(y01)/d(t) = y1$
2. $d(y1)/d(t) = -(1000)*y01+(1)*0.01225*(10^0)*(560+0.27*t-8.41*(10^{-5})*t^2)$

$$(7.10)$$

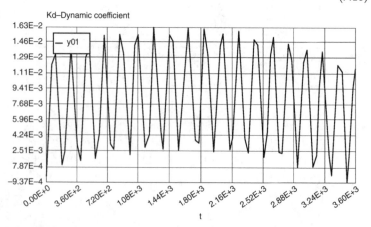

Figure 7.7 Dynamic coefficient.

Case 2: Fast Fire

$K_v = 0.05$ $\omega = 1.0$ Hz

	Variable	Initial Value	Minimal Value	Maximal Value	Final Value
1	t	0	0	3600.	3600.
2	y01	0	–0.4256885	0.7537212	–0.4256885
3	y1	0	–3.08235	3.082485	–0.3114667

TABLE 7.5   Calculated Values of DEQ Variables

Differential Equations

1. $d(y01)/d(t) = y1$
2. $d(y1)/d(t) = -(39.5)*y01+(1)*0.00876*(10^0)*(2212-0.634*t+2.72*(10^{-5})*t^2)$

$$(7.11)$$

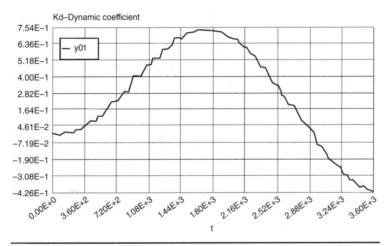

FIGURE 7.8   Dynamic coefficient.

$K_v = 0.05$ $\omega = 1.59$ Hz

	Variable	Initial Value	Minimal Value	Maximal Value	Final Value
1	t	0	0	3600.	3600.
2	y01	0	–0.1661328	0.3769301	0.1958176
3	y1	0	–1.937946	1.938031	–0.9121385

TABLE 7.6   Calculated Values of DEQ Variables

Differential Equations

1. $d(y01)/d(t) = y1$
2. $d(y1)/d(t) = -(100)*y01+(1)*0.00876*(10^0)*(2212-0.634*t+2.72*(10^-5)*t^2)$

$$(7.12)$$

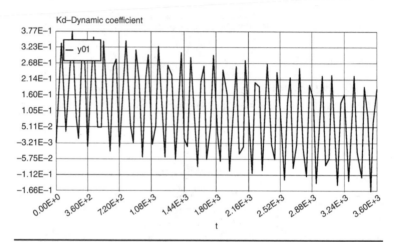

FIGURE 7.9   Dynamic coefficient.

$K_v = 0.05$ $\omega = 5.0$ Hz

	Variable	Initial Value	Minimal Value	Maximal Value	Final Value
1	t	0	0	3600.	3600.
2	y01	0	−0.0160898	0.0370503	0.0217861
3	y1	0	−0.6137264	0.6125758	−0.0628856

TABLE 7.7   Calculated Values of DEQ Variables

Differential Equations

1. $d(y01)/d(t) = y1$
2. $d(y1)/d(t) = -(1000)*y01+(1)*0.00876*(10^0)*(2212-0.634*t+2.72*(10^-5)*t^2)$

$$(7.13)$$

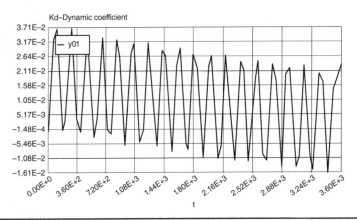

FIGURE 7.10   Dynamic coefficient.

Case 3: Medium Fire
$K_v = 0.05$ $\omega = 1.0$ Hz

	Variable	Initial Value	Minimal Value	Maximal Value	Final Value
1	t	0	0	3600.	3600.
2	y01	0	−0.3185956	0.591878	−0.3185956
3	y1	0	−2.409501	2.40855	−0.2433604

TABLE 7.8   Calculated Values of DEQ Variables

Differential Equations

1. $d(y01)/d(t) = y1$
2. $d(y1)/d(t) = -(39.5)*y01+(1)*0.007125*(10^0)*(2125-0.614*t+3.35*(10^{-5})*t^2)$

$$(7.14)$$

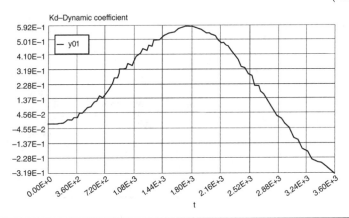

FIGURE 7.11   Dynamic coefficient.

$K_v = 0.05$ $\omega = 1.59$ Hz

	Variable	Initial Value	Minimal Value	Maximal Value	Final Value
1	t	0	0	3600.	3600.
2	y01	0	−0.1244323	0.2942226	0.1585447
3	y1	0	−1.512497	1.514437	−0.7127118

TABLE 7.9   Calculated Values of DEQ Variables

Differential Equations

1. $d(y01)/d(t) = y1$
2. $d(y1)/d(t) = -(100)*y01+(1)*0.007125*(10^0)*(2125-0.614*t+3.35*(10^{-5})*t^2)$

$$(7.15)$$

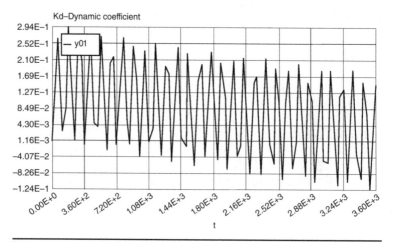

FIGURE 7.12   Dynamic coefficient.

$K_v = 0.05$ $\omega = 5.0$ Hz

	Variable	Initial Value	Minimal Value	Maximal Value	Final Value
1	t	0	0	3600.	3600.
2	y01	0	−0.0121372	0.0289423	0.0175769
3	y1	0	−0.479575	0.4773349	−0.0491363

TABLE 7.10   Calculated Values of DEQ Variables

Differential Equations

1. $d(y01)/d(t) = y1$
2. $d(y1)/d(t) = -(1000)*y01+(1)*0.007125*(10^0)*(2125-0.614*t+3.35*(10^-5)*t^2)$

$$(7.16)$$

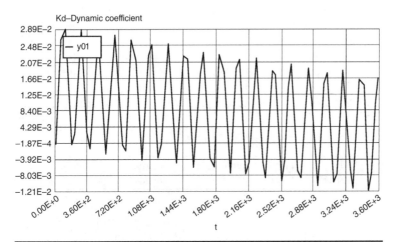

FIGURE 7.13   Dynamic coefficient.

Case 4: Slow Fire

$K_v = 0.05$ $\omega = 1.0$ Hz

	Variable	Initial Value	Minimal Value	Maximal Value	Final Value
1	t	0	0	3600.	3600.
2	y01	0	−0.2481477	0.4654161	−0.2481477
3	y1	0	−1.892469	1.891413	−0.1911095

TABLE 7.11   Calculated Values of DEQ Variables

Differential Equations

1. $d(y01)/d(t) = y1$
2. $d(y1)/d(t) = -(39.5)*y01+(1)*0.00642*(10^0)*(1852-0.532*t+2.93*(10^-5)*t^2)$

$$(7.17)$$

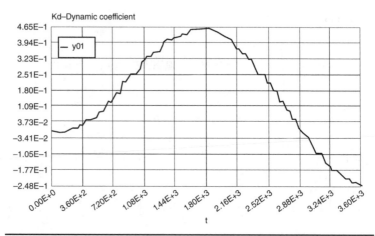

FIGURE 7.14 Dynamic coefficient.

$K_v = 0.05$ $\omega = 1.59$ Hz

	Variable	Initial Value	Minimal Value	Maximal Value	Final Value
1	t	0	0	3600.	3600.
2	y01	0	−0.0971534	0.2312576	0.1253115
3	y1	0	−1.189692	1.189245	−0.559688

TABLE 7.12   Calculated Values of DEQ Variables

Differential Equations

1. $d(y01)/d(t) = y1$
2. $d(y1)/d(t) = -(100)*y01+(1)*0.00642*(10^0)*(1852-0.532*t+2.93*(10^{-5})*t^2)$

$$(7.18)$$

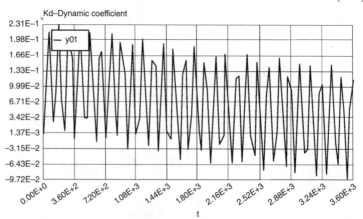

FIGURE 7.15   Dynamic coefficient.

$K_v = 0.05$ $\omega = 5.0$ Hz

	Variable	Initial Value	Minimal Value	Maximal Value	Final Value
1	t	0	0	3600.	3600.
2	y01	0	−0.0094531	0.0227358	0.0138838
3	y1	0	−0.3758828	0.3746159	−0.0385864

TABLE 7.13 Calculated Values of DEQ Variables

Differential Equations

1. $d(y01)/d(t) = y1$
2. $d(y1)/d(t) = -(1000)*y01+(1)*0.00642*(10^0)*(1852-0.532*t+2.93*(10^{-5})*t^2)$

$$(7.19)$$

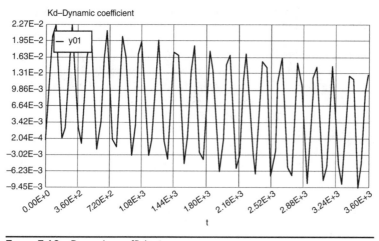

FIGURE 7.16 Dynamic coefficient.

Category	$\omega = $ 0.5 Hz	$\omega = $ 1.0 Hz	$\omega = $ 1.59 Hz	$\omega = $ 5.0 Hz	$\omega > $ 5 Hz
Very fast	1.63	0.414	0.164	0.0163	0
Fast	0.754	0.754	0.377	0.0370	0
Medium	0.592	0.592	0.294	0.0289	0
Slow	0.465	0.465	0.231	0.0227	0
Impact (const. temp.)	2.0	2.0	2.0	2.0	2.0
Beatings and impact	2.3	2.3	2.3	2.3	2.3

TABLE 7.14 Dynamic Coefficient K_d

7.2   Application of General Mechanical Creep Theory

Creep as a time-dependent phenomenon is the progressive accumulation of plastic strain in a structural element under stress at elevated temperature over a period of time. The life of a structural element may be severely limited even for loads less than the design load. In fact, at elevated temperature, a sustained load may produce inelastic strain in a material that increases with time. The material is said to *creep*. The general mechanical theory of creep gives the functional relations among three variables: stress, strain, and time. In this respect, the general creep theory allows the structural engineer to analyze the structural system subjected to impact, static, and dynamic loads at the same time.

Creep, creep failure, and creep fracture of a structural member may occur over a wide range of temperature. Elevated temperature for creep behavior of a metal begins at about one-half the melting temperature; for example, at 205°C for aluminum alloys, 370°C for low-alloy steels, and so on. Creep failure occurs when the accumulated creep strain results in a deformation of the structural member that exceeds the design limits, whereas the creep fracture is an extension where the stressed component actually separates into two parts.

The creep behavior of various materials is often based on a one-dimensional test (tension). In terms of a mathematical representation of a creep curve, for example, in the case of a one-dimensional test (Fig. 7.17), we assume that the creep behavior of steel is a function of stress σ, temperature T, and time t [4,5].

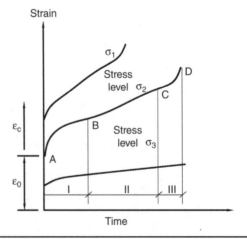

FIGURE 7.17   Creep curve (one-dimensional tension test).

From the figure, three phases can be recognized. Phase one is called *primary* (*initial* or *transient*) *creep*, phase two is called *steady-state creep*, and phase three is *tertiary* (*accelerating*) *creep*. In the figure, ε_o is an instantaneous deformation that may include both elastic and plastic parts. Because of the extreme complexity of creep behavior, the analysis of creep problems very often is based on a curve of experimental data. This representation is referred to as *creep strain* ε_c. In general, to model creep curve (Fig. 7.17), we need an expression of the form [6]

$$\varepsilon_c = f(\sigma, T, t) \tag{7.20}$$

where f is a function of time t, temperature T, and stress σ. It is customary to assume that the effects of σ, T, and t, are separable. Thus, the equation can be written as

$$\varepsilon_c = f_i(\sigma)g_i(T)h_t(t) \tag{7.21}$$

Experiments for one-dimensional creep behavior usually are run allowing only one of the variables (σ, T, or t) to change; for example, σ is varied for constant T and given t. There is no practical interest here for phase three creep. In engineering practice, the big interest is in phase two, especially in the case of isothermal conditions. Many relationships have been proposed to relate stress, strain, time, and temperature in the creep process.

7.3   Elastoviscoplastic Models

Many of the models that have been used to describe viscoelastic or elastoviscoplastic or viscoplastic material response are presented in Refs. 7, 8, and 9. The well-known Voight-Kelvin model (Fig. 7.18a) uses the following viscoelastic equation:

$$\sigma = E\varepsilon + H\dot{\varepsilon} \tag{7.22}$$

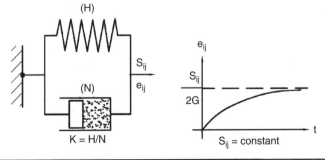

FIGURE 7.18   Voight-Kelvin model and its response at S_{ij} = const.

where σ = stress value
      ε = strain value
      E = Hook's modulus of elastisity (short-term modulus of elastisity)
      H = long-term modulus of elastisity

Initial condition: $\varepsilon_{t=0} = 0$.
    If the stress σ = const. in Eq. (7.22), then the solution to this equation can be presented as:

$$\varepsilon = \frac{\sigma}{E}\left[1 - \exp\left(-\frac{Et}{H}\right)\right] \tag{7.23}$$

This equation describes viscoelastic creep behavior (see Fig. 7.18b).
    The elastic (Hook) and viscose (Newton) elements in Fig. 7.18 have a parallel connection. If these elements were to have a serial connection, then we would have the well-known Maxwell model (Fig. 7.19a).
    The viscoelastic equation of the Maxwell model is

$$\sigma + n\dot{\sigma} = H\dot{\varepsilon} \tag{7.24}$$

where $n = H/E$ is the relaxation time.

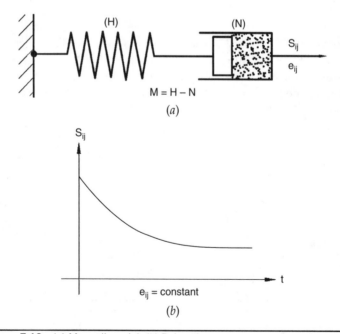

Figure 7.19 (a) Maxwell model. (b) Relaxation curve.

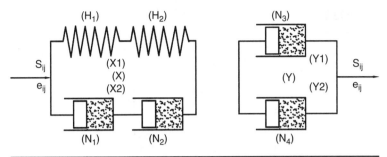

Figure 7.20 Viscoelastic model $[(H_1 - H_2) \mid (N_1 - N_2)] - (N_3 \mid N_4)$, - serial connection, | - parallel connection.

Similar to the preceding case, if the strain ε = const., then the right side of Eq. (7.24) is zero, and the solution is as follows:

$$\sigma = \sigma_o \exp\left(-\frac{t}{n}\right) \tag{7.25}$$

where σ_o is the stress at time $t = 0$ (see Fig. 7.19b).

Obviously, there are many combinations of Hook's and Newton's elements that could be connected in parallel and serially. One of combined viscoelastic model is shown in Fig. 7.20 [10].

The simplest model that will be used here is [11]

$$E\dot{\varepsilon} = \dot{\sigma} + A\sigma \tag{7.26}$$

The most general type of a viscoelastic model can be described by the integral equation

$$E\varepsilon(t) = \sigma(t) + \int_0^t \sigma(t)K(t-\tau)d\tau \tag{7.27}$$

where $K(t-\tau)$ is the kernel of the integral equation (7.27).

The use of creep theory in our case will be limited to two major objectives: (1) to extrapolate the experimental data regarding stiffness of a structural member (system) beyond the testing time (an abnormal fire could last much longer than the prescribed testing time) and to (2) provide an approximate rate of structural element (system) stiffness decrease as a function of temperature only. Since creep theory is very complex, all intermediate mathematical operations are not presented in this study; only final results are provided. Example 7.1 illustrates the application of general creep theory in the case of an abnormal (long-duration) thermal load acting on a structural system.

Example 7.1

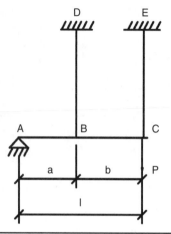

Figure 7. 21   Suspended absolutely rigid beam.

Data   An absolutely rigid beam A-B-C (see Fig. 7.21) is supported by two steel hangers DB and EC. The abnormal fire affects steel hanger EC only. Force P is applied at point C. Both hangers have the same cross-sectional area. H is the height, and $a = b$. After 4 hours of fire, the modulus of elasticity of hanger EC is equal to $0.25E$; therefore, $E_{ec} = 0.25E_{db}$.

Find the distribution of interior forces N_{ec} and N_{db} at this moment, the redistribution of forces at any given moment after 4 hours of fire duration (assume that the fire is continuous with constant temperature $T = 600°C$), and the distribution of forces at $t \to \infty$ (the possibility of progressive collapse).

Solution

The structural system is statically indeterminate system. By separating the structure at point B, two free body diagrams are obtained, and the unknown force is $X_1 = N_{db}$. The corresponding equations (force method) are

$$\delta_{11}X_1 + \Delta_{1p} = 0 \quad \text{and} \quad \delta_{11}\dot{X}_1 + \dot{\Delta}_{1p} = 0 \qquad (7.28)$$

In order to calculate parameters δ_{11}, Δ_{1p}, and $\dot{\Delta}_{1p}$ let's use the creep deformation equation (7.26). After all mathematical simplifications, the following differential equation is obtained:

$$\dot{X}_1 + 0.5X_1 = AP \qquad (7.29)$$

The solution to Eq. (7.29) with initial condition $X_1(0) = P$ is

$$X_1 = N_{bd} = -P[\exp(-0.5At) - 2]$$ (7.30)

and

$$N_{ec} = P - 0.5X_1 = 0.5P\exp(-0.5At)$$ (7.31)

Finally, the answers are

1. At the initial moment (4 hours after the fire started), $N_{bd} = P$ and $N_{ec} = 0.5P$.
2. At any given time after the initial moment, the forces are defined by Eqs. (7.30) and (7.31).
3. $N_{bd} = 2P$ and $N_{ec} = 0$ when $t \to \infty$.

It is interesting to underline that the same results could be obtained by using the general equation (7.27) because the kernel of the integral equation in this case is equal to A. The long-term modulus of elasticity is defined from general linear creep theory as

$$H = \frac{E}{1 + \int\limits_0^\infty K(\theta)d\theta}$$ (7.32)

The stiffness-reduction coefficient n based on Eq. (7.32) for any given time t_0 can be calculated as

$$n = \frac{E}{H} = 1 + \int\limits_0^{t_0} K(\theta)d\theta$$ (7.33)

where H is the long-term modulus of elasticity at a given time t_0. From Eq. (7.32), the long-term stiffness for member EC is zero because the kernel is constant, and therefore, the force $N_{ec} = 0$.

Example 7.2

For a proposed building, the structural and fire protection engineers must collaborate to develop a list of possible locations where the start of a fire could lead to a significant impact on the structural integrity of the building. To do this, the structural engineer must describe the structural system design approach to identify particular structural components that may be critical to building stability. At the same time, the fire protection engineer must, with an understanding of the expected occupancy, determine the areas where fuel loads may be high. Multiple compartments should be analyzed with a view to predicting the range of fire scenarios that reasonably might be expected at any point throughout the building. By doing this, the designers can be assured that when the analysis is complete, the structural system has been designed adequately.

Once various compartments have been selected, the expected fuel load has to be determined. This can be accomplished by using Table 7.15 to estimate the mass of various fuel loads available within the compartments. This includes movable fuel loads such as furniture and book shelves, fixed fuel loads such as doors and window frames if combustible, and protected fuel loads such as wood framing in walls. For noncombustible construction, the fuel load likely will be limited to the furnishings in the room. Care must be taken to account properly for the fuel content of noncellulosic materials such as plastic containers, binders, and so on. Once the mass of the contents in a room is totaled, it is converted to an energy value based on 16.8 MJ/kg for cellulosic products, keeping in mind that petroleum-based materials are to be adjusted by a factor of 2 prior to adding the mass to the cellulosic-based materials to account for the higher heat energy content of those materials. This total fuel load then is divided by either the compartment floor area or the total surface area to yield a per-unit area.

Care should be taken to ensure that the fuel load (MJ/m^2) is calculated correctly for the model chosen.

General Data Four-story steel construction with 0.30-m-thick reinforced masonry exterior and interior malls. Typical floor plan (footprint): 18 m × 24 m. Total height: 14 m. Wind resistance system: reinforced-masonry shear walls. Structural design loads: dead load: 244 kg/m^2 (50 psf); live load: 244 kg/m^2 (50 psf).

Structural steel framing is noncombustible and complies with the requirements of type I and type II construction. Occupancy type: R-2. Allowable maximum height and building areas per IBC Table 503: $H = 19.8$ m and $A = 2230$ m^2. In accordance with IBC Table 601, for type IIA construction, a 1-hour fire resistance rating is required.

Building Description

- Four-story hotel building consisting of sleeping rooms with kitchens, general offices, file storage areas, and meeting rooms, with a floor area of approximately 1,250 m^2

- Noncombustible construction steel-framed building containing column and beams of primary supporting steel and open-web steel joist construction supporting a composite floor-ceiling assembly consisting of metal lathe and 100 mm of poured concrete

- Exterior wall construction consisting of reinforced-masonry walls and fixed glazing

- Interior wall construction consisting of steel stud framing and 13 mm gwb (gypsum wallboard) on either side of the studs

- Exit stair and other shaft walls of 300-mm-thick reinforced masonry
- Combination of suspended ceiling and gypsum wallboard ceiling throughout

Fire Load Design (Kitchens + Living Rooms)

This part of the hotel has dimensions of approximately 6 m × 4.0 m. It has one ventilation opening to outside air with dimensions ranging from 0.6 m width to 2.0 m height. All the glass on the openings is assumed to be broken during a postflashover fire. Although the area has two doors with dimensions of approximately 0.9 m × 2 m each, these are not considered to be a source of ventilation because they are considered to be in the closed position during the modeling. The room contains all sorts of fuel load, ranging from wood materials to plastic materials, which are assumed to be ignited all at once during a postflashover fire.

The mass, dimensions, and exposed surface areas of each item are carefully assessed. Table 7.15 provides a summary descriptions of each item.

During the modeling for each item available inside this room, besides the thinner portions of the cellulosic materials, it is found that noncellulosic materials will have a shorter burning period than the thicker portions of the cellulosic materials. Fire load $377/24 = 15.7$. Fuel load $15.7 \times 16.8 = 264$ MJ/m^2. Use 300 MJ/m^2. Fire severity: Case 3 (medium). $T_{max} = 882$ K. Thus $K_v = 0.05$, and $\theta_{max} = 4.70778$ (see Table 6.27). $T_{max} = 4.71(60) + 600 = 882$ K $= 609°C$.

From Table 5.5, find the value of τ that gives maximum value of θ_{max1}: $\tau = 0.0467$. Find t^* based on Eq. (6.70):

$$\tau = \frac{a_2}{h^2}(3600)t^*$$

$$0.0467 = \frac{1.38 \times 10^{-4}}{9}(3,600)t^* \tag{7.33}$$

$$t^* = 0.846 \text{ (hour)}$$

Note: Thermal diffusivity $a_2 = 1.38 \times 10^{-4}$ is taken from Eq. (6.73) (Case 3). Modify Eq. (5.12) as follows:

$$\theta = A \exp[-(\tau/a - 1)^2/2(\sigma/a)^2]$$

In this case, $A = 4.71$, $\sigma/a = 0.0598/0.0802 = 0.746$, and $\tau_m = a = 0.846$, the dimensionless time at the maximum value of temperature.

$$A = 4.55, \qquad a = 0.0802, \qquad \sigma = 0.0598$$

Fire Load	Type	Quantity	Dimensions (m)	Mass (kg) (equivalent wood)
Fixed				
Carpet	P	1	6.03 × 3.9	53 ea (approx.)
Door + frame	W	2	0.89 × 2 × 0.015	16 ea
Built-in cupboard *a*	W	1	1.5 × 0.74 × 0.9 with thickness 0.02	40 ea
Built-in cupboard *b*			1.16 × 0.74 × 0.9 with thickness 0.03	30 ea
Built-in cupboard *c*			1.2 × 0.32 × 0.6 with thickness 0.04	15 ea
Movable				
Couch	P + W	1	1.95 × 0.9	40 ea
Chair	P + S	4	0.51 × 0.44	4 ea
Table	W + S	1	diameter = 1.19	30 ea
Bed, single	P + W + S	1	1.88 × 0.9	90 ea
TV and cabinet	P + W + S + G	1	0.59 × 0.37	20 ea
Telephone	P + S + E	1	0.165 × 0.225	1 ea
Curtain	P	1	1.96 × 1.11	0.5 ea
		1	0.46 × 1.11	
		1	2.36 × 1.98	
Pictures	W	1	0.82 × 0.57 × 0.02	4 ea
Pillow	P	4	0.36 × 0.36	1 ea
Other plastic things	P	1	0.19 × 0.52	3 ea
			Total:	377 kg

Note: N/A = steel materials are considered to be low fuel load in fire. Thus surfaces are not involved in the fire. W = wood; P = plastic; S = steel; E = electrical; G = glass.

TABLE 7.15   Summary Descriptions of Each Fuel Load in the Hotel (Kitchen + Living Rooms)

Structural fire load (SFL) is as follows:

$$T = 60(4.71) \exp\left[-\frac{\left(\frac{t}{0.846} - 1\right)^2}{2(0.557)} \right] + 600$$

$$= 282 \exp\left[-\frac{\left(\frac{t}{0.846} - 1\right)^2}{2(0.557)} \right] + 600 \qquad (7.34)$$

where t is the time in hours, and T is the gas temperature in kelvins.

Finally, from Table 6.47 using the time-equivalence method (Case 3), the total duration of the fire (including the decay period) is $t = 2.6$ hours for a 1-hour fire rating. Therefore, the decay period is $2.6 - 0.846 = 1.754$ hours.

Since the main objective in all the following examples is the structural analysis and design of structural elements and systems subjected to the SFL, the fuel load calculations will remain the same (see above). These examples should be viewed as an approximate conservative design method that will allow one to weed out the unpractical and non-critical fire scenarios during the preliminary stage of a building design process, on the one hand, and on the other hand, they could be viewed as a final design methods if there is no change in the "original" structural design (members or system as a whole) owing to SFL. The approximations used in all the following examples are based on maximum values of temperature and fire duration, an assumed opening factor parameter, curtain-type passive insulation material, and so on. For any given fire scenario, these parameters obviously should be readjusted. If detailed computer-generated dynamic analysis and structural design are required, then the temperature-time function (7.33) (or similar) can be used as an SFL.

Example 7.3 (Fig. 7.22)

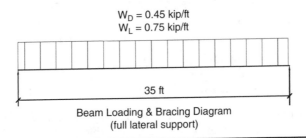

$W_D = 0.45$ kip/ft
$W_L = 0.75$ kip/ft

35 ft

Beam Loading & Bracing Diagram
(full lateral support)

FIGURE 7.22 Beam loading diagram.

Data   live load = 0.5 klf; dead load = 0.5 klf; span $L = 40$ ft, 0 in; $F_y =$ 50 ksi; $A = 18.2$ in^2; $F_u = 65$ ksi; $S_x = 131$ in^3; $I_x = 1,550$ in^4; $Z_x = 153$ in^3; W 24 × 62 (ASTM A992); $T_m = 609°C$

Fire rating: 1 hour
Fire severity: Case 3

Original Structural Design (LRFD Method)   Beam is continuously braced.

1. Ultimate design load:

$$w_u = 1.2(0.5) + 1.6(0.5) = 1.4 \text{ klf}$$

2. Ultimate bending moment:

$$M_u = 1.4(40)^2/8 = 280 \text{ kip-ft}$$

Per the user note in Sec. F2 (AISC), the section is compact. Since the beam is continuously braced and compact, only the yielding limit state applies:

3. $\phi_b M_n = \dfrac{0.9(50)153}{12} = 573.8 > 280$ kip-ft   okay!

4. Calculate the required moment of inertia for live load deflection criterion of $L/360$:

$$\Delta_{max} = \frac{L}{360} = \frac{40(12)}{360} = 1.33 \text{ in}$$

$$I_{x(req)} = \frac{5wL^4}{384E\Delta_{max}} = \frac{5(0.5)(40^4)1728}{384(29,000)1.33}$$

$$= 747 \text{ in}^4 < 1,550 \text{ in}^4 \qquad \text{okay!}$$

The beam is okay.

Beam Design (SFL)

Data   Beam is restrained at both ends; span $L = 40$ ft; $T_m = 609°C$; Case 3; fire rating: 1 hour.

$$\text{Natural frequency } \omega = \sqrt{\frac{g}{\delta_{11}W}} = \sqrt{\frac{32.2(12)}{0.05125(40)1}}$$

$$= 13.73 \text{ rad/s} = 2.18 \text{ Hz}$$

where g = gravitational acceleration
        W = total gravity load
        δ_{11} = deformation of the beam from unit force applied at midspan

Maximum temperature reduction owing to passive fire protection of the beam (assume 10 percent; see Table 6.57): $T_m = 0.9(609) = 548°C$. Total elongation of the beam $\Delta L = \alpha_o T_m L = 0.00117(5.48)480 = (0.00641) \times 480 = 3.08$ in.

1. Since the deformed length of the beam is known $[L_{total} = L + \Delta L =$ as $L(1 + \alpha_o T_{max})]$, the maximum deflection of the beam can be approximated as (large deformations)

$$\Delta_b = L\sqrt{\frac{\alpha_o T_{max}}{2}} = 480\sqrt{\frac{0.00641}{2}} = 27.2 \text{ in} = 2.26 \text{ ft}$$

2. The maximum trust force in this case can be approximated as

$$H = W\sqrt{\frac{1}{24(\alpha_o T_{max})}} = 30\sqrt{\frac{1}{24(0.00641)}} = 76.5 \text{ kip}$$

where $W = 40[0.5 + (0.5)0.5] = 30$ kip—50 percent live load reduction.

3. Additional bending moment and axial force in this case owing to SFL are

$$M = 76.5(2.26) = 172.9 \text{ ft-kip} \qquad N = -76.5 \text{ kip}$$

4. Dynamic coefficient K_d from Table 7.3 (based on linear interpolation) is

$$K_d = \frac{2.82}{3.41}(0.264) + 0.03 = 0.248$$

5. Design load combination in this case [per AISC's, *Manual of Steel Construction*, 13th ed., Appendix 4, Eq. (A-4-1)]: 1.2D + 0.5 L + T. Therefore, for dead load and live load:

$$w_u = 1.2(0.5) + 1.6(0.5)0.5 = 1.0 \text{ klf}$$

6. Finally, the ultimate design forces are

$$M_u = 1.0(1,600)/8 + 172.9(1.248) = 415.8 \text{ ft-kip}$$
$$N_u = -76.5(1.248) = -95.5 \text{ kip}$$
$$V_u = 1.0(40/2) = 20 \text{ kip}$$

7. Check stresses (deformations obviously are not limited in the case of fire).

   a. $C_{mx} = B_{mx} = 1.0$—okay by inspection; $K = 1.0$ (beam is braced)

   b. $P_c = \phi_c P_n = 0.9(50)18.2 = 819 \text{ kip}$
   $M_{cx} = 1.0(50)153/12 = 637.5 \text{ ft-kip}$

   c. $P_r/P_c = 127.3/819 = 0.155 < 0.2$; use equation H1 – 1b:

   d. $\dfrac{P_r}{P_c(2)} + \dfrac{M_{rx}}{M_{cx}} = \dfrac{95.5}{2(819)} + \dfrac{415.8}{637.5} = 0.71 < 1.0$ Beam is okay.

8. Now let's check the catenary's action (large deformations). The kernel of the integral equation (7.27) is as follows: $K(\theta) = \exp(-b\theta)$.

Now, from Eq. (7.32), if $b = 0.333$,

$$n = \frac{E}{H} = 1 + \int_0^\infty \exp(-0.333\theta)d\theta = 4$$

The reduced $F_u = (0.9)65/4 = 14.6$ ksi, and the required cross-sectional area is $A_{req} = 95.5/14.6 = 6.54 << 18.2$, provided. (The progressive collapse requirement is satisfied in this case.)

Example 7.4   (Fig. 7.23)

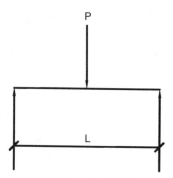

FIGURE 7.23 Steel girder design model.

Data $P_{D.L.} = 100$ kip; $P_{L.L.} = 100$ kip; span $L = 20$ ft, 0 in; $F_y = 50$ ksi; $W36 \times 160$ (ASTM A992); $A = 47.0$ in^2; $F_u = 65$ ksi; $S_x = 542$ in^3; $I_x = 9750$ in^4; $Z_x = 624$ in^3; $T_m = 609°C$

Fire rating: 1 hour
Fire severity: Case 3

Original Structural Design (LRFD Method)   Beam is continuously braced.

1. Ultimate design load:

$$P_u = 1.2(100) + 1.6(100) = 280 \text{ kip}$$

2. Ultimate bending moment:

$$M_u = 280(20)/4 = 1{,}400 \text{ kip-ft}$$

Per the user note in Sec. F2 (AISC), the section is compact. Since the beam is continuously braced and compact, only the yielding limit state applies:

3. $\phi_b M_n = \dfrac{0.9(50)624}{12} = 2{,}340 > 1{,}400$ kip-ft   okay.

4. Calculate the required moment of inertia for live load deflection criterion of $L/360$:

$$\Delta_{max} = \frac{L}{360} = \frac{20(12)}{360} = 0.67 \text{ in}$$

$$I_{x(req)} = \frac{PL^3}{48E\Delta_{max}} = \frac{100(20^3)1728}{48(29,000)0.67}$$

$$= 1482 \text{ in}^4 < 9,750 \text{ in}^4 \qquad \text{okay.}$$

Beam design is okay.

Beam Design (SFL)

Data   Beam is restrained at both ends, span $L = 40$ ft; $T_m = 609°C$; Case 3; fire rating: 1 hour.

$$\text{Natural frequency } \omega = \sqrt{\frac{g}{\delta_{11}W}} = \sqrt{\frac{32.2(12)}{0.001019(200)}}$$

$$= 43.55 \text{ rad/s} = 6.93 \text{ Hz}$$

where g = gravitational acceleration
W = total gravity load
δ_{11} = deformation of the beam from unit force applied at midspan

Maximum temperature reduction owing to passive fire protection of the beam (assume 10 percent see Table 6.56): $T_m = 0.9(609) = 548°C$. Total elongation of the beam $\Delta L = \alpha_o T_m L = 0.00117(5.48)240 = (0.00641)240 = 1.54$ in.

1. Since the deformed length of the beam is known $[L_{total} = L + \Delta L = L(1 + \alpha_o T_{max})]$, the maximum deflection of the beam can be approximated as

$$\Delta_b = L\sqrt{\frac{\alpha_o T_{max}}{2}} = 240\sqrt{\frac{0.00641}{2}} = 13.59 \text{ in} = 1.13 \text{ ft}$$

2. The maximum trust force in this case can be approximated as

$$H = W\sqrt{\frac{1}{24(\alpha_o T_{max})}} = 150\sqrt{\frac{1}{24(0.00641)}} = 382.4 \text{ kip}$$

where $W = 100(1.0) + (0.5)100 = 150$ kip.

3. Additional bending moment and axial force in this case owing to SFL are

$$M = 382.4(1.13) = 432.2 \text{ ft-kip} \qquad N = -382.4 \text{ kip}$$

4. Dynamic coefficient K_d from Table 7.3 ($\omega = 6.93$ Hz > 5.0 Hz) is $K_d = 0$.

5. Design load combination in this case [per AISC's *Manual of Steel Construction*, 13th ed., Appendix 4, Eq. (A-4-1)] is $1.2D + 0.5L + T$. Therefore, for dead load and live load,

$$P_u = 1.2(100) + 1.6(100)0.5 = 200 \text{ kip}$$

6. Finally, the ultimate design forces are

$$M_u = 200(20)/4 + 432.2(1.0) = 1{,}432.2 \text{ ft-kip}$$
$$N_u = -382.4(1.0) = -382.4 \text{ kip}$$
$$V_u = 1.0(200/2) = 100 \text{ kip}$$

7. Unity check (deformations obviously are not limited in case of fire).

   a. $C_{mx} = B_{mx} = 1.0$—okay by inspection; $K = 1.0$ (beam is braced).

   b. $P_c = \phi_c P_n = 0.9(50)47.0 = 2{,}115$ kip.
   $M_{cx} = 1.0(50)624/12 = 2{,}600$ ft-kip.

   c. $P_r/P_c = 382.4/2{,}115 = 0.181 < 0.2$; use equation $H1 - 1b$:

   d. $\dfrac{P_r}{P_c(2)} + \dfrac{M_{rx}}{M_{cx}} = \dfrac{382.4}{2(2{,}115)} + \dfrac{1{,}432.2}{2600}$

$$= 0.641 < 1.0 \qquad \text{Beam is okay.}$$

8. Now, let's check the catenary's action. Again, similar to the preceding example, from Eq. (7.31); $n = 4$. The reduced $F_u = (0.9)65/4 = 14.6$ ksi, and the required cross-sectional area is $A_{req} = 382.4/14.6 = 26.2 < 47.0$, provided. (The progressive collapse requirement is satisfied in this case.)

Example 7.5

The girder from Example 7.4 is supported now by two columns; therefore, it is partially restrained at both ends (owing to bending of columns; see Fig. 7.24).

Data $P_{D.L.} = 100$ kip; $P_{L.L.} = 100$ kip; span $L = 20$ ft, 0 in; $F_y = 50$ ksi

W 36 × 160 (ASTM A992), beam	W 14 × 99, column
$A = 47.0$ in^2	$A = 29.1$ in^2
$F_u = 65$ ksi	$F_u = 65$ ksi
$S_x = 542$ in^3	$S_x = 157$ in^3
$I_x = 9{,}750$ in^4	$I_x = 1{,}110$ in^4
$Z_x = 624$ in^3	$Z_x = 173$ in^3
$T_m = 609°C$	

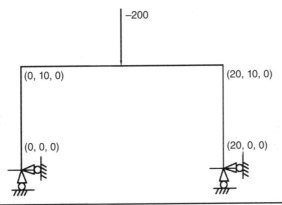

FIGURE 7.24   Transfer girder frame.

Fire rating: 1 hour
Fire severity: Case 3

Computer input and output data are presented below (see Example 7A.1 in the appendix at the end of this chapter). Simple computer analysis provides the following results:

1. From force $P = 200$ kip; moment at midspan $M = 875.68$ ft-kip; deflection at the same point $\Delta_{1st} = 0.024$ ft

$$\text{Natural frequency } \omega = \sqrt{\frac{g}{\Delta_{1st}}} = \sqrt{\frac{32.2}{0.024}}$$

$$= 36.6 \text{ rad/s} = 5.83 \text{ Hz}$$

2. Dynamic coefficient ($\omega = 5.83$ Hz. ≈ 5.0 Hz.): Use $K_d = 0.03$.
3. From temperature $T = 548°C$ (beam only), $M = 269.33$ ft-kip; and deflection $\Delta_t = 0.007$ ft (downward).
4. Dynamic bending moment ($K_d = 0.03$): $M_d = (0.03)656.8(0.007)/(0.024) = 5.75$ ft-kip.
5. Dynamic axial force ($K_d = 0.03$): $N_d = -(0.03)(0.75)12.43(0.007)/(0.024) = -0.082$.
6. Total moment (with 50 percent live load reduced): $M_{total} = 656.8 + 269.33 + 5.75 = 931.9$ ft-kip.
7. Total axial force (with 50 percent live load reduced): $N_{total} = -0.75(12.43) + 26.93 + 0.082 = -36.33$ kip.
8. Total effect from fire (temperature load): $K = 931.9/875.68 = 1.06$ (bending moment).
9. Total effect from fire (temperature load): $K = 36.33/12.43 = 2.92$ (axial force).

Similar calculations are provided for the negative moment at the face of the column:

1. From force $P = 200$ kip; $M^{sup} = 124.32$ ft-kip.
2. From temperature load (the same), $M = 269.33$ ft-kip.
3. Dynamic bending moment $M_d = (0.03)124.32(0.007)/(0.024) = 1.09$ ft-kip.
4. Total moment (with 50 percent live load reduced): $M_{total} = 93.24 + 269.33 + 1.09 = 363.66$ ft-kip.
5. Total effect from fire (temperature load): $K = 2.93$ (bending moment).

Conclusion   The positive bending moment remains practically unchanged (see Example 7.2). However, the negative moment has been increased drastically owing to high temperature.

1. Finally, the design forces (beam, member 3) are as follows [use the allowable stress design (ASD) method]:

$$M = 931.9 \text{ ft-kip}$$
$$N = -36.33 \text{ kip}$$
$$V = 0.75(100) = 75 \text{ kip}$$

2. Unity check (deformations obviously are not limited in the case of fire).

   a. $C_{mx} = 1.0$—okay by inspection; $K = 1.0$ (beam is braced); $kl/r = 20(12)/14.4 = 16.7$.

   b. $F_a = 28.61$ ksi; $f_a = 36.33/47 = 0.773$ ksi; $f_a/F_a = 0.027 < 0.15$.

   c. $F_b = 33$ ksi; $f_b = 931.9(12)/542 = 20.63$ ksi.

   d. $\dfrac{f_a}{F_a} + \dfrac{f_{bx}}{F_{bx}} = \dfrac{0.773}{28.61} + \dfrac{20.63}{33}$

   $$= 0.652 < 1.0 \qquad \text{Beam is okay.}$$

3. Now, let's check the catenary's action. Per Eq. (7.31), $n = 4$. The reduced $F_u = (0.9)65/4 = 14.6$ ksi, and the required cross-sectional area is $A_{req} = 36.33/14.6 = 2.49 < 47.0$, provided. (The progressive collapse requirement is satisfied in this case.)

One would think that the beam subjected directly to the high temperature load is the most critical element in the frame structure of a fire compartment. However, this is not the case here. The "cold" column, which was not intended to be designed for a high level of bending moments in the "original" structural design, is the most critical element. Let's check the column stresses now.

Column W 14 × 99

Data $A = 29.1$ in²; $F_y = 50$ ksi; $F_u = 65$ ksi; $S_x = 157$ in³; $I_x = 1,110$ in⁴; $Z_x = 173$ in³; $r_x = 6.17$ in

1. In the original design, the design forces are
$$M = 124.32 \text{ ft-kip}; N = -100 \text{ kip}$$

   a. $C_{mx} = 1.0$—okay by inspection; $K = 1.0$ (frame is braced); $kl/r = 10(12)/6.17 = 19.45$.

   b. $F_a = 28.31$ ksi; $f_a = 100/29.1 = 3.44$ ksi; $f_a/F_a = 0.12 < 0.15$.

   c. $F_b = 33$ ksi; $f_b = 124.32(12)/157 = 9.5$ ksi.

   d. $\dfrac{f_a}{F_a} + \dfrac{f_{bx}}{F_{bx}} = \dfrac{3.44}{28.31} + \dfrac{9.5}{33} = 0.41 < 1.0.$

2. The new design forces (column, member 2) are as follows (use the ASD method):
$$M = 363.66 \text{ ft-kip}$$
$$N = -75.0 \text{ kip}$$
$$V = 36.33 \text{ kip}$$

3. Unity check (deformations obviously are not limited in the case of fire).

   a. $C_{mx} = 1.0$—okay by inspection; $K = 1.0$ (frame is braced); $kl/r = 10(12)/6.17 = 19.45$.

   b. $F_a = 28.31$ ksi; $f_a = 75/29.1 = 2.58$ ksi; $f_a/F_a = 0.091 < 0.15$.

   c. $F_b = 33$ ksi; $f_b = 363.66(12)/157 = 27.8$ ksi.

   d. $\dfrac{f_a}{F_a} + \dfrac{f_{bx}}{F_{bx}} = \dfrac{2.58}{28.31} + \dfrac{27.8}{33} = 0.934 < 1.0.$

Column is still okay, but $0.934 \gg 0.41$.

Example 7.6   (Fig. 7.25)

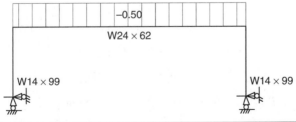

Figure 7.25   Frame computer model.

Computer input and output data are presented below (see Example 7A.2 in the appendix at the end of this chapter).

Beam Design (SFL)

Data   Beam is restrained at both ends; span $L = 40$ ft; $T_m = 609°C$; Case 3; Fire rating: 1 hour.

$$\text{Natural frequency } \omega = \sqrt{\frac{g}{\delta_{11}W}} = \sqrt{\frac{32.2}{0.003(40)1}}$$

$$= 16.38 \text{ rad/s} = 2.61 \text{ Hz}$$

where g = gravitational acceleration
W = total gravity load
δ_{11} = deformation of the beam from unit force applied at mid-span of the beam

1. Dynamic coefficient K_d from Table 7.3 (based on linear interpolation) is

$$K_d = \frac{2.39}{3.41}(0.264) + 0.03 = 0.215$$

2. Design load combination in this case [per AISC's *Manual of Steel Construction*, 13th ed., Appendix 4, Eq. (A-4-1)] is $1.2D + 0.5L + T$. Therefore, for dead load and live load, $w_u = 1.2(0.5) + 1.6(0.5)0.5 = 1.0$ klf

3. Displacement, bending moment, and axial force from unit force applied at the center of the beam:

$$\delta_{11} = 0.003 \text{ ft} \qquad M = 5.98 \text{ ft-kip} \qquad \text{and} \qquad N = -0.4 \text{ kip}$$

4. Finally, the ultimate design forces are (from computer output)

$M_u = 1.2(53.6) + 1.6(0.5)53.6 + 138.14 + 5.98(0.088)(0.215)/(0.003)$
     $= 283.0$ ft-kip

$N_u = -[1.2(5.36) + 1.6(0.5)5.36 + 13.81 + 0.4(0.088)(0.215)/(0.003)]$
     $= -27.0$ kip

5. Check stresses (deformations obviously are not limited in the case of fire).

    a. $C_{mx} = B_{mx} = 1.0$—okay by inspection; $K = 1.0$ (beam is braced).

    b. $P_c = \phi_c P_n = 0.9(50)18.2 = 819$ kip.
        $M_{cx} = 1.0(50)153/12 = 637.5$ ft-kip.

    c. $P_r/P_c = 27.0/819 = 0.033 < 0.2$; use equation $H1 - 1b$:

    d. $\dfrac{P_r}{P_c(2)} + \dfrac{M_{rx}}{M_{cx}} = \dfrac{27}{2(819)} + \dfrac{283}{637.5}$

$$= 0.46 < 1.0 \qquad \text{Beam is okay.}$$

6. Now, let's check the catenary's action. Per Eq. (7.31); $n = 4$. The reduced $F_u = (0.9)65/4 = 14.6$ ksi, and the required cross-sectional area is $A_{req} = 27/14.6 = 1.85 << 18.2$, provided. (The progressive collapse requirement is satisfied in this case.)

Example 7.7 (Fig. 7.26)

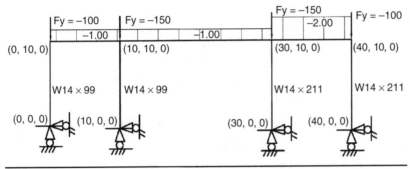

FIGURE 7.26   Frame loading diagram.

Computer input and output data are presented below (see Example 7A.3 in the appendix at the end of this chapter). Simple computer analysis provides the following results:

1. Internal forces from dead load and live load (element 5): maximum negative moment $M = 295$ ft-kip; axial force $N = -29.5$ kip; shear $V = 38.9$ kip; total gravity load $P = 550$ kip.

2. Static internal forces from temperature (SFL) load: maximum negative moment $M = 506$ ft-kip; axial force $N = -50.64$ kip; shear $V = 95$ kip; deflections at center of mass (ODOF, center of element 6): vertical, 0.018 ft; horizontal, 0.031 ft.

3. Deflection from unity vertical (gravity) force $P = 1$ kip, located at center of mass (center of element 6): -0.00034 ft. The corresponding internal forces are as follows: maximum negative moment (element 5): $M = -0.8584$ ft-kip; axial force $N = -0.012$ kip; shear $V = 0.1$ kip.

4. Deflection from unity horizontal force $P = 1$ kip, located at joint 5: -0.00038 ft. The corresponding maximum negative moment (element 5): $M = 1.23$ ft-kip; axial force $N = -0.64$ kip; shear $V = 0.11$ kip.

5. Natural frequency ω (vertical vibrations):

$$\omega = \sqrt{\frac{g}{\Delta_{1st}}} = \sqrt{\frac{32.2}{0.00034(550)}} = 13.1 \text{ rad/s} = 2.1 \text{ Hz}$$

6. Dynamic coefficient ($\omega = 2.1$ Hz, vertical vibrations): Use $K_d = 0.224$.

7. Natural frequency ω (horizontal vibrations):

$$\omega = \sqrt{\frac{g}{\Delta_{1st}}} = \sqrt{\frac{32.2}{0.00038(550)}} = 12.4 \text{ rad/s} = 1.98 \text{ Hz}$$

8. Dynamic coefficient (ω = 1.98 Hz, horizontal vibrations): Use $K_d = 0.234$.

9. Dynamic internal forces (vertical vibrations):

    a. $M_d = (0.8584)(0.009)0.224/(0.00034) = 5.1$ ft-kip.

    b. $N_d = -(0.012)(0.009)0.224/(0.00034) = -0.072$ kip.

    c. $V_d = (0.10)(0.009)0.224/(0.00034) = 0.6$ kip.

10. Dynamic internal forces (horizontal vibrations):

    a. $M_d = (1.23)(0.009)0.234/(0.00038) = 6.8$ ft-kip.

    b. $N_d = -(0.64)(0.009)0.234/(0.00038) = -3.55$ kip.

    c. $V_d = (0.11)(0.009)0.234/(0.00038) = 0.61$ kip.

11. Total moment (with 50 percent live load reduced): $M_{total} = (0.75) \times 12.37 + 432.47 + 5.1 + 6.8 = 453.6$ ft-kip.

12. Total axial force (with 50 percent live load reduced): $N_{total} = -[(0.75)0.67 + 43.25 + 0.072 + 3.55] = -47.4$ kip.

13. Total shear (with 50 percent live load reduced): $V_{total} = (0.75) \times 5.57 + 70.7 + 0.6 + 0.61 = 76.1$ kip.

14. Total effect from fire (temperature): The negative bending moment has been increased owing to the dynamic effect in this case only by 5.4 percent. However, the axial force has been increased by 22.5 percent! It's all dependent on the structural system, relative stiffness ratios (beams and columns), dead and live load distributions, height of the building, and so on. Since the dynamic effect of the SFL cannot be estimated a priori, the structural system has to be checked for static and dynamic portions of the SFL.

15. Finally, the design forces (beam, member 5) are (use the ASD method):

$$M = 453.6 \text{ ft-kip}$$
$$N = -47.4 \text{ kip}$$
$$V = 76.1 \text{ kip}$$

16. Check bending stresses first:

$$f_b = 453.6(12)/117 = 46.5 > 33 \text{ ksi} \qquad \text{N.G.}$$

17. Provide beam flange reinforcement: two cover plates: 1.0×10 in. Calculate new properties:

$$A = 19.1 + 2(1.0)10 = 39.1 \text{ in}^2$$
$$I_x = 1,070 + 2(10)9.5(9.5) = 2,875 \text{ in}^4$$

$$S_x = 2,875/9.5 = 302.6 \text{ in}^3$$

$$r_x = (2,875/39.1)^{1/2} = 8.58 \text{ in}$$

18. Unity check (deformations obviously are not limited in the case of fire).

   a. $C_{mx} = 1.0$—okay by inspection; $K = 1.0$ (beam is braced); $kl/r = 10(12)/8.58 = 14.0$.

   b. $F_a = 28.9 \text{ ksi}; f_a = 47.4/39.1 = 1.21 \text{ ksi}; f_a/F_a = 0.042 < 0.15$.

   c. $F_b = 33 \text{ ksi}; f_b = 453.6(12)/302.6 = 18.0 \text{ ksi}$.

   d. $\dfrac{f_a}{F_a} + \dfrac{f_{bx}}{F_{bx}} = \dfrac{1.21}{28.90} + \dfrac{18.0}{33}$

   $$= 0.587 < 1.0 \qquad \text{Beam is okay.}$$

19. Now, let's check the catenary's action. Per Eq. (7.31), $n = 4$. The reduced $F_u = (0.9)65/4 = 14.6 \text{ ksi}$, and the required cross-sectional area is $A_{\text{req}} = 47.4/14.6 = 3.25 < 39.1$, provided. (The progressive collapse requirement is satisfied in this case.)

   One would think that the short beam ($L = 10$ ft, 0 in) subjected directly to the temperature load is not the most critical element in the frame structure of a fire compartment. However, this is not the case here. The beam, which was not intended to be designed for a high level of bending moments in the "original" structural design, is the most critical element.

Example 7.8   Steel truss (Fig. 7.27)

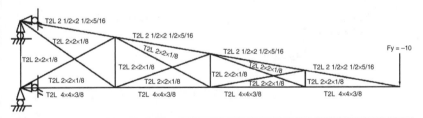

FIGURE 7.27   Steel truss loading diagram.

Data $P_{D.L.} = 10 \text{ kip}$; span $L = 30$ ft, 0 in; $F_y = 50 \text{ ksi}$

Bottom Chord	**Top Chord**
$A = 19.1 \text{ in}^2$	$A = 29.1 \text{ in}^2$
$F_u = 65 \text{ ksi}$	$F_u = 65 \text{ ksi}$
$S_x = 117 \text{ in}^3$	$S_x = 157 \text{ in}^3$
$I_x = 1,070 \text{ in}^4$	$I_x = 1,110 \text{ in}^4$
$Z_x = 133 \text{ in}^3$	$Z_x = 173 \text{ in}^3$

$T_m = 609°C$; fire rating: 1 hour; fire severity: Case 3.

Computer input and output data are presented below (see Example 7A.4 in the appendix at the end of this chapter). Simple computer analysis provides the following results:

1. Deflections from gravity force $P = 10$ kip that is located at the center of mass (joint 2) are

$$\Delta_{st}^h = -0.011 \text{ ft}$$
$$\Delta_{st}^v = -0.191 \text{ ft}$$

2. The corresponding natural frequencies ω are

$$\text{Vertical: } \omega = \sqrt{\frac{32.2}{0.191}} = 12.98 \text{ rad/s} = 2.07 \text{ Hz}$$
$$\text{Horizontal: } \omega = \sqrt{\frac{32.2}{0.011}} = 54.1 \text{ rad/s} = 8.6 \text{ Hz} > 5 \text{ Hz}$$

There is no dynamic effect from horizontal vibrations because the natural frequency $\omega > 5$ Hz.

3. Dynamic coefficient ($\omega = 2.07$ Hz, vertical vibrations): Use $K_d = 0.224$.

4. Maximum axial forces (original design) are as follows:

   a. Bottom chord (member 2): $N = -62.91$ kip.

   b. Top chord (member 8): $N = +57.51$ kip.

   c. Diagonal element (member 13): $N = +2.81$ kip.

   d. Diagonal element (member 16): $N = +3.27$ kip.

5. Maximum axial forces from fire temperature load (SFL) are as follows:

   a. Bottom chord (member 2): $N = -0.48$ kip.

   b. Top chord (member 8): $N = +0.02$ kip.

   c. Diagonal element (member 13): $N = -14.77$ kip.

   d. Diagonal element (member 16): $N = -16.05$ kip.

6. The top and bottom chord elements do not receive any substantial additional forces from temperature (SFL) load; therefore, there is no change in the original structural design of these elements. However, the diagonal elements (members 13 and 16), originally designed as tension members, now became compression members and receive a substantial axial forces.

7. Calculate the additional dynamic axial forces for diagonal members 13 and 16 (compression):

   a. Member 13: $N = -2.81(10)(0.031)(0.224)/(0.191) = -1.02$ kip.

   b. Member 16: $N = -3.27(10)(0.031)(0.224)/(0.191) = -1.19$ kip.

8. Let's check the structural elements (members 13 and 16) for the following load combination:

$$N = 0.9 \text{ D.L. (dead load)} + 1.0 \text{ T.L. (temperature load)}$$

9. Element 13:   *Data*: $2\_L\ 2 \times 2 \times 1/8$ in; $L = 8.38$ ft; $A = 0.96$ in^2; $r_x = 0.626$ in; $k = 1.0$ (it is assumed that the truss is braced at all joints)

10. Calculate axial stresses F_a and f_a:

$$kL/r_x = 8.38(12)/0.626 = 161; \text{ therefore, } F_a = 5.76 \text{ ksi}$$

$$N = 0.9(2.81) - 1.0(14.77 + 1.02) = -13.26 \text{ kip}$$

$$f_a = 13.26/0.96 = 13.81 \text{ ksi} > 5.76 \text{ ksi (compr.)} \qquad \text{Element failed.}$$

11. Element 16:   *Data*: $2\_L\ 2 \times 2 \times 1/8$ in; $L = 9.01$ ft; $A = 0.96$ in^2; $r_x = 0.626$ in; $k = 1.0$ (it is assumed that the truss is braced at all joints)

12. Calculate axial stresses F_a and f_a:

$$kL/r_x = 9.01(12)/0.626 = 173; \qquad \text{therefore,} \quad F_a = 4.99 \text{ ksi}$$

$$N = 0.9(3.27) - 1.0(16.05 + 1.19) = -14.30 \text{ kip}$$

$$f_a = 14.3/0.96 = 14.89 \text{ ksi} > 4.99 \text{ ksi (compr.)} \qquad \text{Element failed.}$$

Conclusion   The structure as a whole might fail in this case owing to structural fire load.

Example 7.9   (Fig. 7.28)

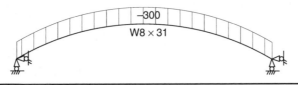

-300
W8 × 31

FIGURE 7.28   Arch loading diagram.

Computer input and output data are presented below (see Example 7A.5 in the appendix at the end of this chapter). It is a well-known fact that a parabolic arch subjected to uniformly distributed load works predominately as a compression (tension) structural element (internal axial force has a much larger effect than bending moment or shear), and such an arch can support a large gravity load [in this case, the total gravity load $W = 3(40) = 120$ kip) with a relatively small section W6 × 20. At the same time, one can see that the parabolic arch represents a "very good" shape when the structure is subjected to the SFL.

Simple computer analysis provides the following results:

1. Deflection from gravity load at the center of mass (member 6) is

$$\Delta_{st}^h = 0 \text{ ft}$$
$$\Delta_{st}^v = -0.03 \text{ ft}$$

2. Corresponding natural frequency ω is

$$\text{Vertical: } \omega = \sqrt{\frac{32.2}{0.03}} = 32.76 \text{ rad/s} = 5.2 \text{ Hz} > 5 \text{ Hz}$$

There is no dynamic effect from horizontal and vertical vibrations ($K_d = 0$).

3. Maximum internal forces (original design) are as follows:

   a. Compression force (member 6): $N = -120.4$ kip.

   b. Bending moment (member 6): $M = 1.1$ ft-kip.

4. Maximum internal forces from fire temperature load (SFL) are as follows:

   a. Compression force (member 6): $N = -9.0$ kip.

   b. Bending moment (member 6): $M = 44.2$ ft-kip.

5. Let's check the structural element (member 6) for the following load internal forces:

$$N = -129.4 \text{ kip} \quad \text{and} \quad M = 45.3 \text{ ft-kip}$$

6. Element 6: Data: W8 × 31; $A = 9.13$ in²; $S = 27.5$ in³; $k = 1.0$ (it is assumed that the arch is continuously braced); $C_{mx} = 0.85$ (compression member)

7. Calculate axial stresses F_a and f_a:

$$F_a = (0.6)50 = 30.0 \text{ ksi}$$
$$f_a = 129.4/9.13 = 14.17 \text{ ksi}$$

8. Calculate bending stresses F_b and f_b:

$$F_b = (0.66)50 = 33.0 \text{ ksi}$$
$$F_b = 45.3(12)/33 = 19.7 \text{ ksi}$$

9. Unity check:

$$\frac{f_a}{F_a} + \frac{f_{bx}}{F_{bx}} = \frac{14.17}{30} + \frac{0.85(19.7)}{33} = 0.98 < 1.0$$

10. The section W8 × 31 is okay.

Example 7.10

Data Continuous reinforced-concrete beam restrained against longitudinal expansion. End span $L_n = 30$ ft; data are taken from CRSI (Concrete Reinforcing Steel Institute) (1984): Service load: dead load = 87 psf; 25 percent live load = 43 psf; spacing, 14 ft, 0 in; total service load: 1.8 klf; total ultimate load: 6.4 klf; concrete: $f_c = 4$ ksi; steel $f_y = 60$ ksi; section: $w = 16$ in; $h = 30$ in; reinforcement: top 4_#11; bottom 4_# with #5 at 13 in o.c. ties; modulus of elasticity $E = 3,600$ ksi; moment of inertia $I_g = 16(30)^3/12 = 36,000$ in⁴; maximum temperature: $T_m = 609°C$; Case 3.

1. Beam design (SFL)

    Data Beam is restrained at both ends; span $L = 30$ ft.

$$\text{Natural frequency } \omega = \sqrt{\frac{g}{\delta_{11}W}} = \sqrt{\frac{32.2(12)}{0.0075(30)1.8}}$$

$$= 30.89 \text{ rad/s} = 4.92 \text{ Hz}$$

   where g = gravitational acceleration
   W = total gravity load
   δ_{11} = deformation of the beam from unit force applied at midspan

   Maximum temperature reduction owing to passive fire protection of the beam (assume 10 percent; see Table 6.56): $T_m = 0.9(609) = 548°C$. Total elongation of the beam $\Delta L = \alpha_o T_m L = 0.00099(5.48)360 = (0.00542)360 = 1.95$ in.

2. Since the deformed length of the beam is known $[L_{total} = L + \Delta L = L(1 + \alpha_o T_{max})]$, the maximum deflection of the beam can be approximated as

$$\Delta_b = L\sqrt{\frac{\alpha_o T_{max}}{2}} = 360\sqrt{\frac{0.00542}{2}} = 18.74 \text{ in} = 1.56 \text{ ft}$$

3. The maximum trust force in this case can be approximated as

$$H = W\sqrt{\frac{1}{24(\alpha_o T_{max})}} = 30(1.52)\sqrt{\frac{1}{24(0.00542)}} = 126.4 \text{ kip}$$

   where $W = 30[0.087 + (0.5)0.043]14 = 30(1.52)$ kip

4. Additional bending moment and axial force in this case owing to the SFL are

$$M = 126.4(1.56) = 197.2 \text{ ft-kip} \quad \text{and} \quad N = -126.4 \text{ kip}$$

5. Dynamic coefficient K_d from Table 7.3 (based on linear interpolation) is

$$K_d = \frac{0.08}{3.41}(0.264) + 0.03 = 0.036$$

6. The design load combination in this case [per AISC's *Manual of Steel Construction*, 13th ed., Appendix 4, Eq. (A-4-1)] is 1.2D + 0.5L + T. Therefore, for dead load and live loads:

$$w_u = (1.2(0.087) + 1.6(0.043)0.5)14 = 1.94 \text{ klf.}$$

7. The ultimate design forces are

$$M_u = 1.94(900)/8 + 197.2(1.036) = 422.5 \text{ ft-kip}$$
$$N_u = -126.4(1.036) = -131.0 \text{ kip}$$
$$V_u = 1.94(30/2) = 29.1 \text{ kip}$$

8. Now, let's calculate the fire endurance rating time. Try $t = 3.5$ hours. The kernel of the integral equation (7.26) is as follows: $K(\theta) = \exp(-b\theta)$. Other kernels' types for different materials are given in Rabotnov [11]. Calculate the reduction coefficient n based on Eq. (7.32) for any given time:

$$n = \frac{E}{H} = 1 + \int_0^{3.5} K(\theta)d\theta \qquad (7.35)$$

where H is the long-term modulus of elasticity for a given time $t = 3.5$. Calculate n as follows: If $b = 0.333$, $n = 2.07$. Therefore, the yielding stress of the steel reinforcement is as follows: $F_y = 60/2.07 = 29.1$ ksi. Compare this result with Fig. 2.9 from ACI216.2.97 [12]: $n = 52$ percent or $F_y = 31.2$ ksi. Now, let's draw the interaction diagram (based on PCACol.software; see Figs. 7.29 and 7.30).

f'c = 4 ksi	fy = 29.1 ksi	Ag = 480 in^2	8 bars
Ec = 3605 ksi	Es = 29000 ksi	As = 11.32 in^2	Rho = 2.36%
fc = 3.4 ksi	e_rup = infinity	Xo = 0.00 in	Ix = 36000 in^4

9. Finally, check the fire endurance of the beam when the duration of fire is much longer than 3.5 hours, that is, $t \gg 3.5$ (catenary's action or progressive collapse arrest). Equation (7.31) in this case is

$$n = 1 + \int_0^\infty \exp(-0.333\theta)d\theta = 4 \qquad (7.36)$$

Therefore, $F_y = 60/4 = 15$ ksi and the maximum catenary's force is

$$T = (4(1.27) + (1.56)(4)(15) = 169.8 \text{ kip} > 126.4 \text{ kip} \qquad \text{okay.}$$

(The progressive collapse requirement is satisfied in this case.)

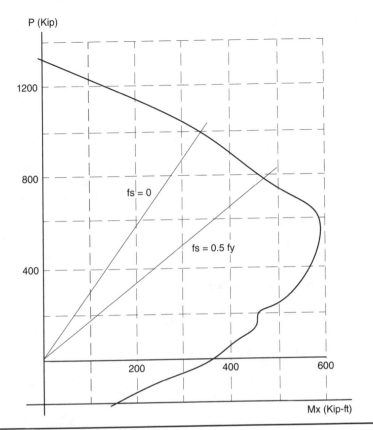

Figure 7.29   Interaction diagram.

Figure 7.30   Cross section.

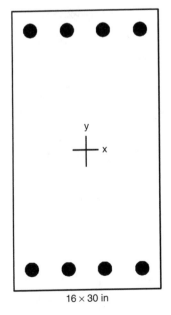

16 × 30 in

379

Example 7.11

Data   Simply supported reinforced-concrete transfer girder restrained against longitudinal expansion; span $L_n = 20$ ft. Fire exposure on three sides; therefore, additional bending moment accrues owing to restrained ends only. Cross section: $w = 24$ in; $h = 48$ in. Main reinforcement: 10_#9 (top and bottom). Concentrated force (service load) at midspan $P = 200$ kip (dead load = live load); $T_m = 609°C$; Case 3; concrete: $f'_c = 4$ ksi; steel: $f_y = 60$ ksi; #5 at 12 in o.c. ties; modulus of elasticity $E = 3,600$ ksi; moment of inertia $I_g = 24(48)^3/12 = 221,184$ in^4.

1. Beam design (SFL).

   Data   Beam is restrained at both ends, span $L = 20$ ft.

   $$\text{Natural frequency } \omega = \sqrt{\frac{g}{\delta_{11}W}} = \sqrt{\frac{32.2(12)}{0.000362(200)}}$$

   $$= 73.05 \text{ rad/s} = 11.6 \text{ Hz} > 5 \text{ Hz}$$

   where g = gravitational acceleration
   W = total gravity load
   δ_{11} = deformation of the beam from unit force applied at midspan

   Maximum temperature reduction owing to passive fire protection of the beam (assume 10 percent; see Table 6.57): $T_m = 0.9(609) = 548°C$. Total elongation of the beam:

   $$\Delta L = \alpha_o T_m L = 0.00099(5.48)240 = (0.00542)240 = 1.30 \text{ in.}$$

2. Since the deformed length of the beam is known [$L_{total} = L + \Delta L = L(1 + \alpha_o T_{max})$], the maximum deflection of the beam can be approximated as

   $$\Delta_b = L\sqrt{\frac{\alpha_o T_{max}}{2}} = 240\sqrt{\frac{0.00542}{2}} = 12.5 \text{ in} = 1.04 \text{ ft}$$

3. The maximum trust force in this case can be approximated as

   $$H = W\sqrt{\frac{1}{24(\alpha_o T_{max})}} = 240\sqrt{\frac{1}{24(0.00542)}} = 665.4 \text{ kip}$$

   where $W = 200$ kip.

4. Additional bending moment and axial force in this case owing to SFL are as follows:

   $$M = 665(1.04) = 692 \text{ ft-kip} \quad \text{and} \quad N = -665.4 \text{ kip}$$

5. Dynamic coefficient K_d from Table 7.3: Use $K_d = 0$.

6. Design load combination in this case [per AISC's *Manual of Steel Construction*, 13th ed., Appendix 4, Eq. (A-4-1)] is $1.2D + 0.5L + T$. Therefore,

$$P_u = (0.5)100(1.6) + 1.2(100) = 200 \text{ kip}$$

7. The ultimate design forces are

$$M_u = 20(200)/4 + 692 = 1{,}692 \text{ ft-kip}$$
$$N_u = -665.4 \text{ kip}$$
$$V_u = 200/2 = 100 \text{ kip}$$

8. Now, let's calculate the fire endurance rating time. Try $t = 3.5$ hours. The kernel of the integral equation (3.2) is as follows: $K(\theta) = \exp(-b\theta)$. Other kernels' types for different materials are given in Rabotnov [11]. Calculate the reduction coefficient n based on Eq. (7.32) for any given time:

$$n = \frac{E}{H} = 1 + \int_0^{3.5} K(\theta)\,d\theta \qquad (7.37)$$

where H is the long-term modulus of elasticity for a given time $t = 3.5$. Calculate n as follows: If $b = 0.333$; $n = 2.07$. Therefore, the yielding stress of the steel reinforcement is as follows: $F_y = 60/2.07 = 29.1$ ksi. Compare this result with Fig. 2.9 from ACI216.1.97 [12]: $n = 52$ percent or $F_y = 31.2$ ksi. Now, let's draw the interaction diagram (based on the PCACol .software; see Figs. 7.31 and 7.32).

f'c = 4 ksi	fy = 29.1 ksi	Ag = 1152 in^2	26 bars
Ec = 3605 ksi	Es = 29000 ksi	As = 21.86 in^2	Rho = 1.90%
fc = 3.4 ksi	e_rup = infinity	Xo = 0.00 in	Ix = 221184 in^4

Section is okay.

9. Finally, check the fire endurance of the beam when the duration of fire is much longer than 3.5 hours, that is, $t \gg 3.5$ (catenary's action or progressive collapse arrest). Equation (7.31) in this case is

$$n = 1 + \int_0^\infty \exp(-0.333\theta)\,d\theta = 4 \qquad (7.38)$$

Therefore, $F_y = 60/4 = 15$ ksi, and the maximum catenary's force is

$$T = (20(1) + (0.31)6)(15) = 327.9 \text{ kip} > 665.4 \text{ kip}$$

(The progressive collapse requirement is not satisfied in this case.)

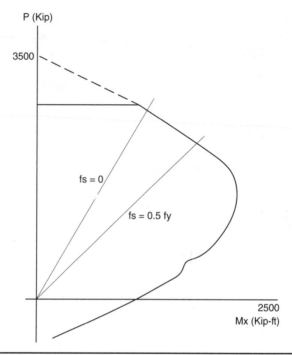

FIGURE 7.31 Interaction diagram.

FIGURE 7.32 Cross section.

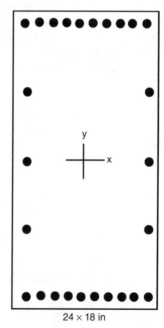

24 × 18 in

Example 7.12   (Fig. 7.33)

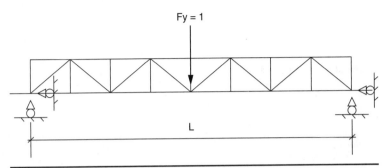

FIGURE 7.33   Floor truss.

1. From the Steel Joist Institute (SJI) table for a 24K12 joist; $L = 40$ ft; total allowable distributed load is 438 plf; total distributed live load is 247 plf.; total load $P = 0.438(40) = 17.52$ kip.

2. Moment of inertia (per SLI): $I = 26.767(w)(L - 0.33)10^{-6} = 26.767(247)(40 - 0.33)10^{-6} = 412.75$ in⁴.

3. Deflection (per SJI): $\Delta_{st} = 1.15(w)L^4/(384EI) = 1.15(0.3145)40^4 \times (1{,}728)/384(29{,}000)412.75 = 1.74$ in.

4. Natural frequency:

$$\omega = \sqrt{\frac{32.2(12)}{1.74}} = 14.9 \text{ rad/s} = 2.37 \text{ Hz}$$

5. Dynamic coefficient:

$$K_d = \frac{2.63}{3.41}0.264 + 0.03 = 0.234$$

6. Maximum internal force (end panel) from unit force $P = 1.0$ kip applied at the center of the truss:

$$N_{11} = 0.5/\tan \alpha = 0.5/0.8 = 0.625 \text{ kip}$$

7. Maximum deflection at the centerline of the truss (center of mass) from SFL (temperature load):

$$\Delta_{st} = L\sqrt{\frac{\alpha_o T_{max}}{2}} = 480\sqrt{\frac{0.00641}{2}} = 27.2 \text{ in} = 2.26 \text{ ft}$$

8. Total internal force from SFL (end panel):

$$N_{11} = 0.625(12.58) + 0.625(12.58)27.2(1.234)/1.74$$
$$= 7.86 + 151.5 = 159.36 \text{ kip}$$

9. Internal force (original design): $N_{11} = 0.625(17.52) = 10.95$ kip $\ll$ 159.36 kip

10. Total original maximum design force (bottom chord):

$$N_{max} = \frac{0.438(40)^2}{2(8)} = 43.8 \text{ kip} \ll 159.36 \text{ kip}$$

Conclusion   End panel is failing.

References

1. American Society of Civil Engineering (ASCE). *Minimum Design Loads for Buildings and Other Structures*, ASCE-7-05. ASCE, New York, 2005.
2. Razdolsky, L., "Extreme Thermal Load and Concrete Structures Design," in *CONSEC-07: Proceedings of the Fifth International Conference on Concrete under Severe Conditions of Environment and Loading*, Tour, France. Taylor & Francis Group, London, U.K., 2007.
3. Razdolsky, L., "Fire Load in a Concrete Building Design," in *Proceedings of the International Conference, Concrete: Construction's Sustainability Option*, Dundee, Scotland, 2008.
4. Owen, D. R. J., and Hinton, E., *Finite Elements in Plasticity*, Pineridge Press, Ltd., Swansea, UK, 1986.
5. Penny, R. K., and Marriot, D. L., *Design for Creep*. Chapman & Hall, London, 1995.
6. Boresi, A. P., Schmidt, R. J., and Sidebottom, O. M., *Advanced Mechanics of Materials*, Wiley, New York, 1993.
7. Drozdov, A. D., *Finite Elasticity and Viscoelasticity*, World Scientific, Princeton, NJ, 1996.
8. Drozdov, A. D., *Mechanics of Viscoelastic Solids*, Wiley, New York, 1998.
9. Findley, W. N., Lai, J. S., and Onaran, K., *Creep and Relaxation of Nonlinear Viscoelastic Materials*. Dover Publications, New York, 1989.
10. Brnić, J., *Elastomechanics and Plastomechanics* [in Croatian], Skolska knjiga, Zagreb, 1996.
11. Rabotnov, Y. N., *Some Problems of the Theory of Creep*. National Advisory Committee for Aeronautics (NACA), Washington, DC, 1953.
12. American Concrete Institute (ACI 216.1-97). Farmington Hills, MI, 1997.

Appendix 7A

Example 7A.1
Computer input and output data are presented below:

LOAD DATA

File: F:\RAM Advanse\Data\Fire 03 Frame.AVW

Units: Kip-ft

CONCENTRATED FORCES ON MEMBERS					
	ConditiBeam	Dir.	Value (kip)	Distance (ft)	%
dl	3	Y	−200	10	0

PRESSURE & TEMPERATURE							
	ConditiBeam	Pressure (kip/ft²)	iPressure (kip/ft²)	iPressure (kip/ft²)	iTemp 1 [°C]	Temp 2 [°F/ft]	Temp 3 [°F/ft]
tl	3	0	0	0	548	0	0

ANALYSIS RESULTS

File: F:\RAM Advanse\Data\Fire 03 Frame.AVW

Units: Kip-ft

TRANSLATIONS						
	TRANSLATIONS (ft)			ROTATIONS (rad)		
Node	TX	TY	TZ	RX	RY	RZ
Condition dl (dead load)						
1	0.00000	0.00000	0.00000	0.00000	0.00000	0.00087
2	0.00000	0.00000	0.00000	0.00000	0.00000	−0.00087
3	0.00009	−0.00118	0.00000	0.00000	0.00000	−0.00190
4	−0.00009	−0.00118	0.00000	0.00000	0.00000	0.00190
Condition tl (temperature load)						
1	0.00000	0.00000	0.00000	0.00000	0.00000	0.00736
2	0.00000	0.00000	0.00000	0.00000	0.00000	−0.00736
3	−0.05460	0.00000	0.00000	0.00000	0.00000	0.00136
4	0.05460	0.00000	0.00000	0.00000	0.00000	−0.00136

REACTIONS						
	FORCES (kip)			**MOMENTS (kip*ft)**		
Node	**FX**	**FY**	**FZ**	**MX**	**MY**	**MZ**
Condition dl (dead load)						
1	12.43211	100.00000	0.00000	0.00000	0.00000	0.00000
2	−12.43211	100.00000	0.00000	0.00000	0.00000	0.00000
SUM	0.00000	200.00000	0.00000	0.00000	0.00000	0.00000
Condition tl (temperature load)						
1	26.93346	0.00000	0.00000	0.00000	0.00000	0.00000
2	−26.93346	0.00000	0.00000	0.00000	0.00000	0.00000
SUM	0.00000	0.00000	0.00000	0.00000	0.00000	0.00000
MEMBER FORCES						
	M33 (kip*ft)	**V2 (kip)**	**M22 (kip*ft)**	**V3 (kip)**	**Axial (kip)**	**Torsion (kip*ft)**
Condition dl (dead load)						
MEMBER 1						
0%	0.00	12.43	0.00	0.00	−100.00	0.00
25%	−31.08	12.43	0.00	0.00	−100.00	0.00
50%	−62.16	12.43	0.00	0.00	−100.00	0.00
75%	−93.24	12.43	0.00	0.00	−100.00	0.00
100%	−124.32	12.43	0.00	0.00	−100.00	0.00
MEMBER 2						
0%	0.00	−12.43	0.00	0.00	−100.00	0.00
25%	31.08	−12.43	0.00	0.00	−100.00	0.00
50%	62.16	−12.43	0.00	0.00	−100.00	0.00
75%	93.24	−12.43	0.00	0.00	−100.00	0.00
100%	124.32	−12.43	0.00	0.00	−100.00	0.00
MEMBER 3						
0%	−124.32	−100.00	0.00	0.00	−12.43	0.00
25%	375.68	−100.00	0.00	0.00	−12.43	0.00
50%	875.68	100.00	0.00	0.00	−12.43	0.00
75%	375.68	100.00	0.00	0.00	−12.43	0.00
100%	−124.32	100.00	0.00	0.00	−12.43	0.00

	M33 (kip*ft)	V2 (kip)	M22 (kip*ft)	V3 (kip)	Axial (kip)	Torsion (kip*ft)
MEMBER FORCES						
Condition tl (temperature load)						
MEMBER 1						
0%	0.00	26.93	0.00	0.00	0.00	0.00
25%	−67.33	26.93	0.00	0.00	0.00	0.00
50%	−134.67	26.93	0.00	0.00	0.00	0.00
75%	−202.00	26.93	0.00	0.00	0.00	0.00
100%	−269.33	26.93	0.00	0.00	0.00	0.00
MEMBER 2						
0%	0.00	−26.93	0.00	0.00	0.00	0.00
25%	67.33	−26.93	0.00	0.00	0.00	0.00
50%	134.67	−26.93	0.00	0.00	0.00	0.00
75%	202.00	−26.93	0.00	0.00	0.00	0.00
100%	269.33	−26.93	0.00	0.00	0.00	0.00
MEMBER 3						
0%	−269.33	0.00	0.00	0.00	−26.93	0.00
25%	−269.33	0.00	0.00	0.00	−26.93	0.00
50%	−269.33	0.00	0.00	0.00	−26.93	0.00
75%	−269.33	0.00	0.00	0.00	−26.93	0.00
100%	−269.33	0.00	0.00	0.00	−26.93	0.00

Station	At 1 (ft)	At 2 (ft)	At 3 (ft)	Rotation11 (rad)	Slopes	
					L/ defl(2)	L/ defl(3)
MEMBER LOCAL DEFLECTIONS						
Condition dl (dead load)						
MEMBER 1						
0%	0.000	0.000	0.000	0.00000	—	—
25%	−0.000	0.002	0.000	0.00000	4672.42	—
50%	−0.001	0.003	0.000	0.00000	2928.02	—
75%	−0.001	0.003	0.000	0.00000	3378.38	—
100%	−0.001	0.000	0.000	0.00000	—	—

	MEMBER LOCAL DEFLECTIONS					
					Slopes	
Station	At 1 (ft)	At 2 (ft)	At 3 (ft)	Rotation 11 (rad)	L/defl (2)	L/defl (3)
Condition dl (dead load)						
MEMBER 2						
0%	0.000	0.000	0.000	0.00000	—	—
25%	−0.000	−0.002	0.000	0.00000	4672.42	—
50%	−0.001	−0.003	0.000	0.00000	2928.02	—
75%	−0.001	−0.003	0.000	0.00000	3378.38	—
100%	−0.001	0.000	0.000	0.00000	—	—
MEMBER 3						
0%	0.000	−0.001	0.000	0.00000	—	—
25%	0.000	−0.015	0.000	0.00000	1351.20	—
50%	0.000	−0.024	0.000	0.00000	844.90	—
75%	0.000	−0.015	0.000	0.00000	1351.20	—
100%	0.000	−0.001	0.000	0.00000	—	—
Condition tl (temperature load)						
MEMBER 1						
0%	0.000	0.000	0.000	0.00000	—	—
25%	0.000	0.018	0.000	0.00000	545.36	—
50%	0.000	0.035	0.000	0.00000	287.36	—
75%	0.000	0.048	0.000	0.00000	210.47	—
100%	0.000	0.055	0.000	0.00000	183.14	—
MEMBER 2						
0%	0.000	0.000	0.000	0.00000	—	—
25%	0.000	−0.018	0.000	0.00000	545.36	—
50%	0.000	−0.035	0.000	0.00000	287.36	—
75%	0.000	−0.048	0.000	0.00000	210.47	—
100%	0.000	−0.055	0.000	0.00000	183.14	—
MEMBER 3						
0%	−0.055	0.000	0.000	0.00000	—	—
25%	−0.027	0.005	0.000	0.00000	3914.20	—
50%	0.000	0.007	0.000	0.00000	2935.65	—
75%	0.027	0.005	0.000	0.00000	3914.20	—
100%	0.055	0.000	0.000	0.00000	—	—

Example 7A.2
LOAD DATA

File: F:\RAM Advanse\Data\Fire, Frame, Dist. Load.AVW
Units: Kip-ft

DISTRIBUTED FORCES ON MEMBERS			
	ConditiBeam	Dir.	Value (kip/ft)
dl	3	Y	−0.5
ll	1	Y	0
	2	Y	0
	3	Y	0
CONCENTRATED FORCES ON MEMBERS				
	ConditiBeam	Dir.	Value (kip)	Distance (ft)	%
ll	3	y	−1	20	0
PRESSURE & TEMPERATURE						
	ConditiBeam	Pressure (kip/ft²)	iPressure (kip/ft²)	iPressure (kip/ft²)	iTemp1 (°F)	Temp 2 (°F/ft)	Temp 3 (°F/ft)
tl	3	0	0	0	548	0	0

ANALYSIS RESULTS

File: F:\RAM Advanse\Data\Fire, Frame, Dist. Load.AVW
Units: Kip-ft
Date: 5/5/2010
Time: 1:39:08 PM

TRANSLATIONS						
	TRANSLATIONS (ft)			ROTATIONS (rad)		
Node	TX	TY	TZ	RX	RY	RZ
Condition dl (dead load)						
1	0.00000	0.00000	0.00000	0.00000	0.00000	0.00036
2	0.00020	−0.00012	0.00000	0.00000	0.00000	−0.00084
3	0.00000	0.00000	0.00000	0.00000	0.00000	−0.00036
4	−0.00020	−0.00012	0.00000	0.00000	0.00000	0.00084
Condition tl (temperature load)						
1	0.00000	0.00000	0.00000	0.00000	0.00000	0.01188
2	−0.10908	0.00000	0.00000	0.00000	0.00000	0.00881
3	0.00000	0.00000	0.00000	0.00000	0.00000	−0.01188
4	0.10908	0.00000	0.00000	0.00000	0.00000	−0.00881

TRANSLATIONS					
	TRANSLATIONS (ft)			ROTATIONS (rad)	
Node	TX	TY	TZ	RX	RY	RZ
Condition II (live load)						
1	0.00000	0.00000	0.00000	0.00000	0.00000	0.00003
2	0.00002	−0.00001	0.00000	0.00000	0.00000	−0.00006
3	0.00000	0.00000	0.00000	0.00000	0.00000	−0.00003
4	−0.00002	−0.00001	0.00000	0.00000	0.00000	0.00006
REACTIONS						
	FORCES (kip)			MOMENTS (kip*ft)		
Node	FX	FY	FZ	MX	MY	MZ
Condition dl (dead load)						
1	5.35655	10.00000	0.00000	0.00000	0.00000	0.00000
3	−5.35655	10.00000	0.00000	0.00000	0.00000	0.00000
SUM	0.00000	20.00000	0.00000	0.00000	0.00000	0.00000
Condition tl (temperature load)						
1	13.81364	0.00000	0.00000	0.00000	0.00000	0.00000
3	−13.81364	0.00000	0.00000	0.00000	0.00000	0.00000
SUM	0.00000	0.00000	0.00000	0.00000	0.00000	0.00000
Condition II (live load)						
1	0.40174	0.50000	0.00000	0.00000	0.00000	0.00000
3	−0.40174	0.50000	0.00000	0.00000	0.00000	0.00000
SUM	0.00000	1.00000	0.00000	0.00000	0.00000	0.00000
MEMBER FORCES						
	M33 (kip*ft)	V2 (kip)	M22 (kip*ft)	V3 (kip)	Axial (kip)	Torsion (kip*ft)
Condition dl (dead load)						
MEMBER 1						
0%	0.00	5.36	0.00	0.00	−10.00	0.00
25%	−13.39	5.36	0.00	0.00	−10.00	0.00
50%	−26.78	5.36	0.00	0.00	−10.00	0.00
75%	−40.17	5.36	0.00	0.00	−10.00	0.00
100%	−53.57	5.36	0.00	0.00	−10.00	0.00

MEMBER FORCES						
	M33 (kip*ft)	V2 (kip)	M22 (kip*ft)	V3 (kip)	Axial (kip)	Torsion (kip*ft)
Condition dl (dead load)						
MEMBER 2						
0%	0.00	−5.36	0.00	0.00	−10.00	0.00
25%	13.39	−5.36	0.00	0.00	−10.00	0.00
50%	26.78	−5.36	0.00	0.00	−10.00	0.00
75%	40.17	−5.36	0.00	0.00	−10.00	0.00
100%	53.57	−5.36	0.00	0.00	−10.00	0.00
MEMBER 3						
0%	−53.57	−10.00	0.00	0.00	−5.36	0.00
25%	21.43	−5.00	0.00	0.00	−5.36	0.00
50%	46.43	0.00	0.00	0.00	−5.36	0.00
75%	21.43	5.00	0.00	0.00	−5.36	0.00
100%	−53.57	10.00	0.00	0.00	−5.36	0.00
Condition tl (temperature load)						
MEMBER 1						
0%	0.00	13.81	0.00	0.00	0.00	0.00
25%	−34.53	13.81	0.00	0.00	0.00	0.00
50%	−69.07	13.81	0.00	0.00	0.00	0.00
75%	−103.60	13.81	0.00	0.00	0.00	0.00
100%	−138.14	13.81	0.00	0.00	0.00	0.00
MEMBER 2						
0%	0.00	−13.81	0.00	0.00	0.00	0.00
25%	34.53	−13.81	0.00	0.00	0.00	0.00
50%	69.07	−13.81	0.00	0.00	0.00	0.00
75%	103.60	−13.81	0.00	0.00	0.00	0.00
100%	138.14	−13.81	0.00	0.00	0.00	0.00
MEMBER 3						
0%	−138.14	0.00	0.00	0.00	−13.81	0.00
25%	−138.14	0.00	0.00	0.00	−13.81	0.00
50%	−138.14	0.00	0.00	0.00	−13.81	0.00
75%	−138.14	0.00	0.00	0.00	−13.81	0.00
100%	−138.14	0.00	0.00	0.00	−13.81	0.00

			MEMBER FORCES			
	M33 (kip*ft)	V2 (kip)	M22 (kip*ft)	V3 (kip)	Axial (kip)	Torsion (kip*ft)
			Condition II (live load)			
MEMBER 1						
0%	0.00	0.40	0.00	0.00	−0.50	0.00
25%	−1.00	0.40	0.00	0.00	−0.50	0.00
50%	−2.01	0.40	0.00	0.00	−0.50	0.00
75%	−3.01	0.40	0.00	0.00	−0.50	0.00
100%	−4.02	0.40	0.00	0.00	−0.50	0.00
MEMBER 2						
0%	0.00	−0.40	0.00	0.00	−0.50	0.00
25%	1.00	−0.40	0.00	0.00	−0.50	0.00
50%	2.01	−0.40	0.00	0.00	−0.50	0.00
75%	3.01	−0.40	0.00	0.00	−0.50	0.00
100%	4.02	−0.40	0.00	0.00	−0.50	0.00
MEMBER 3						
0%	−4.02	−0.50	0.00	0.00	−0.40	0.00
25%	0.98	−0.50	0.00	0.00	−0.40	0.00
50%	5.98	0.50	0.00	0.00	−0.40	0.00
75%	0.98	0.50	0.00	0.00	−0.40	0.00
100%	−4.02	0.50	0.00	0.00	−0.40	0.00
		MEMBER LOCAL DEFLECTIONS				
Station	At 1 (ft)	At 2 (ft)	At 3 (ft)	Rotation 11 (rad)	SLOPES	
					L/defl(2)	L/defl(3)
			Condition dl (dead load)			
MEMBER 1						
0%	0.000	0.000	0.000	0.00000	—	—
25%	0.000	0.001	0.000	0.00000	—	—
50%	0.000	0.001	0.000	0.00000	7194.61	—
75%	0.000	0.001	0.000	0.00000	8673.23	—
100%	−0.000	−0.000	0.000	0.00000	—	—

Station	At 1 (ft)	At 2 (ft)	At 3 (ft)	Rotation 11 (rad)	SLOPES L/defl(2)	L/defl(3)
MEMBER LOCAL DEFLECTIONS						
Condition dl (dead load)						
MEMBER 2						
0%	0.000	0.000	0.000	0.00000	—	—
25%	0.000	−0.001	0.000	0.00000	—	—
50%	0.000	−0.001	0.000	0.00000	7194.61	—
75%	0.000	−0.001	0.000	0.00000	8673.23	—
100%	−0.000	0.000	0.000	0.00000	—	—
MEMBER 3						
0%	0.000	−0.000	0.000	0.00000	—	—
25%	0.000	−0.013	0.000	0.00000	3124.90	—
50%	0.000	−0.020	0.000	0.00000	2032.09	—
75%	−0.000	−0.013	0.000	0.00000	3124.90	—
100%	−0.000	−0.000	0.000	0.00000	—	—
Condition tl (temperature load)						
MEMBER 1						
0%	0.000	0.000	0.000	0.00000	—	—
25%	0.000	0.030	0.000	0.00000	337.01	—
50%	0.000	0.058	0.000	0.00000	171.28	—
75%	0.000	0.085	0.000	0.00000	117.41	—
100%	0.000	0.109	0.000	0.00000	91.68	—
MEMBER 2						
0%	0.000	0.000	0.000	0.00000	—	—
25%	0.000	−0.030	0.000	0.00000	337.01	—
50%	0.000	−0.058	0.000	0.00000	171.28	—
75%	0.000	−0.085	0.000	0.00000	117.41	—
100%	0.000	−0.109	0.000	0.00000	91.68	—
MEMBER 3						
0%	−0.109	0.000	0.000	0.00000	—	—
25%	−0.055	0.066	0.000	0.00000	605.64	—
50%	0.000	0.088	0.000	0.00000	454.23	—
75%	0.055	0.066	0.000	0.00000	605.64	—
100%	0.109	0.000	0.000	0.00000	—	—

					SLOPES	
Station	**At 1 (ft)**	**At 2 (ft)**	**At 3 (ft)**	**Rotation 11 (rad)**	**L/defl(2)**	**L/defl(3)**
\multicolumn{7}{c}{Condition II (live load)}						

MEMBER LOCAL DEFLECTIONS

Station	At 1 (ft)	At 2 (ft)	At 3 (ft)	Rotation 11 (rad)	L/defl(2)	L/defl(3)
\multicolumn{7}{c}{Condition II (live load)}						
MEMBER 1						
0%	0.000	0.000	0.000	0.00000	—	—
25%	0.000	0.000	0.000	0.00000	—	—
50%	0.000	0.000	0.000	0.00000	—	—
75%	0.000	0.000	0.000	0.00000	—	—
100%	0.000	0.000	0.000	0.00000	—	—
MEMBER 2						
0%	0.000	0.000	0.000	0.00000	—	—
25%	0.000	0.000	0.000	0.00000	—	—
50%	0.000	−0.000	0.000	0.00000	—	—
75%	0.000	0.000	0.000	0.00000	—	—
100%	0.000	0.000	0.000	0.00000	—	—
MEMBER 3						
0%	0.000	0.000	0.000	0.00000	—	—
25%	0.000	−0.002	0.000	0.00000	—	—
50%	0.000	−0.003	0.000	0.00000	—	—
75%	0.000	−0.002	0.000	0.00000	—	—
100%	0.000	0.000	0.000	0.00000	—	—

Example 7A.3

Data

$P_{D.L.} = P_{L.L.}$

Span $L = 10$ ft, 0 in

$F_y = 50$ ksi

W 18 × 65 (ASTM A992)—Beam	**W 14 × 99—Column**
$A = 19.1$ in^2	$A = 29.1$ in^2
$F_u = 65$ ksi	$F_u = 65$ ksi
$S_x = 117$ in^3	$S_x = 157$ in^3
$I_x = 1{,}070$ in^4	$I_x = 1{,}110$ in^4
$Z_x = 133$ in^3	$Z_x = 173$ in^3
$T_m = 609°C$	

Fire rating: 1 hour

Fire severity: Case 3

Computer input and output data are presented below:

LOAD DATA

File: F:\RAM Advanse\Data\Fire 02.AVW

Units: Kip-ft

NODAL FORCES							
	ConditiNode	FX (kip)	FY (kip)	FZ (kip)	MX (kip*ft)	MY (kip*ft)	MZ (kip*ft)
dl	5	0	−100	0	0	0	0
	6	0	−150	0	0	0	0
	7	0	−150	0	0	0	0
	8	0	−100	0	0	0	0
wl	5	100	0	0	0	0	0

DISTRIBUTED FORCES ON MEMBERS			
	ConditiBeam	Dir.	Value (kip/ft)
dl	5	Y	−1
	6	Y	−1
	7	Y	−2

CONCENTRATED FORCES ON MEMBERS				
	ConditiBeam	Dir.	Value (kip)	Distance (ft)	%
ll	6	y	−100	10	0

PRESSURE & TEMPERATURE						
	ConditiBeam	Pressure (kip/ft^2)	iPressure (kip/ft^2)	iPressure (kip/ft^2)	iTemp1 (°F)	Temp 2 (°F/ft)	Temp 3 (°F/ft)
tl	5	0	0	0	548	0	0
	6	0	0	0	548	0	0

ANALYSIS RESULTS

File: F:\RAM Advanse\Data\Fire 02.AVW
Units: Kip-ft
Date: 5/5/2010
Time: 3:46:13 PM

MEMBER FORCES						
	M33 (kip*ft)	V2 (kip)	M22 (kip*ft)	V3 (kip)	Axial (kip)	Torsion (kip*ft)
Condition dl (dead load)						
MEMBER 1						
0%	0.00	0.67	0.00	0.00	−104.43	0.00
25%	−1.67	0.67	0.00	0.00	−104.43	0.00
50%	−3.33	0.67	0.00	0.00	−104.43	0.00
75%	−5.00	0.67	0.00	0.00	−104.43	0.00
100%	−6.66	0.67	0.00	0.00	−104.43	0.00

MEMBER FORCES						
	M33 (kip*ft)	V2 (kip)	M22 (kip*ft)	V3 (kip)	Axial (kip)	Torsion (kip*ft)
Condition dl (dead load)						
MEMBER 2						
0%	0.00	1.03	0.00	0.00	−164.91	0.00
25%	−2.57	1.03	0.00	0.00	−164.91	0.00
50%	−5.15	1.03	0.00	0.00	−164.91	0.00
75%	−7.72	1.03	0.00	0.00	−164.91	0.00
100%	−10.29	1.03	0.00	0.00	−164.91	0.00
MEMBER 3						
0%	0.00	−0.94	0.00	0.00	−172.55	0.00
25%	2.35	−0.94	0.00	0.00	−172.55	0.00
50%	4.70	−0.94	0.00	0.00	−172.55	0.00
75%	7.04	−0.94	0.00	0.00	−172.55	0.00
100%	9.39	−0.94	0.00	0.00	−172.55	0.00
MEMBER 4						
0%	0.00	−0.76	0.00	0.00	−108.11	0.00
25%	1.89	−0.76	0.00	0.00	−108.11	0.00
50%	3.78	−0.76	0.00	0.00	−108.11	0.00
75%	5.67	−0.76	0.00	0.00	−108.11	0.00
100%	7.56	−0.76	0.00	0.00	−108.11	0.00
MEMBER 5						
0%	−6.66	−4.43	0.00	0.00	−0.67	0.00
25%	1.29	−1.93	0.00	0.00	−0.67	0.00
50%	2.98	0.57	0.00	0.00	−0.67	0.00
75%	−1.57	3.07	0.00	0.00	−0.67	0.00
100%	−12.37	5.57	0.00	0.00	−0.67	0.00
MEMBER 6						
0%	−22.66	−9.34	0.00	0.00	−1.70	0.00
25%	11.54	−4.34	0.00	0.00	−1.70	0.00
50%	20.75	0.66	0.00	0.00	−1.70	0.00
75%	4.96	5.66	0.00	0.00	−1.70	0.00
100%	−35.83	10.66	0.00	0.00	−1.70	0.00

MEMBER FORCES						
	M33 (kip*ft)	V2 (kip)	M22 (kip*ft)	V3 (kip)	Axial (kip)	Torsion (kip*ft)
Condition dl (dead load)						
MEMBER 7						
0%	−26.44	−11.89	0.00	0.00	−0.76	0.00
25%	−2.97	−6.89	0.00	0.00	−0.76	0.00
50%	8.00	−1.89	0.00	0.00	−0.76	0.00
75%	6.47	3.11	0.00	0.00	−0.76	0.00
100%	−7.56	8.11	0.00	0.00	−0.76	0.00
Condition tl (temperature load)						
MEMBER 1						
0%	0.00	43.25	0.00	0.00	−70.69	0.00
25%	−108.12	43.25	0.00	0.00	−70.69	0.00
50%	−216.23	43.25	0.00	0.00	−70.69	0.00
75%	−324.35	43.25	0.00	0.00	−70.69	0.00
100%	−432.47	43.25	0.00	0.00	−70.69	0.00
MEMBER 2						
0%	0.00	31.46	0.00	0.00	72.92	0.00
25%	−78.65	31.46	0.00	0.00	72.92	0.00
50%	−157.31	31.46	0.00	0.00	72.92	0.00
75%	−235.96	31.46	0.00	0.00	72.92	0.00
100%	−314.61	31.46	0.00	0.00	72.92	0.00
MEMBER 3						
0%	0.00	−40.48	0.00	0.00	64.02	0.00
25%	101.21	−40.48	0.00	0.00	64.02	0.00
50%	202.42	−40.48	0.00	0.00	64.02	0.00
75%	303.63	−40.48	0.00	0.00	64.02	0.00
100%	404.84	−40.48	0.00	0.00	64.02	0.00
MEMBER 4						
0%	0.00	−34.22	0.00	0.00	−66.25	0.00
25%	85.56	−34.22	0.00	0.00	−66.25	0.00
50%	171.12	−34.22	0.00	0.00	−66.25	0.00
75%	256.68	−34.22	0.00	0.00	−66.25	0.00
100%	342.24	−34.22	0.00	0.00	−66.25	0.00

MEMBER FORCES					
M33 (kip*ft)	V2 (kip)	M22 (kip*ft)	V3 (kip)	Axial (kip)	Torsion (kip*ft)
Condition tl (temperature load)					
MEMBER 5					
0% −432.47	−70.69	0.00	0.00	−43.25	0.00
25% −255.73	−70.69	0.00	0.00	−43.25	0.00
50% −78.99	−70.69	0.00	0.00	−43.25	0.00
75% 97.74	−70.69	0.00	0.00	−43.25	0.00
100% 274.48	−70.69	0.00	0.00	−43.25	0.00
MEMBER 6					
0% −40.13	2.22	0.00	0.00	−74.71	0.00
25% −51.25	2.22	0.00	0.00	−74.71	0.00
50% −62.38	2.22	0.00	0.00	−74.71	0.00
75% −73.50	2.22	0.00	0.00	−74.71	0.00
100% −84.62	2.22	0.00	0.00	−74.71	0.00
MEMBER 7					
0% 320.22	66.25	0.00	0.00	−34.22	0.00
25% 154.61	66.25	0.00	0.00	−34.22	0.00
50% −11.01	66.25	0.00	0.00	−34.22	0.00
75% −176.62	66.25	0.00	0.00	−34.22	0.00
100% −342.24	66.25	0.00	0.00	−34.22	0.00
Condition II (live load)					
MEMBER 1					
0% 0.00	−1.15	0.00	0.00	9.74	0.00
25% 2.89	−1.15	0.00	0.00	9.74	0.00
50% 5.77	−1.15	0.00	0.00	9.74	0.00
75% 8.66	−1.15	0.00	0.00	9.74	0.00
100% 11.55	−1.15	0.00	0.00	9.74	0.00
MEMBER 2					
0% 0.00	9.99	0.00	0.00	−57.62	0.00
25% −24.97	9.99	0.00	0.00	−57.62	0.00
50% −49.95	9.99	0.00	0.00	−57.62	0.00
75% −74.92	9.99	0.00	0.00	−57.62	0.00
100% −99.89	9.99	0.00	0.00	−57.62	0.00
MEMBER 3					
0% 0.00	−12.50	0.00	0.00	−66.09	0.00
25% 31.25	−12.50	0.00	0.00	−66.09	0.00
50% 62.50	−12.50	0.00	0.00	−66.09	0.00
75% 93.75	−12.50	0.00	0.00	−66.09	0.00
100% 125.00	−12.50	0.00	0.00	−66.09	0.00

MEMBER FORCES					
M33 (kip*ft)	V2 (kip)	M22 (kip*ft)	V3 (kip)	Axial (kip)	Torsion (kip*ft)
Condition II (live load)					
MEMBER 4					
0.00	3.67	0.00	0.00	13.97	0.00
−9.16	3.67	0.00	0.00	13.97	0.00
−18.33	3.67	0.00	0.00	13.97	0.00
−27.49	3.67	0.00	0.00	13.97	0.00
−36.65	3.67	0.00	0.00	13.97	0.00
MEMBER 5					
11.55	9.74	0.00	0.00	1.15	0.00
−12.80	9.74	0.00	0.00	1.15	0.00
−37.15	9.74	0.00	0.00	1.15	0.00
−61.50	9.74	0.00	0.00	1.15	0.00
−85.84	9.74	0.00	0.00	1.15	0.00
MEMBER 6					
−185.74	−47.88	0.00	0.00	−8.83	0.00
53.68	−47.88	0.00	0.00	−8.83	0.00
293.10	52.12	0.00	0.00	−8.83	0.00
32.52	52.12	0.00	0.00	−8.83	0.00
−228.07	52.12	0.00	0.00	−8.83	0.00
MEMBER 7					
−103.06	−13.97	0.00	0.00	3.67	0.00
−68.13	−13.97	0.00	0.00	3.67	0.00
−33.21	−13.97	0.00	0.00	3.67	0.00
1.72	−13.97	0.00	0.00	3.67	0.00
36.65	−13.97	0.00	0.00	3.67	0.00
Condition wl (wind in X)					
MEMBER 1					
0.00	−15.32	0.00	0.00	26.31	0.00
38.30	−15.32	0.00	0.00	26.31	0.00
76.59	−15.32	0.00	0.00	26.31	0.00
114.89	−15.32	0.00	0.00	26.31	0.00
153.18	−15.32	0.00	0.00	26.31	0.00
MEMBER 2					
0.00	−20.82	0.00	0.00	−15.26	0.00
52.04	−20.82	0.00	0.00	−15.26	0.00
104.08	−20.82	0.00	0.00	−15.26	0.00
156.13	−20.82	0.00	0.00	−15.26	0.00
208.17	−20.82	0.00	0.00	−15.26	0.00

Note: Each MEMBER block's rows correspond to positions 0%, 25%, 50%, 75%, and 100%.

MEMBER FORCES					
M33 (kip*ft)	V2 (kip)	M22 (kip*ft)	V3 (kip)	Axial (kip)	Torsion (kip*ft)
Condition wl (wind in X)					
MEMBER 3					
0%	0.00	−36.39	0.00	0.00	40.53	0.00
25%	90.98	−36.39	0.00	0.00	40.53	0.00
50%	181.97	−36.39	0.00	0.00	40.53	0.00
75%	272.95	−36.39	0.00	0.00	40.53	0.00
100%	363.93	−36.39	0.00	0.00	40.53	0.00
MEMBER 4					
0%	0.00	−27.47	0.00	0.00	−51.58	0.00
25%	68.68	−27.47	0.00	0.00	−51.58	0.00
50%	137.36	−27.47	0.00	0.00	−51.58	0.00
75%	206.04	−27.47	0.00	0.00	−51.58	0.00
100%	274.72	−27.47	0.00	0.00	−51.58	0.00
MEMBER 5					
0%	153.18	26.31	0.00	0.00	−84.68	0.00
25%	87.40	26.31	0.00	0.00	−84.68	0.00
50%	21.62	26.31	0.00	0.00	−84.68	0.00
75%	−44.16	26.31	0.00	0.00	−84.68	0.00
100%	−109.95	26.31	0.00	0.00	−84.68	0.00
MEMBER 6					
0%	98.22	11.05	0.00	0.00	−63.87	0.00
25%	42.97	11.05	0.00	0.00	−63.87	0.00
50%	−12.29	11.05	0.00	0.00	−63.87	0.00
75%	−67.55	11.05	0.00	0.00	−63.87	0.00
100%	−122.80	11.05	0.00	0.00	−63.87	0.00
MEMBER 7					
0%	241.13	51.58	0.00	0.00	−27.47	0.00
25%	112.17	51.58	0.00	0.00	−27.47	0.00
50%	−16.79	51.58	0.00	0.00	−27.47	0.00
75%	−145.75	51.58	0.00	0.00	−27.47	0.00
100%	−274.72	51.58	0.00	0.00	−27.47	0.00

	MEMBER LOCAL DEFLECTIONS					
Station	At 1 (ft)	At 2 (ft)	At 3 (ft)	Rotation 11 (rad)	Slopes L/defl(2)	L/defl(3)
Condition dl (dead load)						
MEMBER 1						
0%	0.000	0.000	0.000	0.00000	—	—
25%	–0.000	0.000	0.000	0.00000	—	—
50%	–0.001	0.000	0.000	0.00000	—	—
75%	–0.001	0.000	0.000	0.00000	—	—
100%	–0.001	0.000	0.000	0.00000	—	—
MEMBER 2						
0%	0.000	0.000	0.000	0.00000	—	—
25%	–0.000	0.000	0.000	0.00000	—	—
50%	–0.001	0.000	0.000	0.00000	—	—
75%	–0.001	0.001	0.000	0.00000	—	—
100%	–0.002	0.000	0.000	0.00000	—	—
MEMBER 3						
0%	0.000	0.000	0.000	0.00000	—	—
25%	–0.000	0.000	0.000	0.00000	—	—
50%	–0.000	0.000	0.000	0.00000	—	—
75%	–0.001	0.000	0.000	0.00000	—	—
100%	–0.001	0.000	0.000	0.00000	—	—
MEMBER 4						
0%	0.000	0.000	0.000	0.00000	—	—
25%	–0.000	0.000	0.000	0.00000	—	—
50%	–0.000	0.000	0.000	0.00000	—	—
75%	–0.000	0.000	0.000	0.00000	—	—
100%	–0.001	0.000	0.000	0.00000	—	—
MEMBER 5						
0%	–0.000	–0.001	0.000	0.00000	8111.82	—
25%	–0.000	–0.002	0.000	0.00000	6666.01	—
50%	–0.000	–0.002	0.000	0.00000	5845.04	—
75%	–0.000	–0.002	0.000	0.00000	5505.43	—
100%	–0.000	–0.002	0.000	0.00000	5136.72	—
MEMBER 6						
0%	–0.000	–0.002	0.000	0.00000	—	—
25%	–0.000	–0.003	0.000	0.00000	6541.64	—
50%	–0.000	–0.003	0.000	0.00000	5893.48	—
75%	–0.000	–0.002	0.000	0.00000	8575.96	—
100%	–0.000	–0.001	0.000	0.00000	—	—

					Slopes	
Station	**At 1 (ft)**	**At 2 (ft)**	**At 3 (ft)**	**Rotation 11 (rad)**	**L/defl(2)**	**L/defl(3)**

MEMBER LOCAL DEFLECTIONS

Condition dl (dead load)

MEMBER 7

Station	At 1 (ft)	At 2 (ft)	At 3 (ft)	Rotation 11 (rad)	L/defl(2)	L/defl(3)
0%	−0.000	−0.001	0.000	0.00000	—	—
25%	−0.000	−0.001	0.000	0.00000	—	—
50%	−0.000	−0.001	0.000	0.00000	9637.59	—
75%	−0.000	−0.001	0.000	0.00000	—	—
100%	−0.000	−0.001	0.000	0.00000	—	—

Condition tl (temperature load)

MEMBER 1

Station	At 1 (ft)	At 2 (ft)	At 3 (ft)	Rotation 11 (rad)	L/defl(2)	L/defl(3)
0%	0.000	0.000	0.000	0.00000	—	—
25%	−0.000	0.037	0.000	0.00000	273.04	—
50%	−0.000	0.070	0.000	0.00000	142.37	—
75%	−0.001	0.098	0.000	0.00000	102.21	—
100%	−0.001	0.116	0.000	0.00000	85.91	—

MEMBER 2

Station	At 1 (ft)	At 2 (ft)	At 3 (ft)	Rotation 11 (rad)	L/defl(2)	L/defl(3)
0%	0.000	0.000	0.000	0.00000	—	—
25%	0.000	0.021	0.000	0.00000	474.67	—
50%	0.000	0.040	0.000	0.00000	250.35	—
75%	0.001	0.054	0.000	0.00000	183.68	—
100%	0.001	0.062	0.000	0.00000	160.32	—

MEMBER 3

Station	At 1 (ft)	At 2 (ft)	At 3 (ft)	Rotation 11 (rad)	L/defl(2)	L/defl(3)
0%	0.000	0.000	0.000	0.00000	—	—
25%	0.000	−0.014	0.000	0.00000	711.24	—
50%	0.000	−0.027	0.000	0.00000	370.89	—
75%	0.000	−0.038	0.000	0.00000	266.32	—
100%	0.000	−0.045	0.000	0.00000	223.91	—

MEMBER 4

Station	At 1 (ft)	At 2 (ft)	At 3 (ft)	Rotation 11 (rad)	L/defl(2)	L/defl(3)
0%	0.000	0.000	0.000	0.00000	—	—
25%	0.000	−0.013	0.000	0.00000	742.07	—
50%	−0.000	−0.026	0.000	0.00000	385.02	—
75%	−0.000	−0.037	0.000	0.00000	273.88	—
100%	−0.000	−0.044	0.000	0.00000	226.68	—

		MEMBER LOCAL DEFLECTIONS				
Station	At 1 (ft)	At 2 (ft)	At 3 (ft)	Rotation 11 (rad)	Slopes L/defl(2)	L/defl(3)
Condition tl (temperature load)						
MEMBER 5						
0%	−0.116	−0.001	0.000	0.00000	—	—
25%	−0.103	0.005	0.000	0.00000	1819.10	—
50%	−0.089	0.005	0.000	0.00000	2209.83	—
75%	−0.076	0.001	0.000	0.00000	7708.80	—
100%	−0.062	0.001	0.000	0.00000	—	—
MEMBER 6						
0%	−0.062	0.001	0.000	0.00000	—	—
25%	−0.036	0.007	0.000	0.00000	2989.43	—
50%	−0.009	0.009	0.000	0.00000	2209.93	—
75%	0.018	0.007	0.000	0.00000	2782.63	—
100%	0.045	0.000	0.000	0.00000	—	—
MEMBER 7						
0%	0.045	0.000	0.000	0.00000	—	—
25%	0.045	−0.001	0.000	0.00000	9085.11	—
50%	0.044	0.000	0.000	0.00000	—	—
75%	0.044	0.002	0.000	0.00000	5806.17	—
100%	0.044	−0.000	0.000	0.00000	—	—
Condition II (live load)						
MEMBER 1						
0%	0.000	0.000	0.000	0.00000	—	—
25%	0.000	0.000	0.000	0.00000	—	—
50%	0.000	0.001	0.000	0.00000	—	—
75%	0.000	0.001	0.000	0.00000	—	—
100%	0.000	0.002	0.000	0.00000	5936.45	—
MEMBER 2						
0%	0.000	0.000	0.000	0.00000	—	—
25%	−0.000	0.002	0.000	0.00000	4642.77	—
50%	−0.000	0.004	0.000	0.00000	2768.08	—
75%	−0.001	0.004	0.000	0.00000	2716.66	—
100%	−0.001	0.002	0.000	0.00000	6010.03	—
MEMBER 3						
0%	0.000	0.000	0.000	0.00000	—	—
25%	0.000	−0.000	0.000	0.00000	—	—
50%	−0.000	−0.000	0.000	0.00000	—	—
75%	−0.000	0.000	0.000	0.00000	—	—
100%	−0.000	0.002	0.000	0.00000	5083.63	—

	MEMBER LOCAL DEFLECTIONS					
					Slopes	
Station	At 1 (ft)	At 2 (ft)	At 3 (ft)	Rotation 11 (rad)	L/defl(2)	L/defl(3)
Condition II (live load)						
MEMBER 4						
0%	0.000	0.000	0.000	0.00000	—	—
25%	0.000	0.001	0.000	0.00000	—	—
50%	0.000	0.001	0.000	0.00000	7280.11	—
75%	0.000	0.002	0.000	0.00000	5560.68	—
100%	0.000	0.002	0.000	0.00000	5239.45	—
MEMBER 5						
0%	−0.002	0.000	0.000	0.00000	—	—
25%	−0.002	0.001	0.000	0.00000	8621.17	—
50%	−0.002	0.002	0.000	0.00000	5436.85	—
75%	−0.002	0.001	0.000	0.00000	6860.11	—
100%	−0.002	−0.001	0.000	0.00000	—	—
MEMBER 6						
0%	−0.002	−0.001	0.000	0.00000	—	—
25%	−0.002	−0.019	0.000	0.00000	1036.57	—
50%	−0.002	−0.034	0.000	0.00000	583.52	—
75%	−0.002	−0.018	0.000	0.00000	1085.88	—
100%	−0.002	−0.000	0.000	0.00000	—	—
MEMBER 7						
0%	−0.002	−0.000	0.000	0.00000	—	—
25%	−0.002	0.001	0.000	0.00000	9618.81	—
50%	−0.002	0.001	0.000	0.00000	8831.31	—
75%	−0.002	0.001	0.000	0.00000	—	—
100%	−0.002	0.000	0.000	0.00000	—	—
Condition wl (wind in X)						
MEMBER 1						
0%	0.000	0.000	0.000	0.00000	—	—
25%	0.000	−0.013	0.000	0.00000	786.13	—
50%	0.000	−0.024	0.000	0.00000	410.26	—
75%	0.000	−0.034	0.000	0.00000	295.01	—
100%	0.000	−0.040	0.000	0.00000	248.62	—

	At 1 (ft)	At 2 (ft)	At 3 (ft)	Rotation 11 (rad)	Slopes L/defl(2)	L/defl(3)
MEMBER LOCAL DEFLECTIONS						
					Slopes	
Station	At 1 (ft)	At 2 (ft)	At 3 (ft)	Rotation 11 (rad)	L/defl(2)	L/defl(3)
Condition wl (wind in X)						
MEMBER 2						
0%	0.000	0.000	0.000	0.00000	—	—
25%	0.000	–0.013	0.000	0.00000	751.93	—
50%	0.000	–0.025	0.000	0.00000	397.62	—
75%	–0.000	–0.034	0.000	0.00000	293.23	—
100%	–0.000	–0.039	0.000	0.00000	258.33	—
MEMBER 3						
0%	0.000	0.000	0.000	0.00000	—	—
25%	0.000	–0.012	0.000	0.00000	852.41	—
50%	0.000	–0.022	0.000	0.00000	445.99	—
75%	0.000	–0.031	0.000	0.00000	322.26	—
100%	0.000	–0.037	0.000	0.00000	273.84	—
MEMBER 4						
0%	0.000	0.000	0.000	0.00000	—	—
25%	0.000	–0.011	0.000	0.00000	910.42	—
50%	–0.000	–0.021	0.000	0.00000	472.09	—
75%	–0.000	–0.030	0.000	0.00000	335.47	—
100%	–0.000	–0.036	0.000	0.00000	277.17	—
MEMBER 5						
0%	0.040	0.000	0.000	0.00000	—	—
25%	0.040	–0.002	0.000	0.00000	5961.01	—
50%	0.039	–0.001	0.000	0.00000	8549.98	—
75%	0.039	0.000	0.000	0.00000	—	—
100%	0.039	–0.000	0.000	0.00000	—	—
MEMBER 6						
0%	0.039	–0.000	0.000	0.00000	—	—
25%	0.038	–0.001	0.000	0.00000	—	—
50%	0.038	0.002	0.000	0.00000	—	—
75%	0.037	0.003	0.000	0.00000	6174.47	—
100%	0.037	0.000	0.000	0.00000	—	—

MEMBER LOCAL DEFLECTIONS					
Station	At 1 (ft)	At 2 (ft)	At 3 (ft)	Rotation 11 (rad)	Slopes
					L/defl(2)	L/defl(3)
Condition wl (wind in X)						
MEMBER 7						
0%	0.037	0.000	0.000	0.00000	—	—
25%	0.036	−0.001	0.000	0.00000	—	—
50%	0.036	0.001	0.000	0.00000	—	—
75%	0.036	0.002	0.000	0.00000	6389.32	—
100%	0.036	−0.000	0.000	0.00000	—	—

Example 7A.4

LOAD DATA

File:
Units: Kip-ft
Date: 5/7/2010
Time: 10:25:44 AM

NODAL FORCES							
	ConditiNode	FX (kip)	FY (kip)	FZ (kip)	MX (kip*ft)	MY (kip*ft)	MZ (kip*ft)
dl	2	0	−10	0	0	0	0
PRESSURE & TEMPERATURE							
	ConditiBeam	Pressure (kip/ft²)	iPressure (kip/ft²)	iPressure (kip/ft²)	iTemp1 (°F)	Temp 2 (°F/ft)	Temp 3 (°F/ft)
TL	1	0	0	0	548	0	0
	2	0	0	0	548	0	0
	3	0	0	0	548	0	0
	4	0	0	0	548	0	0
	5	0	0	0	548	0	0
	6	0	0	0	548	0	0
	7	0	0	0	548	0	0
	8	0	0	0	548	0	0
	9	0	0	0	548	0	0
	10	0	0	0	548	0	0
	11	0	0	0	548	0	0
	12	0	0	0	548	0	0
	13	0	0	0	548	0	0
	14	0	0	0	548	0	0
	15	0	0	0	548	0	0
	16	0	0	0	548	0	0
	17	0	0	0	548	0	0
	18	0	0	0	548	0	0

MODEL DATA ECHO

File:
Units: Kip-ft

NODES				
Node	**X (ft)**	**Y (ft)**	**Z (ft)**	**Floor**
1	0	0	0	0
2	30	0	0	0
3	0	5	0	0
4	7.5	0	0	0
5	15	0	0	0
6	22.5	0	0	0
7	7.5	3.75	0	0
8	15	2.5	0	0
9	22.5	1.25	0	0
MEMBERS					
Beam	**NJ**	**NK**	**Description**	**Section**	**Material**
1	1	4	g1	T2L $4 \times 4 \times 3\_8$	A50
2	4	5	g1	T2L $4 \times 4 \times 3\_8$	A50
3	5	6	g1	T2L $4 \times 4 \times 3\_8$	A50
4	6	2	g1	T2L $4 \times 4 \times 3\_8$	A50
5	3	7	g2	T2L $2\text{-}1\_2 \times 2\text{-}1\_2 \times 5\_16$	A50
6	7	8	g2	T2L $2\text{-}1\_2 \times 2\text{-}1\_2 \times 5\_16$	A50
7	8	9	g2	T2L $2\text{-}1\_2 \times 2\text{-}1\_2 \times 5\_16$	A50
8	9	2	g2	T2L $2\text{-}1\_2 \times 2\text{-}1\_2 \times 5\_16$	A50
9	1	3	g3	T2L $2 \times 2 \times 1\_8$	A50
10	4	7	g3	T2L $2 \times 2 \times 1\_8$	A50
11	5	8	g3	T2L $2 \times 2 \times 1\_8$	A50
12	6	9	g3	T2L $2 \times 2 \times 1\_8$	A50
13	1	7	g4	T2L $2 \times 2 \times 1\_8$	A50
14	4	8	g4	T2L $2 \times 2 \times 1\_8$	A50
15	5	9	g4	T2L $2 \times 2 \times 1\_8$	A50
16	4	3	g5	T2L $2 \times 2 \times 1\_8$	A50
17	5	7	g5	T2L $2 \times 2 \times 1\_8$	A50
18	6	8	g5	T2L $2 \times 2 \times 1\_8$	A50

ANALYSIS RESULTS

File:

Units: Kip-ft

	M33 (kip*ft)	V2 (kip)	M22 (kip*ft)	V3 (kip)	Axial (kip)	Torsion (kip*ft)
MEMBER FORCES						
Condition dl (dead load)						
MEMBER 1						
0%	−0.02	0.09	0.00	0.00	−62.50	0.00
25%	−0.18	0.09	0.00	0.00	−62.50	0.00
50%	−0.35	0.09	0.00	0.00	−62.50	0.00
75%	−0.51	0.09	0.00	0.00	−62.50	0.00
100%	−0.67	0.09	0.00	0.00	−62.50	0.00
MEMBER 2						
0%	−0.66	−0.04	0.00	0.00	−62.91	0.00
25%	−0.58	−0.04	0.00	0.00	−62.91	0.00
50%	−0.51	−0.04	0.00	0.00	−62.91	0.00
75%	−0.43	−0.04	0.00	0.00	−62.91	0.00
100%	−0.35	−0.04	0.00	0.00	−62.91	0.00
MEMBER 3						
0%	−0.24	0.34	0.00	0.00	−62.19	0.00
25%	−0.89	0.34	0.00	0.00	−62.19	0.00
50%	−1.53	0.34	0.00	0.00	−62.19	0.00
75%	−2.17	0.34	0.00	0.00	−62.19	0.00
100%	−2.82	0.34	0.00	0.00	−62.19	0.00
MEMBER 4						
0%	−3.33	−0.44	0.00	0.00	−56.71	0.00
25%	−2.51	−0.44	0.00	0.00	−56.71	0.00
50%	−1.69	−0.44	0.00	0.00	−56.71	0.00
75%	−0.87	−0.44	0.00	0.00	−56.71	0.00
100%	−0.05	−0.44	0.00	0.00	−56.71	0.00

MEMBER FORCES					
M33 (kip*ft)	V2 (kip)	M22 (kip*ft)	V3 (kip)	Axial (kip)	Torsion (kip*ft)
Condition dl (dead load)					
MEMBER 5					
0% −0.02	0.01	0.00	0.00	58.06	0.00
25% −0.04	0.01	0.00	0.00	58.06	0.00
50% −0.07	0.01	0.00	0.00	58.06	0.00
75% −0.09	0.01	0.00	0.00	58.06	0.00
100% −0.12	0.01	0.00	0.00	58.06	0.00
MEMBER 6					
0% −0.09	0.00	0.00	0.00	57.45	0.00
25% −0.10	0.00	0.00	0.00	57.45	0.00
50% −0.11	0.00	0.00	0.00	57.45	0.00
75% −0.12	0.00	0.00	0.00	57.45	0.00
100% −0.13	0.00	0.00	0.00	57.45	0.00
MEMBER 7					
0% −0.06	0.05	0.00	0.00	55.67	0.00
25% −0.15	0.05	0.00	0.00	55.67	0.00
50% −0.24	0.05	0.00	0.00	55.67	0.00
75% −0.33	0.05	0.00	0.00	55.67	0.00
100% −0.42	0.05	0.00	0.00	55.67	0.00
MEMBER 8					
0% −0.78	−0.11	0.00	0.00	57.51	0.00
25% −0.57	−0.11	0.00	0.00	57.51	0.00
50% −0.37	−0.11	0.00	0.00	57.51	0.00
75% −0.16	−0.11	0.00	0.00	57.51	0.00
100% 0.05	−0.11	0.00	0.00	57.51	0.00
MEMBER 9					
0% 0.02	0.01	0.00	0.00	0.00	0.00
25% 0.01	0.01	0.00	0.00	0.00	0.00
50% 0.00	0.01	0.00	0.00	0.00	0.00
75% −0.01	0.01	0.00	0.00	0.00	0.00
100% −0.02	0.01	0.00	0.00	0.00	0.00

MEMBER FORCES					
M33 (kip*ft)	V2 (kip)	M22 (kip*ft)	V3 (kip)	Axial (kip)	Torsion (kip*ft)
Condition dl (dead load)					
MEMBER 10					
0%	−0.01	−0.01	0.00	0.00	−2.73	0.00
25%	−0.01	−0.01	0.00	0.00	−2.73	0.00
50%	0.00	−0.01	0.00	0.00	−2.73	0.00
75%	0.01	−0.01	0.00	0.00	−2.73	0.00
100%	0.02	−0.01	0.00	0.00	−2.73	0.00
MEMBER 11					
0%	−0.09	−0.07	0.00	0.00	−2.34	0.00
25%	−0.05	−0.07	0.00	0.00	−2.34	0.00
50%	−0.01	−0.07	0.00	0.00	−2.34	0.00
75%	0.04	−0.07	0.00	0.00	−2.34	0.00
100%	0.08	−0.07	0.00	0.00	−2.34	0.00
MEMBER 12					
0%	0.39	0.53	0.00	0.00	−0.85	0.00
25%	0.22	0.53	0.00	0.00	−0.85	0.00
50%	0.06	0.53	0.00	0.00	−0.85	0.00
75%	−0.11	0.53	0.00	0.00	−0.85	0.00
100%	−0.27	0.53	0.00	0.00	−0.85	0.00
MEMBER 13					
0%	−0.01	0.00	0.00	0.00	2.81	0.00
25%	−0.01	0.00	0.00	0.00	2.81	0.00
50%	−0.01	0.00	0.00	0.00	2.81	0.00
75%	−0.01	0.00	0.00	0.00	2.81	0.00
100%	−0.02	0.00	0.00	0.00	2.81	0.00
MEMBER 14					
0%	−0.03	−0.00	0.00	0.00	3.30	0.00
25%	−0.03	−0.00	0.00	0.00	3.30	0.00
50%	−0.02	−0.00	0.00	0.00	3.30	0.00
75%	−0.02	−0.00	0.00	0.00	3.30	0.00
100%	−0.02	−0.00	0.00	0.00	3.30	0.00

MEMBER FORCES						
	M33 (kip*ft)	V2 (kip)	M22 (kip*ft)	V3 (kip)	Axial (kip)	Torsion (kip*ft)
Condition dl (dead load)						
MEMBER 15						
0%	−0.03	0.01	0.00	0.00	2.36	0.00
25%	−0.04	0.01	0.00	0.00	2.36	0.00
50%	−0.06	0.01	0.00	0.00	2.36	0.00
75%	−0.07	0.01	0.00	0.00	2.36	0.00
100%	−0.08	0.01	0.00	0.00	2.36	0.00
MEMBER 16						
0%	−0.03	−0.00	0.00	0.00	3.27	0.00
25%	−0.02	−0.00	0.00	0.00	3.27	0.00
50%	−0.01	−0.00	0.00	0.00	3.27	0.00
75%	−0.01	−0.00	0.00	0.00	3.27	0.00
100%	−0.00	−0.00	0.00	0.00	3.27	0.00
MEMBER 17						
0%	−0.02	0.00	0.00	0.00	3.49	0.00
25%	−0.02	0.00	0.00	0.00	3.49	0.00
50%	−0.02	0.00	0.00	0.00	3.49	0.00
75%	−0.02	0.00	0.00	0.00	3.49	0.00
100%	−0.02	0.00	0.00	0.00	3.49	0.00
MEMBER 18						
0%	−0.13	−0.02	0.00	0.00	5.21	0.00
25%	−0.09	−0.02	0.00	0.00	5.21	0.00
50%	−0.06	−0.02	0.00	0.00	5.21	0.00
75%	−0.03	−0.02	0.00	0.00	5.21	0.00
100%	0.00	−0.02	0.00	0.00	5.21	0.00
Condition tl (temperature load)						
MEMBER 1						
0%	−0.10	−0.11	0.00	0.00	13.21	0.00
25%	0.11	−0.11	0.00	0.00	13.21	0.00
50%	0.32	−0.11	0.00	0.00	13.21	0.00
75%	0.53	−0.11	0.00	0.00	13.21	0.00
100%	0.73	−0.11	0.00	0.00	13.21	0.00

MEMBER FORCES					
M33 (kip*ft)	V2 (kip)	M22 (kip*ft)	V3 (kip)	Axial (kip)	Torsion (kip*ft)
Condition tl (temperature load)					
MEMBER 2					
0%	0.62	0.11	0.00	0.00	−0.48	0.00
25%	0.41	0.11	0.00	0.00	−0.48	0.00
50%	0.21	0.11	0.00	0.00	−0.48	0.00
75%	0.00	0.11	0.00	0.00	−0.48	0.00
100%	−0.21	0.11	0.00	0.00	−0.48	0.00
MEMBER 3					
0%	−0.16	−0.03	0.00	0.00	0.02	0.00
25%	−0.11	−0.03	0.00	0.00	0.02	0.00
50%	−0.06	−0.03	0.00	0.00	0.02	0.00
75%	−0.01	−0.03	0.00	0.00	0.02	0.00
100%	0.04	−0.03	0.00	0.00	0.02	0.00
MEMBER 4					
0%	0.03	0.00	0.00	0.00	−0.02	0.00
25%	0.02	0.00	0.00	0.00	−0.02	0.00
50%	0.01	0.00	0.00	0.00	−0.02	0.00
75%	0.00	0.00	0.00	0.00	−0.02	0.00
100%	−0.00	0.00	0.00	0.00	−0.02	0.00
MEMBER 5					
0%	0.04	0.04	0.00	0.00	13.51	0.00
25%	−0.03	0.04	0.00	0.00	13.51	0.00
50%	−0.09	0.04	0.00	0.00	13.51	0.00
75%	−0.16	0.04	0.00	0.00	13.51	0.00
100%	−0.23	0.04	0.00	0.00	13.51	0.00
MEMBER 6					
0%	−0.12	−0.02	0.00	0.00	−0.41	0.00
25%	−0.08	−0.02	0.00	0.00	−0.41	0.00
50%	−0.03	−0.02	0.00	0.00	−0.41	0.00
75%	0.01	−0.02	0.00	0.00	−0.41	0.00
100%	0.06	−0.02	0.00	0.00	−0.41	0.00

	M33 (kip*ft)	V2 (kip)	M22 (kip*ft)	V3 (kip)	Axial (kip)	Torsion (kip*ft)
MEMBER FORCES						
Condition tl (temperature load)						
MEMBER 7						
0%	0.02	0.00	0.00	0.00	−0.01	0.00
25%	0.02	0.00	0.00	0.00	−0.01	0.00
50%	0.01	0.00	0.00	0.00	−0.01	0.00
75%	−0.00	0.00	0.00	0.00	−0.01	0.00
100%	−0.01	0.00	0.00	0.00	−0.01	0.00
MEMBER 8						
0%	−0.01	−0.00	0.00	0.00	0.02	0.00
25%	−0.00	−0.00	0.00	0.00	0.02	0.00
50%	0.00	−0.00	0.00	0.00	0.02	0.00
75%	0.00	−0.00	0.00 ·	0.00	0.02	0.00
100%	0.00	−0.00	0.00	0.00	0.02	0.00
MEMBER 9						
0%	0.10	0.01	0.00	0.00	−153.95	0.00
25%	0.09	0.01	0.00	0.00	−153.95	0.00
50%	0.07	0.01	0.00	0.00	−153.95	0.00
75%	0.06	0.01	0.00	0.00	−153.95	0.00
100%	0.04	0.01	0.00	0.00	−153.95	0.00
MEMBER 10						
0%	0.03	0.00	0.00	0.00	8.56	0.00
25%	0.03	0.00	0.00	0.00	8.56	0.00
50%	0.03	0.00	0.00	0.00	8.56	0.00
75%	0.03	0.00	0.00	0.00	8.56	0.00
100%	0.03	0.00	0.00	0.00	8.56	0.00
MEMBER 11						
0%	−0.02	−0.00	0.00	0.00	−0.15	0.00
25%	−0.01	−0.00	0.00	0.00	−0.15	0.00
50%	−0.01	−0.00	0.00	0.00	−0.15	0.00
75%	−0.01	−0.00	0.00	0.00	−0.15	0.00
100%	−0.00	−0.00	0.00	0.00	−0.15	0.00

	M33 (kip*ft)	V2 (kip)	M22 (kip*ft)	V3 (kip)	Axial (kip)	Torsion (kip*ft)
MEMBER FORCES						
Condition tl (temperature load)						
MEMBER 12						
0%	0.01	0.01	0.00	0.00	−0.02	0.00
25%	0.01	0.01	0.00	0.00	−0.02	0.00
50%	0.00	0.01	0.00	0.00	−0.02	0.00
75%	0.00	0.01	0.00	0.00	−0.02	0.00
100%	0.00	0.01	0.00	0.00	−0.02	0.00
MEMBER 13						
0%	−0.00	−0.01	0.00	0.00	−14.77	0.00
25%	0.01	−0.01	0.00	0.00	−14.77	0.00
50%	0.02	−0.01	0.00	0.00	−14.77	0.00
75%	0.04	−0.01	0.00	0.00	−14.77	0.00
100%	0.05	−0.01	0.00	0.00	−14.77	0.00
MEMBER 14						
0%	0.03	0.01	0.00	0.00	0.37	0.00
25%	0.02	0.01	0.00	0.00	0.37	0.00
50%	0.01	0.01	0.00	0.00	0.37	0.00
75%	−0.01	0.01	0.00	0.00	0.37	0.00
100%	−0.02	0.01	0.00	0.00	0.37	0.00
MEMBER 15						
0%	−0.01	−0.00	0.00	0.00	0.04	0.00
25%	−0.00	−0.00	0.00	0.00	0.04	0.00
50%	−0.00	−0.00	0.00	0.00	0.04	0.00
75%	0.00	−0.00	0.00	0.00	0.04	0.00
100%	0.00	−0.00	0.00	0.00	0.04	0.00
MEMBER 16						
0%	−0.06	−0.01	0.00	0.00	−16.05	0.00
25%	−0.04	−0.01	0.00	0.00	−16.05	0.00
50%	−0.03	−0.01	0.00	0.00	−16.05	0.00
75%	−0.01	−0.01	0.00	0.00	−16.05	0.00
100%	0.00	−0.01	0.00	0.00	−16.05	0.00

MEMBER FORCES					
M33 (kip*ft)	V2 (kip)	M22 (kip*ft)	V3 (kip)	Axial (kip)	Torsion (kip*ft)
Condition tl (temperature load)					
MEMBER 17					
0%	0.02	0.01	0.00	0.00	0.61	0.00
25%	0.01	0.01	0.00	0.00	0.61	0.00
50%	−0.00	0.01	0.00	0.00	0.61	0.00
75%	−0.02	0.01	0.00	0.00	0.61	0.00
100%	−0.03	0.01	0.00	0.00	0.61	0.00
MEMBER 18					
0%	−0.00	−0.00	0.00	0.00	−0.05	0.00
25%	−0.00	−0.00	0.00	0.00	−0.05	0.00
50%	0.00	−0.00	0.00	0.00	−0.05	0.00
75%	0.00	−0.00	0.00	0.00	−0.05	0.00
100%	0.01	−0.00	0.00	0.00	−0.05	0.00

MEMBER LOCAL DEFLECTIONS					
Station	At 1 (ft)	At 2 (ft)	At 3 (ft)	Rotation11 (rad)	Slopes	
					L/defl(2)	L/defl(3)
Condition dl (dead load)					
MEMBER 1					
0%	0.000	0.000	0.000	0.00000	—	—
25%	−0.001	−0.001	0.000	0.00000	—	—
50%	−0.001	−0.002	0.000	0.00000	4467.27	—
75%	−0.002	−0.003	0.000	0.00000	2210.00	—
100%	−0.003	−0.006	0.000	0.00000	1223.08	—
MEMBER 2					
0%	−0.003	−0.006	0.000	0.00000	1223.08	—
25%	−0.004	−0.010	0.000	0.00000	740.16	—
50%	−0.004	−0.015	0.000	0.00000	490.09	—
75%	−0.005	−0.021	0.000	0.00000	349.01	—
100%	−0.006	−0.029	0.000	0.00000	262.83	—

MEMBER LOCAL DEFLECTIONS

Station	At 1 (ft)	At 2 (ft)	At 3 (ft)	Rotation 11 (rad)	Slopes L/defl(2)	L/defl(3)
			Condition dl (dead load)			
MEMBER 3						
0%	−0.006	−0.029	0.000	0.00000	262.83	—
25%	−0.006	−0.036	0.000	0.00000	205.99	—
50%	−0.007	−0.046	0.000	0.00000	162.81	—
75%	−0.008	−0.059	0.000	0.00000	127.55	—
100%	−0.008	−0.076	0.000	0.00000	98.82	—
MEMBER 4						
0%	−0.008	−0.076	0.000	0.00000	98.82	—
25%	−0.009	−0.099	0.000	0.00000	75.98	—
50%	−0.010	−0.127	0.000	0.00000	59.26	—
75%	−0.010	−0.158	0.000	0.00000	47.53	—
100%	−0.011	−0.191	0.000	0.00000	39.31	—
MEMBER 5						
0%	0.000	0.000	0.000	0.00000	—	—
25%	0.001	−0.000	0.000	0.00000	—	—
50%	0.003	−0.001	0.000	0.00000	5355.12	—
75%	0.004	−0.003	0.000	0.00000	2468.38	—
100%	0.005	−0.006	0.000	0.00000	1329.40	—
MEMBER 6						
0%	0.005	−0.006	0.000	0.00000	1329.40	—
25%	0.006	−0.009	0.000	0.00000	805.23	—
50%	0.008	−0.014	0.000	0.00000	533.92	—
75%	0.009	−0.020	0.000	0.00000	376.36	—
100%	0.010	−0.027	0.000	0.00000	277.32	—
MEMBER 7						
0%	0.010	−0.027	0.000	0.00000	277.32	—
25%	0.012	−0.036	0.000	0.00000	212.51	—
50%	0.013	−0.046	0.000	0.00000	166.15	—
75%	0.014	−0.058	0.000	0.00000	130.36	—
100%	0.015	−0.074	0.000	0.00000	102.16	—

				MEMBER LOCAL DEFLECTIONS		
	At 1	At 2	At 3	Rotation 11	Slopes	
Station	(ft)	(ft)	(ft)	(rad)	L/defl(2)	L/defl(3)
			Condition dl (dead load)			
MEMBER 8						
0%	0.015	−0.074	0.000	0.00000	102.16	—
25%	0.017	−0.096	0.000	0.00000	78.88	—
50%	0.018	−0.124	0.000	0.00000	61.10	—
75%	0.019	−0.156	0.000	0.00000	48.62	—
100%	0.020	−0.190	0.000	0.00000	40.02	—
MEMBER 9						
0%	0.000	0.000	0.000	0.00000	—	—
25%	0.000	−0.000	0.000	0.00000	—	—
50%	0.000	0.000	0.000	0.00000	—	—
75%	0.000	0.000	0.000	0.00000	—	—
100%	0.000	0.000	0.000	0.00000	—	—
MEMBER 10						
0%	−0.006	0.003	0.000	0.00000	1326.72	—
25%	−0.006	0.001	0.000	0.00000	3465.36	—
50%	−0.006	−0.001	0.000	0.00000	5158.82	—
75%	−0.006	−0.003	0.000	0.00000	1492.82	—
100%	−0.006	−0.004	0.000	0.00000	896.14	—
MEMBER 11						
0%	−0.029	0.006	0.000	0.00000	440.79	—
25%	−0.029	0.003	0.000	0.00000	836.70	—
50%	−0.029	0.000	0.000	0.00000	—	—
75%	−0.029	−0.003	0.000	0.00000	856.17	—
100%	−0.029	−0.006	0.000	0.00000	439.47	—
MEMBER 12						
0%	−0.076	0.008	0.000	0.00000	147.34	—
25%	−0.076	0.005	0.000	0.00000	230.30	—
50%	−0.076	0.003	0.000	0.00000	470.52	—
75%	−0.076	0.000	0.000	0.00000	—	—
100%	−0.076	−0.003	0.000	0.00000	434.89	—

	MEMBER LOCAL DEFLECTIONS					
					Slopes	
Station	At 1 (ft)	At 2 (ft)	At 3 (ft)	Rotation 11 (rad)	L/defl(2)	L/defl(3)
	Condition dl (dead load)					
MEMBER 13						
0%	0.000	0.000	0.000	0.00000	—	—
25%	0.000	−0.001	0.000	0.00000	9259.16	—
50%	0.000	−0.002	0.000	0.00000	3498.43	—
75%	0.001	−0.005	0.000	0.00000	1818.92	—
100%	0.001	−0.008	0.000	0.00000	1091.63	—
MEMBER 14						
0%	−0.005	−0.005	0.000	0.00000	1605.69	—
25%	−0.004	−0.009	0.000	0.00000	863.46	—
50%	−0.004	−0.015	0.000	0.00000	538.67	—
75%	−0.004	−0.021	0.000	0.00000	370.18	—
100%	−0.004	−0.029	0.000	0.00000	271.97	—
MEMBER 15						
0%	−0.010	−0.027	0.000	0.00000	279.38	—
25%	−0.010	−0.036	0.000	0.00000	213.81	—
50%	−0.010	−0.046	0.000	0.00000	165.43	—
75%	−0.010	−0.059	0.000	0.00000	128.81	—
100%	−0.010	−0.075	0.000	0.00000	100.88	—
MEMBER 16						
0%	−0.001	−0.007	0.000	0.00000	1351.40	—
25%	−0.001	−0.003	0.000	0.00000	2628.69	—
50%	−0.001	−0.002	0.000	0.00000	5957.65	—
75%	−0.000	−0.001	0.000	0.00000	—	—
100%	0.000	0.000	0.000	0.00000	—	—
MEMBER 17						
0%	−0.008	−0.028	0.000	0.00000	298.83	—
25%	−0.007	−0.020	0.000	0.00000	413.34	—
50%	−0.007	−0.014	0.000	0.00000	615.57	—
75%	−0.007	−0.008	0.000	0.00000	1029.38	—
100%	−0.007	−0.004	0.000	0.00000	2128.99	—

	MEMBER LOCAL DEFLECTIONS					
					Slopes	
Station	At 1 (ft)	At 2 (ft)	At 3 (ft)	Rotation 11 (rad)	L/defl(2)	L/defl(3)
Condition dl (dead load)						
MEMBER 18						
0%	−0.016	−0.075	0.000	0.00000	105.85	—
25%	−0.016	−0.057	0.000	0.00000	139.10	—
50%	−0.015	−0.044	0.000	0.00000	180.60	—
75%	−0.015	−0.034	0.000	0.00000	233.46	—
100%	−0.014	−0.025	0.000	0.00000	310.39	—
Condition tl (temperature load)						
MEMBER 1						
0%	0.000	0.000	0.000	0.00000	—	—
25%	0.010	−0.005	0.000	0.00000	1482.70	—
50%	0.021	−0.010	0.000	0.00000	758.14	—
75%	0.031	−0.014	0.000	0.00000	532.43	—
100%	0.042	−0.017	0.000	0.00000	435.47	—
MEMBER 2						
0%	0.042	−0.017	0.000	0.00000	435.47	—
25%	0.052	−0.019	0.000	0.00000	391.99	—
50%	0.062	−0.020	0.000	0.00000	370.95	—
75%	0.073	−0.021	0.000	0.00000	359.02	—
100%	0.083	−0.022	0.000	0.00000	347.83	—
MEMBER 3						
0%	0.083	−0.022	0.000	0.00000	347.83	—
25%	0.093	−0.023	0.000	0.00000	333.06	—
50%	0.103	−0.024	0.000	0.00000	316.57	—
75%	0.114	−0.025	0.000	0.00000	300.23	—
100%	0.124	−0.026	0.000	0.00000	285.35	—
MEMBER 4						
0%	0.124	−0.026	0.000	0.00000	285.35	—
25%	0.134	−0.028	0.000	0.00000	272.41	—
50%	0.144	−0.029	0.000	0.00000	261.00	—
75%	0.155	−0.030	0.000	0.00000	250.73	—
100%	0.165	−0.031	0.000	0.00000	241.29	—

					Slopes	
Station	At 1 (ft)	At 2 (ft)	At 3 (ft)	Rotation 11 (rad)	L/defl(2)	L/defl(3)

MEMBER LOCAL DEFLECTIONS

Condition tl (temperature load)

MEMBER 5

Station	At 1 (ft)	At 2 (ft)	At 3 (ft)	Rotation 11 (rad)	L/defl(2)	L/defl(3)
0%	0.000	0.000	0.000	0.00000	—	—
25%	0.011	0.004	0.000	0.00000	1875.48	—
50%	0.021	0.008	0.000	0.00000	969.98	—
75%	0.032	0.011	0.000	0.00000	715.72	—
100%	0.043	0.012	0.000	0.00000	651.10	—

MEMBER 6

0%	0.043	0.012	0.000	0.00000	651.10	—
25%	0.053	0.011	0.000	0.00000	688.23	—
50%	0.064	0.010	0.000	0.00000	792.70	—
75%	0.074	0.008	0.000	0.00000	976.74	—
100%	0.085	0.006	0.000	0.00000	1246.34	—

MEMBER 7

0%	0.085	0.006	0.000	0.00000	1246.34	—
25%	0.095	0.005	0.000	0.00000	1600.19	—
50%	0.105	0.004	0.000	0.00000	2131.39	—
75%	0.116	0.002	0.000	0.00000	3095.74	—
100%	0.126	0.001	0.000	0.00000	5734.06	—

MEMBER 8

0%	0.126	0.001	0.000	0.00000	5734.06	—
25%	0.137	0.000	0.000	0.00000	—	—
50%	0.147	−0.001	0.000	0.00000	6929.17	—
75%	0.157	−0.002	0.000	0.00000	3264.36	—
100%	0.168	−0.004	0.000	0.00000	2149.21	—

MEMBER 9

0%	0.000	0.000	0.000	0.00000	—	—
25%	0.000	−0.002	0.000	0.00000	2108.89	—
50%	0.000	−0.003	0.000	0.00000	1682.22	—
75%	0.000	−0.002	0.000	0.00000	2395.24	—
100%	0.000	0.000	0.000	0.00000	—	—

		MEMBER LOCAL DEFLECTIONS				
	At 1	At 2	At 3	Rotation 11	Slopes	
Station	(ft)	(ft)	(ft)	(rad)	L/defl(2)	L/defl(3)
		Condition tl (temperature load)				
MEMBER 10						
0%	−0.017	−0.042	0.000	0.00000	89.93	—
25%	−0.012	−0.043	0.000	0.00000	87.66	—
50%	−0.006	−0.044	0.000	0.00000	86.09	—
75%	−0.001	−0.044	0.000	0.00000	85.16	—
100%	0.004	−0.044	0.000	0.00000	84.82	—
MEMBER 11						
0%	−0.022	−0.083	0.000	0.00000	30.20	—
25%	−0.018	−0.083	0.000	0.00000	30.09	—
50%	−0.015	−0.083	0.000	0.00000	29.95	—
75%	−0.011	−0.084	0.000	0.00000	29.80	—
100%	−0.008	−0.084	0.000	0.00000	29.64	—
MEMBER 12						
0%	−0.026	−0.124	0.000	0.00000	10.09	—
25%	−0.025	−0.124	0.000	0.00000	10.07	—
50%	−0.023	−0.124	0.000	0.00000	10.06	—
75%	−0.021	−0.124	0.000	0.00000	10.04	—
100%	−0.019	−0.125	0.000	0.00000	10.03	—
MEMBER 13						
0%	0.000	0.000	0.000	0.00000	—	—
25%	0.010	−0.006	0.000	0.00000	1493.26	—
50%	0.021	−0.011	0.000	0.00000	786.15	—
75%	0.031	−0.014	0.000	0.00000	585.25	—
100%	0.042	−0.016	0.000	0.00000	531.57	—
MEMBER 14						
0%	0.034	−0.030	0.000	0.00000	267.76	—
25%	0.045	−0.031	0.000	0.00000	251.59	—
50%	0.056	−0.032	0.000	0.00000	244.19	—
75%	0.067	−0.033	0.000	0.00000	239.15	—
100%	0.078	−0.034	0.000	0.00000	231.51	—

					Slopes	
Station	**At 1 (ft)**	**At 2 (ft)**	**At 3 (ft)**	**Rotation 11 (rad)**	**L/defl(2)**	**L/defl(3)**

MEMBER LOCAL DEFLECTIONS

Condition tl (temperature load)

Station	At 1 (ft)	At 2 (ft)	At 3 (ft)	Rotation 11 (rad)	L/defl(2)	L/defl(3)
MEMBER 15						
0%	0.078	−0.035	0.000	0.00000	218.01	—
25%	0.089	−0.036	0.000	0.00000	211.97	—
50%	0.099	−0.037	0.000	0.00000	204.95	—
75%	0.109	−0.038	0.000	0.00000	197.93	—
100%	0.120	−0.040	0.000	0.00000	191.69	—
MEMBER 16						
0%	−0.044	0.009	0.000	0.00000	1024.36	—
25%	−0.033	0.010	0.000	0.00000	904.39	—
50%	−0.022	0.008	0.000	0.00000	1096.83	—
75%	−0.011	0.005	0.000	0.00000	1976.61	—
100%	0.000	0.000	0.000	0.00000	—	—
MEMBER 17						
0%	−0.084	0.018	0.000	0.00000	472.87	—
25%	−0.072	0.019	0.000	0.00000	437.27	—
50%	−0.061	0.021	0.000	0.00000	396.00	—
75%	−0.049	0.023	0.000	0.00000	365.18	—
100%	−0.038	0.024	0.000	0.00000	352.76	—
MEMBER 18						
0%	−0.126	0.014	0.000	0.00000	555.25	—
25%	−0.115	0.015	0.000	0.00000	510.52	—
50%	−0.104	0.017	0.000	0.00000	475.27	—
75%	−0.093	0.018	0.000	0.00000	443.48	—
100%	−0.083	0.019	0.000	0.00000	411.63	—

Example 7A.5
MODEL DATA ECHO

File:

Units: Kip-ft

NODES				
Node	X (ft)	Y (ft)	Z (ft)	Floor
1	0	0	0	0
2	4	1.8	0	0
3	8	3.2	0	0
4	12	4.2	0	0
5	16	4.8	0	0
6	20	5	0	0
7	24	4.8	0	0
8	28	4.2	0	0
9	32	3.2	0	0
10	36	1.8	0	0
11	40	0	0	0

RESTRAINTS						
Node	TX	TY	TZ	RX	RY	RZ
1	1	1	1	0	0	0
11	1	1	1	0	0	0

MEMBERS					
Beam	NJ	NK	Description	Section	Material
1	1	2	BEAM 1	W 8 × 31	A50
2	2	3	BEAM 1	W 8 × 31	A50
3	3	4	BEAM 1	W 8 × 31	A50
4	4	5	BEAM 1	W 8 × 31	A50
5	5	6	BEAM 1	W 8 × 31	A50
6	6	7	BEAM 1	W 8 × 31	A50
7	7	8	BEAM 1	W 8 × 31	A50
8	8	9	BEAM 1	W 8 × 31	A50
9	9	10	BEAM 1	W 8 × 31	A50
10	10	11	BEAM 1	W 8 × 31	A50

LOAD DATA

File:

Units: Kip-ft

NODAL FORCES						
ConditiNode	FX (kip)	FY (kip)	FZ (kip)	MX (kip*ft)	MY (kip*ft)	MZ (kip*ft)
ll 6	0	−100	0	0	0	0
DISTRIBUTED FORCES ON MEMBERS						
ConditiBeam	Dir.	Value (kip/ft)				
dl 1	Y	−3				
2	Y	−3				
3	Y	−3				
4	Y	−3				
5	Y	−3				
6	Y	−3				
7	Y	−3				
8	Y	−3				
9	Y	−3				
10	Y	−3				
PRESSURE & TEMPERATURE						
ConditiBeam	Pressure (kip/ft²)	iPressure (kip/ft²)	iPressure (kip/ft²)	iTemp1 (°F)	Temp 2 (°F/ft)	Temp 3 (°F/ft)
tl 1	0	0	0	548	0	0
2	0	0	0	548	0	0
3	0	0	0	548	0	0
4	0	0	0	548	0	0
5	0	0	0	548	0	0
6	0	0	0	548	0	0
7	0	0	0	548	0	0
8	0	0	0	548	0	0
9	0	0	0	548	0	0
10	0	0	0	548	0	0

ANALYSIS RESULTS

File:
Units: Kip-ft
Date: 5/10/2010
Time: 9:10:51 AM

	TRANSLATIONS					
	TRANSLATIONS (ft)			ROTATIONS (rad)		
Node	TX	TY	TZ	RX	RY	RZ
	Condition dl (dead load)					
1	0.00000	0.00000	0.00000	0.00000	0.00000	−0.00306
2	0.00282	−0.01160	0.00000	0.00000	0.00000	−0.00231
3	0.00366	−0.02017	0.00000	0.00000	0.00000	−0.00165
4	0.00314	−0.02606	0.00000	0.00000	0.00000	−0.00106
5	0.00177	−0.02950	0.00000	0.00000	0.00000	−0.00052
6	0.00000	−0.03063	0.00000	0.00000	0.00000	0.00000
7	−0.00177	−0.02950	0.00000	0.00000	0.00000	0.00052
8	−0.00314	−0.02606	0.00000	0.00000	0.00000	0.00106
9	−0.00366	−0.02017	0.00000	0.00000	0.00000	0.00165
10	−0.00282	−0.01160	0.00000	0.00000	0.00000	0.00231
11	0.00000	0.00000	0.00000	0.00000	0.00000	0.00306
	Condition tl (temperature load)					
1	0.00000	0.00000	0.00000	0.00000	0.00000	0.02756
2	−0.02693	0.11808	0.00000	0.00000	0.00000	0.02595
3	−0.03878	0.22182	0.00000	0.00000	0.00000	0.02163
4	−0.03568	0.30203	0.00000	0.00000	0.00000	0.01542
5	−0.02098	0.35252	0.00000	0.00000	0.00000	0.00800
6	0.00000	0.36974	0.00000	0.00000	0.00000	0.00000
7	0.02098	0.35252	0.00000	0.00000	0.00000	−0.00800
8	0.03568	0.30203	0.00000	0.00000	0.00000	−0.01542
9	0.03878	0.22182	0.00000	0.00000	0.00000	−0.02163
10	0.02693	0.11808	0.00000	0.00000	0.00000	−0.02595
11	0.00000	0.00000	0.00000	0.00000	0.00000	−0.02756

REACTIONS					
FORCES (kip)			**MOMENTS (kip*ft)**		
Node **FX**	**FY**	**FZ**	**MX**	**MY**	**MZ**
Condition dl (dead load)					
1 120.22848	60.00000	0.00000	0.00000	0.00000	0.00000
11 −120.22848	60.00000	0.00000	0.00000	0.00000	0.00000
SUM 0.00000	120.00000	0.00000	0.00000	0.00000	0.00000
Condition tl (temperature load)					
1 9.02092	0.00000	0.00000	0.00000	0.00000	0.00000
11 −9.02092	0.00000	0.00000	0.00000	0.00000	0.00000
SUM 0.00000	0.00000	0.00000	0.00000	0.00000	0.00000

MEMBER FORCES					
M33 **(kip*ft)**	**V2** **(kip)**	**M22** **(kip*ft)**	**V3** **(kip)**	**Axial** **(kip)**	**Torsion** **(kip*ft)**
Condition dl (dead load)					
MEMBER 1					
0% 0.00	−5.38	0.00	0.00	−134.26	0.00
25% 4.40	−2.64	0.00	0.00	−133.03	0.00
50% 5.79	0.09	0.00	0.00	−131.80	0.00
75% 4.19	2.83	0.00	0.00	−130.57	0.00
100% −0.41	5.57	0.00	0.00	−129.34	0.00
MEMBER 2					
0% −0.41	−5.59	0.00	0.00	−129.34	0.00
25% 4.01	−2.76	0.00	0.00	−128.34	0.00
50% 5.43	0.08	0.00	0.00	−127.35	0.00
75% 3.85	2.91	0.00	0.00	−126.36	0.00
100% −0.73	5.74	0.00	0.00	−125.37	0.00
MEMBER 3					
0% −0.73	−5.77	0.00	0.00	−125.37	0.00
25% 3.71	−2.86	0.00	0.00	−124.64	0.00
50% 5.15	0.06	0.00	0.00	−123.91	0.00
75% 3.60	2.97	0.00	0.00	−123.19	0.00
100% −0.96	5.88	0.00	0.00	−122.46	0.00

MEMBER FORCES					
M33 (kip*ft)	V2 (kip)	M22 (kip*ft)	V3 (kip)	Axial (kip)	Torsion (kip*ft)
Condition dl (dead load)					
MEMBER 4					
0% -0.96	-5.90	0.00	0.00	-122.46	0.00
25% 3.51	-2.93	0.00	0.00	-122.01	0.00
50% 4.97	0.03	0.00	0.00	-121.57	0.00
75% 3.44	3.00	0.00	0.00	-121.12	0.00
100% -1.10	5.97	0.00	0.00	-120.68	0.00
MEMBER 5					
0% -1.10	-5.98	0.00	0.00	-120.68	0.00
25% 3.39	-2.98	0.00	0.00	-120.53	0.00
50% 4.88	0.01	0.00	0.00	-120.38	0.00
75% 3.37	3.01	0.00	0.00	-120.23	0.00
100% -1.14	6.00	0.00	0.00	-120.08	0.00
MEMBER 6					
0% -1.14	-6.00	0.00	0.00	-120.08	0.00
25% 3.37	-3.01	0.00	0.00	-120.23	0.00
50% 4.88	-0.01	0.00	0.00	-120.38	0.00
75% 3.39	2.98	0.00	0.00	-120.53	0.00
100% -1.10	5.98	0.00	0.00	-120.68	0.00
MEMBER 7					
0% -1.10	-5.97	0.00	0.00	-120.68	0.00
25% 3.44	-3.00	0.00	0.00	-121.12	0.00
50% 4.97	-0.03	0.00	0.00	-121.57	0.00
75% 3.51	2.93	0.00	0.00	-122.01	0.00
100% -0.96	5.90	0.00	0.00	-122.46	0.00
MEMBER 8					
0% -0.96	-5.88	0.00	0.00	-122.46	0.00
25% 3.60	-2.97	0.00	0.00	-123.19	0.00
50% 5.15	-0.06	0.00	0.00	-123.91	0.00
75% 3.71	2.86	0.00	0.00	-124.64	0.00
100% -0.73	5.77	0.00	0.00	-125.37	0.00

MEMBER FORCES					
M33 (kip*ft)	V2 (kip)	M22 (kip*ft)	V3 (kip)	Axial (kip)	Torsion (kip*ft)
Condition dl (dead load)					
MEMBER 9					
0%	−0.73	−5.74	0.00	0.00	−125.37	0.00
25%	3.85	−2.91	0.00	0.00	−126.36	0.00
50%	5.43	−0.08	0.00	0.00	−127.35	0.00
75%	4.01	2.76	0.00	0.00	−128.34	0.00
100%	−0.41	5.59	0.00	0.00	−129.34	0.00
MEMBER 10					
0%	−0.41	−5.57	0.00	0.00	−129.34	0.00
25%	4.19	−2.83	0.00	0.00	−130.57	0.00
50%	5.79	−0.09	0.00	0.00	−131.80	0.00
75%	4.40	2.64	0.00	0.00	−133.03	0.00
100%	0.00	5.38	0.00	0.00	−134.26	0.00
Condition tl (temperature load)					
MEMBER 1					
0%	0.00	3.70	0.00	0.00	−8.23	0.00
25%	−4.06	3.70	0.00	0.00	−8.23	0.00
50%	−8.12	3.70	0.00	0.00	−8.23	0.00
75%	−12.18	3.70	0.00	0.00	−8.23	0.00
100%	−16.24	3.70	0.00	0.00	−8.23	0.00
MEMBER 2					
0%	−16.24	2.98	0.00	0.00	−8.51	0.00
25%	−19.39	2.98	0.00	0.00	−8.51	0.00
50%	−22.55	2.98	0.00	0.00	−8.51	0.00
75%	−25.71	2.98	0.00	0.00	−8.51	0.00
100%	−28.87	2.98	0.00	0.00	−8.51	0.00
MEMBER 3					
0%	−28.87	2.19	0.00	0.00	−8.75	0.00
25%	−31.12	2.19	0.00	0.00	−8.75	0.00
50%	−33.38	2.19	0.00	0.00	−8.75	0.00
75%	−35.63	2.19	0.00	0.00	−8.75	0.00
100%	−37.89	2.19	0.00	0.00	−8.75	0.00

MEMBER FORCES					
M33 (kip*ft)	V2 (kip)	M22 (kip*ft)	V3 (kip)	Axial (kip)	Torsion (kip*ft)
Condition tl (temperature load)					
MEMBER 4					
0% −37.89	1.34	0.00	0.00	−8.92	0.00
25% −39.24	1.34	0.00	0.00	−8.92	0.00
50% −40.59	1.34	0.00	0.00	−8.92	0.00
75% −41.95	1.34	0.00	0.00	−8.92	0.00
100% −43.30	1.34	0.00	0.00	−8.92	0.00
MEMBER 5					
0% −43.30	0.45	0.00	0.00	−9.01	0.00
25% −43.75	0.45	0.00	0.00	−9.01	0.00
50% −44.20	0.45	0.00	0.00	−9.01	0.00
75% −44.65	0.45	0.00	0.00	−9.01	0.00
100% −45.10	0.45	0.00	0.00	−9.01	0.00
MEMBER 6					
0% −45.10	−0.45	0.00	0.00	−9.01	0.00
25% −44.65	−0.45	0.00	0.00	−9.01	0.00
50% −44.20	−0.45	0.00	0.00	−9.01	0.00
75% −43.75	−0.45	0.00	0.00	−9.01	0.00
100% −43.30	−0.45	0.00	0.00	−9.01	0.00
MEMBER 7					
0% −43.30	−1.34	0.00	0.00	−8.92	0.00
25% −41.95	−1.34	0.00	0.00	−8.92	0.00
50% −40.59	−1.34	0.00	0.00	−8.92	0.00
75% −39.24	−1.34	0.00	0.00	−8.92	0.00
100% −37.89	−1.34	0.00	0.00	−8.92	0.00
MEMBER 8					
0% −37.89	−2.19	0.00	0.00	−8.75	0.00
25% −35.63	−2.19	0.00	0.00	−8.75	0.00
50% −33.38	−2.19	0.00	0.00	−8.75	0.00
75% −31.12	−2.19	0.00	0.00	−8.75	0.00
100% −28.87	−2.19	0.00	0.00	−8.75	0.00

MEMBER FORCES						
	M33 (kip*ft)	V2 (kip)	M22 (kip*ft)	V3 (kip)	Axial (kip)	Torsion (kip*ft)
Condition tl (temperature load)						
MEMBER 9						
0%	–28.87	–2.98	0.00	0.00	–8.51	0.00
25%	–25.71	–2.98	0.00	0.00	–8.51	0.00
50%	–22.55	–2.98	0.00	0.00	–8.51	0.00
75%	–19.39	–2.98	0.00	0.00	–8.51	0.00
100%	–16.24	–2.98	0.00	0.00	–8.51	0.00
MEMBER 10						
0%	–16.24	–3.70	0.00	0.00	–8.23	0.00
25%	–12.18	–3.70	0.00	0.00	–8.23	0.00
50%	–8.12	–3.70	0.00	0.00	–8.23	0.00
75%	–4.06	–3.70	0.00	0.00	–8.23	0.00
100%	0.00	–3.70	0.00	0.00	–8.23	0.00

MEMBER LOCAL DEFLECTIONS						
					Slopes	
Station	At 1 (ft)	At 2 (ft)	At 3 (ft)	Rotation 11 (rad)	L/defl(2)	L/defl(3)
Condition dl (dead load)						
MEMBER 1						
0%	0.000	0.000	0.000	0.00000	—	—
25%	–0.001	–0.003	0.000	0.00000	1304.99	—
50%	–0.001	–0.006	0.000	0.00000	678.83	—
75%	–0.002	–0.009	0.000	0.00000	475.46	—
100%	–0.002	–0.012	0.000	0.00000	373.63	—
MEMBER 2						
0%	–0.001	–0.012	0.000	0.00000	356.59	—
25%	–0.002	–0.014	0.000	0.00000	295.34	—
50%	–0.002	–0.017	0.000	0.00000	255.49	—
75%	–0.003	–0.019	0.000	0.00000	228.73	—
100%	–0.003	–0.020	0.000	0.00000	209.27	—

	MEMBER LOCAL DEFLECTIONS					
	At 1	At 2	At 3	Rotation 11	Slopes	
Station	(ft)	(ft)	(ft)	(rad)	L/defl(2)	L/defl(3)
	Condition dl (dead load)					
MEMBER 3						
0%	−0.001	−0.020	0.000	0.00000	201.52	—
25%	−0.002	−0.022	0.000	0.00000	185.80	—
50%	−0.002	−0.024	0.000	0.00000	173.82	—
75%	−0.003	−0.025	0.000	0.00000	165.06	—
100%	−0.003	−0.026	0.000	0.00000	158.33	—
MEMBER 4						
0%	−0.001	−0.026	0.000	0.00000	154.17	—
25%	−0.001	−0.027	0.000	0.00000	147.90	—
50%	−0.002	−0.028	0.000	0.00000	143.06	—
75%	−0.002	−0.029	0.000	0.00000	139.75	—
100%	−0.003	−0.029	0.000	0.00000	137.42	—
MEMBER 5						
0%	0.000	−0.030	0.000	0.00000	135.54	—
25%	−0.000	−0.030	0.000	0.00000	133.02	—
50%	−0.001	−0.030	0.000	0.00000	131.35	—
75%	−0.001	−0.031	0.000	0.00000	130.76	—
100%	−0.002	−0.031	0.000	0.00000	130.93	—
MEMBER 6						
0%	0.002	−0.031	0.000	0.00000	130.93	—
25%	0.001	−0.031	0.000	0.00000	130.76	—
50%	0.001	−0.030	0.000	0.00000	131.35	—
75%	0.000	−0.030	0.000	0.00000	133.02	—
100%	−0.000	−0.030	0.000	0.00000	135.54	—
MEMBER 7						
0%	0.003	−0.029	0.000	0.00000	137.42	—
25%	0.002	−0.029	0.000	0.00000	139.75	—
50%	0.002	−0.028	0.000	0.00000	143.06	—
75%	0.001	−0.027	0.000	0.00000	147.90	—
100%	0.001	−0.026	0.000	0.00000	154.17	—

					MEMBER LOCAL DEFLECTIONS	
	At 1	At 2	At 3	Rotation 11	Slopes	
Station	(ft)	(ft)	(ft)	(rad)	L/defl(2)	L/defl(3)
Condition dl (dead load)						
MEMBER 8						
0%	0.003	−0.026	0.000	0.00000	158.33	—
25%	0.003	−0.025	0.000	0.00000	165.06	—
50%	0.002	−0.024	0.000	0.00000	173.82	—
75%	0.002	−0.022	0.000	0.00000	185.80	—
100%	0.001	−0.020	0.000	0.00000	201.52	—
MEMBER 9						
0%	0.003	−0.020	0.000	0.00000	209.27	—
25%	0.003	−0.019	0.000	0.00000	228.73	—
50%	0.002	−0.017	0.000	0.00000	255.49	—
75%	0.002	−0.014	0.000	0.00000	295.34	—
100%	0.001	−0.012	0.000	0.00000	356.59	—
MEMBER 10						
0%	0.002	−0.012	0.000	0.00000	373.63	—
25%	0.002	−0.009	0.000	0.00000	475.46	—
50%	0.001	−0.006	0.000	0.00000	678.83	—
75%	0.001	−0.003	0.000	0.00000	1304.99	—
100%	0.000	0.000	0.000	0.00000	—	—
Condition tl (temperature load)						
MEMBER 1						
0%	0.000	0.000	0.000	0.00000	—	—
25%	0.006	0.030	0.000	0.00000	145.08	—
50%	0.012	0.060	0.000	0.00000	72.81	—
75%	0.018	0.090	0.000	0.00000	48.84	—
100%	0.024	0.119	0.000	0.00000	36.94	—
MEMBER 2						
0%	0.014	0.120	0.000	0.00000	35.21	—
25%	0.019	0.147	0.000	0.00000	28.74	—
50%	0.025	0.174	0.000	0.00000	24.42	—
75%	0.031	0.199	0.000	0.00000	21.35	—
100%	0.037	0.222	0.000	0.00000	19.07	—

	MEMBER LOCAL DEFLECTIONS					
	At 1	At 2	At 3	Rotation 11	Slopes	
Station	(ft)	(ft)	(ft)	(rad)	L/defl(2)	L/defl(3)
	Condition tl (temperature load)					
MEMBER 3						
0%	0.016	0.225	0.000	0.00000	18.36	—
25%	0.022	0.246	0.000	0.00000	16.75	—
50%	0.027	0.266	0.000	0.00000	15.48	—
75%	0.033	0.285	0.000	0.00000	14.47	—
100%	0.039	0.302	0.000	0.00000	13.67	—
MEMBER 4						
0%	0.010	0.304	0.000	0.00000	13.31	—
25%	0.015	0.319	0.000	0.00000	12.69	—
50%	0.021	0.332	0.000	0.00000	12.20	—
75%	0.026	0.343	0.000	0.00000	11.80	—
100%	0.032	0.352	0.000	0.00000	11.50	—
MEMBER 5						
0%	−0.003	0.353	0.000	0.00000	11.34	—
25%	0.002	0.360	0.000	0.00000	11.12	—
50%	0.008	0.365	0.000	0.00000	10.97	—
75%	0.013	0.368	0.000	0.00000	10.88	—
100%	0.018	0.369	0.000	0.00000	10.85	—
MEMBER 6						
0%	−0.018	0.369	0.000	0.00000	10.85	—
25%	−0.013	0.368	0.000	0.00000	10.88	—
50%	−0.008	0.365	0.000	0.00000	10.97	—
75%	−0.002	0.360	0.000	0.00000	11.12	—
100%	0.003	0.353	0.000	0.00000	11.34	—
MEMBER 7						
0%	−0.032	0.352	0.000	0.00000	11.50	—
25%	−0.026	0.343	0.000	0.00000	11.80	—
50%	−0.021	0.332	0.000	0.00000	12.20	—
75%	−0.015	0.319	0.000	0.00000	12.69	—
100%	−0.010	0.304	0.000	0.00000	13.31	—

		MEMBER LOCAL DEFLECTIONS				
Station	At 1 (ft)	At 2 (ft)	At 3 (ft)	Rotation 11 (rad)	Slopes L/defl(2)	L/defl(3)
		Condition tl (temperature load)				
MEMBER 8						
0%	−0.039	0.302	0.000	0.00000	13.67	—
25%	−0.033	0.285	0.000	0.00000	14.47	—
50%	−0.027	0.266	0.000	0.00000	15.48	—
75%	−0.022	0.246	0.000	0.00000	16.75	—
100%	−0.016	0.225	0.000	0.00000	18.36	—
MEMBER 9						
0%	−0.037	0.222	0.000	0.00000	19.07	—
25%	−0.031	0.199	0.000	0.00000	21.35	—
50%	−0.025	0.174	0.000	0.00000	24.42	—
75%	−0.019	0.147	0.000	0.00000	28.74	—
100%	−0.014	0.120	0.000	0.00000	35.21	—
MEMBER 10						
0%	−0.024	0.119	0.000	0.00000	36.94	—
25%	−0.018	0.090	0.000	0.00000	48.84	—
50%	−0.012	0.060	0.000	0.00000	72.81	—
75%	−0.006	0.030	0.000	0.00000	145.08	—
100%	0.000	0.000	0.000	0.00000	—	—

Index

Note: Page numbers followed by *f* denote figures; page numbers followed by *t* denote tables.